Scientific Computing and Algorithms in Industrial Simulations

Michael Griebel • Anton Schüller •
Marc Alexander Schweitzer

Editors

Scientific Computing and Algorithms in Industrial Simulations

Projects and Products of Fraunhofer SCAI

 Springer

Editors

Michael Griebel
Schloss Birlinghoven
Fraunhofer-Institut SCAI
Sankt Augustin, Germany

Anton Schüller
Schloss Birlinghoven
Fraunhofer-Institut SCAI
Sankt Augustin, Germany

Marc Alexander Schweitzer
Schloss Birlinghoven
Fraunhofer-Institut SCAI
Sankt Augustin, Germany

ISBN 978-3-319-87317-6   ISBN 978-3-319-62458-7   (eBook)
DOI 10.1007/978-3-319-62458-7

Mathematics Subject Classification (2010): 74S99, 65M99, 65N99, 65F99, 68N99, 68P99

Printed on acid-free paper

This Springer imprint is published by Springer Nature
The registered company is Springer International Publishing AG
The registered company address is: Gewerbestrasse 11, 6330 Cham, Switzerland

Preface

Applied research is the foundation of the Fraunhofer-Gesellschaft (FhG). The 69 Fraunhofer institutes team up with companies to transform original ideas into innovations that benefit society and strengthen both the German and the European economy. To this end, one of the key missions of a Fraunhofer institute is to bridge the gap between fundamental research, usually conducted at universities, and the requirements of industry.

The Fraunhofer Institute for Algorithms and Scientific Computing SCAI combines excellent research and application-oriented development to achieve this goal and to provide added value for our partners. SCAI develops numerical techniques, parallel algorithms, and specialized software tools to support and optimize industrial simulations. Moreover, we implement custom software solutions for production and logistics, and offer calculations on high-performance computers. Our services and products are based on state-of-the-art methods from applied mathematics and information technology. Currently, the main areas of research and development at SCAI are:

- Bioinformatics
- Fast solvers
- High performance computing
- Multiphysics
- Optimization

- Computational finance
- High performance analytics
- Meshfree multiscale methods
- Numerical data driven prediction
- Virtual material design

Within these fields of application, SCAI has developed a number of software products that focus on specific industrial requirements:

- AutoNester is a software package for automatic marker making on fabrics, leather, sheet metal, wood, or other materials. It is widely used in the garment and upholstery manufacturing industry.
- DesParO is a toolbox for the intuitive exploration, automatic analysis, and optimization of parameterized problems in production processes. It can be coupled with simulation programs or data from physical experiments.
- ModelCompare is a data analysis plug-in for finite element pre- and post-processing tools which allows for an easy comparison of similar simulation

results identifying the differences between models based on the geometry described by the mesh.

- The MpCCI software suite enables the coupling of various simulation codes to enable the speedy implementation of multiphysics applications such as fluid-structure-interaction problems. It is used in a wide range of applications from mechanical engineering, and aeroelasticity to the simulation of micro-mechanical components.
- MYNTS is a multiphysical network simulator for electrical circuits, gas and energy transport, and water distribution. It models networks as systems of differential-algebraic equations and helps users to manage, analyze, and optimize their networks.
- PackAssistant is utilized worldwide especially in the automotive industry by numerous manufacturers and suppliers to automatically compute optimal filling of transport containers with identical parts.
- The text-mining tool ProMiner supports users in the identification of genes and proteins and respective illnesses and treatments by analyzing scientific texts and libraries.
- The library SAMG provides highly efficient algorithms for the parallel solution of large linear systems of equations with sparse matrices. It is used in different industries ranging from manufacturing and automotive to oil and gas exploration.
- Tremolo-X is a massively parallel software package for molecular dynamics simulations. It is used in various projects and industrial applications in nanotechnology, materials science, biochemistry, and biophysics.

Besides the development of software tools of industrial relevance, SCAI is involved in many collaborative research projects funded by the EU and the BMBF. For instance, SCAI's bioinformatics group is coordinating AETIONOMY which is one of the flagship projects of the current Innovative Medicines Initiative; see

https://www.scai.fraunhofer.de/en/projects.html

for details.

The articles in this book give an overview of current research projects and selected software products of the Fraunhofer Institute for Algorithms and Scientific Computing SCAI. They show the wide range of challenges and solutions in scientific computing and its important role in applications for industry. This exciting field of applied collaborative research and development is surely attractive for scientists, practitioners, and students alike.

Sankt Augustin, Germany   Michael Griebel
April 2017   Anton Schüller
                                                                     Marc Alexander Schweitzer

Contents

Part I
Methods

Calculation of Chemical Equilibria in Multi-Phase: Multicomponent Systems

Marco Hülsmann, Bernhard Klaassen, Andreas Krämer,
Ottmar Krämer-Fuhrmann, Tosawi Pangalela, Dirk Reith, Klaus Hack,
and Johannes Linden

1 Introduction

The calculation of chemical equilibria for systems with multiple phases and components is relevant for both scientific and industrial applications, especially for the development of new materials. Important physical and chemical properties can be derived from them. They can be used to predict behavior of materials, e. g. breaking resistance and corrosion susceptibility, and much more.

In general, chemical equilibria of a thermodynamic system are computed by minimizing its Gibbs free energy, a nonlinear multidimensional constrained optimization task. In the late 1950s, White et al. [26] applied a variant of the Newton-Raphson method for the minimization of the free enthalpy of an ideal gas mixture. Their method, known as the RAND algorithm, was the fundamental basis for a number of computer programs developed over the years. The method

M. Hülsmann (✉) • D. Reith
Fraunhofer Institute for Algorithms and Scientific Computing SCAI, Schloss Birlinghoven, 53757 Sankt Augustin, Germany

Bonn-Rhein-Sieg University of Applied Sciences, Grantham-Allee 20, 53757 Sankt Augustin, Germany
e-mail: marco.huelsmann@h-brs.de; dirk.reith@scai.fraunhofer.de

O. Krämer-Fuhrmann • B. Klaassen • T. Pangalela • J. Linden
Fraunhofer Institute for Algorithms and Scientific Computing SCAI, Schloss Birlinghoven, 53757 Sankt Augustin, Germany

A. Krämer
Bonn-Rhein-Sieg University of Applied Sciences, Grantham-Allee 20, 53757 Sankt Augustin, Germany

K. Hack
GTT-Technologies, Kaiserstraße 100, 52134 Herzogenrath, Germany

© Springer International Publishing AG 2017
M. Griebel et al. (eds.), *Scientific Computing and Algorithms in Industrial Simulations*, DOI 10.1007/978-3-319-62458-7_1

was extended to solid stoichiometric phases [1, 7, 14, 23] and liquid mixture phases [2, 5, 8, 10].

An important part of the problem is the correct selection of phases which will finally occur in the equilibrium. These phases are called *active phases* and they are typically not known in advance. If one simply involves all possible phases in a nonlinear optimization process including the non-active ones the process could fail. This is due to the fact that the mole numbers of the inactive phases are zero and their mole fractions are not determined.

Eriksson and Thompson [11] developed a method for both selecting the active phases and providing initial guesses for the phase compositions before starting the nonlinear local Gibbs energy minimization.

Nichita et al. [20, 21] suggest an approach in which the complete thermodynamic system including the non-active phases was considered from the beginning. Their algorithm is based on a combination of a local Quasi-Newton search with the tunneling method [16] and represents a global search approach. It successively finds intermediate local minima, maybe arising from a preliminary and probably incorrect phase selection. It then tries to move away from them afterwards. Although the tunneling method turns out to be successful and robust in some cases as demonstrated by Nichita et al., its convergency is not guaranteed, and it often leads to a new local solution in the close neighborhood without finding the desired global minimum, cf. [12].

A different method was proposed by Emelianenko et al. [6], which utilizes geometric properties of the energy surfaces in combination with effective sampling techniques for phase selection. The underlying restriction in their paper, however, is that the phase constituents are identical to the system components.

In this paper, we follow the principal approach by Eriksson and Thompson [11], i. e. we first select the potentially correct phases from a set of possible candidates using a discretized linear model and then we enter into nonlinear optimizations.

Our developments have been realized through a new software named BePhaSys, which is capable of efficiently and robustly predicting phase equilibria in arbitrary systems without limitations in regard to the number of system components or the number of phases and their constituents. It handles non-ideal solid and liquid mixture phases, gas phases, and stoichiometric phases. BePhaSys combines an estimation method predicting the relevant phases in the chemical equilibrium with a Newton-Lagrange method solving the resulting constraint Gibbs energy minimization problem. In automated Gibbs energy minimizing software special attention must be paid to the case when phases with partially concave Gibbs energy curves or surfaces occur. In such cases miscibility gaps may develop at equilibrium. Chen et al. [3] for example proposed a method to automatically detect such miscibility gaps based on a discretization of the continuous problem. A similar method is described in [6]. In BePhaSys we use a method for detecting concavity domains of the objective function which allows an identification of possible miscibility gaps.

For the generation of complete phase diagrams, BePhaSys follows a parallelization approach on many core systems to reduce response times significantly: For
shared memory systems, an OpenMP [24] based distribution of work is used, and
MPI [19] for network connected clusters. Both strategies work very well, since
the computation of phase diagrams can be split into (nearly) independent tasks,
while the synchronization and communication effort necessary to steer the parallel
computations is low. Another speed-up is delivered by the interpolation of the phase
boundaries arising from a special treatment of two-phase and three-phase areas. The
methodology described here is applicable for two-dimensional isothermal sections
in ternary systems only. However, the methods presented here can be used as a basis
and be extended for the calculation of phase diagrams of higher dimensions.

The paper is organized as follows: In Sect. 2, the mathematical Gibbs energy
minimization problem is introduced and analyzed. Section 3 presents the algorithms
and methodologies for the solution of the minimization problem. The method for the
automated detection of miscibility gaps is described in Sect. 4. Section 5 contains
results for a ternary example system at different temperatures and combinations
of system components. We then present results and some algorithmic details for
the (parallel) computation of ternary phase diagrams. The results were validated
by comparing them to results produced by the well-known reference software
ChemApp [9, 25].

2 Problem Formulation

In the following, the notations listed in Fig. 1 are used. We consider a chemical
system with L system components and N phases at a given temperature and pressure.

2.1 Non-Ideal Gibbs Function

The calculation of chemical equilibria leads to an optimization problem, which is
specified in this section. The objective function is the total Gibbs energy function,
which is given in the following as a linear combination of the phase specific molar
Gibbs energy functions, which are weighted by their molar amounts.

In a system with N phases, let (f, x) be the vector containing all mole numbers
and fractions, i. e. $(f, x) := \left(f^{(i)}, x^{(i)} \right)_{i=1,\dots,N} \in \mathbb{R}^{N+\kappa}$. The Gibbs energy function is
then given by:

$$G((f,x)) := \sum_{i=1}^{N} f^{(i)} G^{(i)}(x^{(i)}). \tag{1}$$

- L, the number of system components $j = 1, ..., L$,
- N, the number of phases $i = 1, .., N$,
- K^i, the number of phase constituents $k = 1, ..., K^i$ for each phase $i \in \{1, ..., N\}$, i. e. there are $\kappa := \sum_{i=1}^{N} K^i$ constituents in total,
- $f^{(i)}$, $i = 1, ..., N$, the mole numbers of the phases and f_k^i, the mole number of a phase constituent $k \in \{1, ..., K^i\}$ in a phase $i \in \{1, ..., N\}$, where

$$\forall_{i=1,...,N} \sum_{k=1}^{K^i} f_k^i = f^{(i)},$$

- $\mathbb{E}^{(i)} := \{x \in \mathbb{R}^{K^i} | \langle x^{(i)}, e^{K^i} \rangle_{\mathbb{R}^{K^i}} = 1, 0 \leq x^{(i)} \leq 1 \text{ (component − wise)}\}$ the unit simplex of $\mathbb{R}^{K^i}$, $e^{K^i} := (1, ..., 1)^T \in \mathbb{R}^{K^i}$, $i \in \{1, ..., N\}$,
- $x^{(i)} \in \mathbb{E}^{(i)}$ the vector of mole fractions of the phase constituents $x_k^i := \left(\frac{f_k^i}{f^{(i)}} \right)_{k=1,...,K^i}$, of phase i,
- $\zeta := \left(f^{(i)} x^{(i)} \right)_{i=1,...,N} \in \mathbb{R}^{\kappa}$,
- $b \in \mathbb{E}^{(L)}$ the vector of amounts of substance, given in mole, of the system components,
- $A_i := \begin{pmatrix} (a_1^i)_1 & \cdots & (a_1^i)_L \\ \vdots & \ddots & \vdots \\ (a_{K^i}^i)_1 & \cdots & (a_{K^i}^i)_L \end{pmatrix} \in \mathbb{R}^{K^i \times L}$, the phase specific stoichiometric matrix of a phase i,
- $s^{(i)} := A_i e^L \in \mathbb{R}^{K^i}$ the vector of row sums of the matrix A_i, $e^L := (1, ..., 1)^T \in \mathbb{R}^L$, and
- $G^{(i)}(x^{(i)})$ the phase specific Gibbs function of phase i.

Fig. 1 Notations

Here, $G^{(i)}(x^{(i)})$ describes the molar Gibbs free energy of a phase $i \in \{1, ..., N\}$ *(phase specific Gibbs function)*, consisting of a reference part $G_{\text{ref}}^{(i)}$, an ideal part $G_{\text{id}}^{(i)}$, and an excess part $G_{\text{ex}}^{(i)}$ describing the deviation of the real mixture from the ideal one. More details are given in the appendix.

2.2 Stoichiometric Constraints

Phases and phase constituents are formed within chemical processes, e. g. dissolution, precipitation etc. The amounts of substance of the system components are distributed stoichiometrically over the phase constituents due to mass conservation. The mass conservation is expressed by the *stoichiometric constraints*:

$$\sum_{i=1}^{N} f^{(i)} A_i^T x^{(i)} = b \tag{2}$$

$$A^T \zeta = b, \tag{3}$$

where

$$A := \begin{pmatrix} A_1 \\ \vdots \\ A_N \end{pmatrix} \in \mathbb{R}^{\kappa \times L}$$

is the stoichiometric matrix considering all phases $i = 1, \ldots, N$. Without loss of generality, the amounts of substance are normalized, i. e. $\sum_{j=1}^{L} b_j = 1$.

Stoichiometric matrices represent the stoichiometric relation between system components and phase constituents. Table 1 shows the example of a stoichiometric matrix for a thermodynamic system consisting of seven system components and three phases with 11, 5, and 7 phase constituents, respectively. These phase constituents are either identical to the system components or they are species consisting of well defined amounts of the system components.

Table 1 Example of a stoichiometric matrix for seven system components and three phases

Phase	Components	System components						
		Fe	N	O	C	Ca	Si	Mg
Gas	Fe	1	0	0	0	0	0	0
	N_2	0	2	0	0	0	0	0
	O_2	0	0	2	0	0	0	0
	C	0	0	0	1	0	0	0
	CO	0	0	1	1	0	0	0
	CO_2	0	0	2	1	0	0	0
	Ca	0	0	0	0	1	0	0
	CaO	0	0	1	0	1	0	0
	Si	0	0	0	0	0	1	0
	SiO	0	0	1	0	0	1	0
	Mg	0	0	0	0	0	0	1
Slag	SiO_2	0	0	2	0	0	1	0
	Fe_2O_3	2	0	3	0	0	0	0
	CaO	0	0	1	0	1	0	0
	FeO	1	0	1	0	0	0	0
	MgO	0	0	1	0	0	0	1
Liq. Fe	Fe	1	0	0	0	0	0	0
	N	0	1	0	0	0	0	0
	O	0	0	1	0	0	0	0
	C	0	0	0	1	0	0	0
	Ca	0	0	0	0	1	0	0
	Si	0	0	0	0	0	1	0
	Mg	0	0	0	0	0	0	1

2.3   The Optimization Problem

For the determination of phase equilibria, the following mathematical optimization problem has to be solved:

$$\min_{(f,x)} \left\{ G((f,x)) := \sum_{i=1}^{N} f^{(i)} G^{(i)}(x^{(i)}) \right\},$$

$$\text{where } \langle x^{(i)}, e^{K^i} \rangle_{\mathbb{R}^{K^i}} = 1, \ i = 1, \ldots, N,$$

$$A^T \zeta = b,$$

$$f^{(i)} \geq 0, \ x_k^i \geq 0, \ i = 1, \ldots, N, \ k = 1, \ldots, K^i. \tag{4}$$

Problem (4) is denoted as the *G-problem* in this work. It is a general nonlinear minimization problem. Due to the general form of the phase specific Gibbs functions, assumptions concerning specific properties of the objective function like convexity cannot be made in general.

The corresponding Lagrangian is given by

$$\mathcal{L} : \mathbb{R}^{N+\kappa} \times \mathbb{R}^N \times \mathbb{R}^L \to \mathbb{R}$$

$$((f,x), \lambda, \eta) \mapsto \sum_{i=1}^{N} f^{(i)} G^{(i)}(x^{(i)}) - \sum_{i=1}^{N} \lambda_i \left(\langle x^{(i)}, e^{K^i} \rangle_{\mathbb{R}^{K^i}} - 1 \right)$$

$$- \sum_{j=1}^{L} \eta_j \left(A^T \zeta - b \right),$$

with the Lagrangian multipliers λ_i, $i = 1, \ldots, n$, and η_j, $j = 1, \ldots, L$.

The corresponding Lagrangian equations are:

$$G^{(i)}(x^{(i)}) = \langle \eta, A_i^T x^{(i)} \rangle_{\mathbb{R}^L}, \ i = 1, \ldots, N, \tag{5}$$

$$\lambda_i \cdot e^{\mathbb{R}^{K^i}} = f^{(i)} \left(\nabla G^{(i)}(x^{(i)}) - A_i \eta \right), \ i = 1, \ldots, N. \tag{6}$$

Based on this, the following properties hold:

(i) If there is a phase $i \in \{1, \ldots, N\}$ so that $f^{(i)} \neq 0$ and the column vectors $a_j^i \in \mathbb{R}^{K^i}$, $j = 1, \ldots, L$, of A_i are linearly independent, then each feasible point of the G-problem is regular. This is also true for a weaker condition, i. e. if $\bigcap_{f^{(i)} \neq 0} \ker A_i = \{0\}$.

(ii) The points

$$P_i := \left((A^i)^T x^{(i)}, G^{(i)}(x^{(i)}) \right)^T \in \mathbb{R}^{L+1}$$

lie on the hyperplane given by

$$H_\eta := \{z \in \mathbb{R}^{L+1} | \langle \hat{\eta}, z \rangle = 0\}, \ \hat{\eta} := \begin{pmatrix} \eta \\ -1 \end{pmatrix}.$$

(iii) It holds $(b, G((f, x))) \in H_\eta$ and

$$\forall_{i,j \in \{1,\ldots,N\}} \ G^{(i)}(x^{(i)}) - G^{(j)}(x^{(j)}) = \langle \eta, A_i^T x^{(i)} - A_j^T x^{(j)} \rangle. \tag{7}$$

H_η is the common tangent hyperplane between the phase specific Gibbs energy functions $G^{(i)}$ and $G^{(j)}$.

(iv) The Lagrangian parameters η_j, $j = 1, \ldots, L$ as a part of the solution of the G-problem are unique if and only if $\bigcap_{f^{(i)} \neq 0} \ker A^{(i)} = \{0\}$.

3 Methodology for the Calculation of Chemical Equilibria

In practice, an additional problem arises: It is usually not clear from the beginning which phases from a set of possible candidates will really occur in an equilibrium, i. e. have a positive mole number $f^{(i)}$. Such phases are called active. If one simply starts the nonlinear minimization process by involving all possible phases, one will possibly run into trouble. It is clear from the Gibbs phase rule that at most L phases can be active, where L is the number of system components. Therefore, the mole numbers of some of the phases must vanish. Formally in this case, the minimum is never strict, cf. Fig. 2. That could lead to numerical difficulties, e. g. because iteration matrices could become singular and the iterative processes may diverge. Apart from that, the numerical effort increases with the number of phases involved.

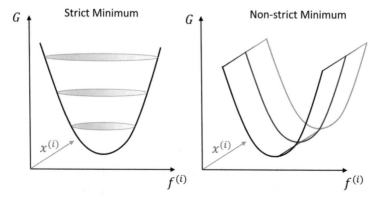

Fig. 2 Draft of a strict and non-strict minimum of $G((f^{(i)}, x^{(i)}))$. Published with kind permission of ©Fraunhofer SCAI 2016. All Rights Reserved

It is therefore crucial to identify the active phases in a pre-processing step and to discard the others before starting a nonlinear minimization procedure. The method used for the selection of the active phases described here is based on a discretization and linearization of the G-problem. It depends on the choice of sampling points within a search space. The linear method is then combined with a corrector step based on the consideration of the chemical activities of all phases. This additional step allows to reconsider the whole set of possible phases and possibly finds better suited active candidates. For more details see Sect. 3.3.

3.1 Reformulation of the Minimization Problem

The phase selection is principally based on a comparison of the phase specific Gibbs functions [8, 11]. A phase whose Gibbs energy surface is situated above the remaining ones can be excluded, i. e. it does not become active. Here, a discretization approach is used to identify the active phases. It is based on the following fully equivalent re-formulation of the G-problem:

$$\min_{(\phi, x)} \left\{ H((\phi, x)) = \sum_{i=1}^{N} \phi^{(i)} H^i(x^{(i)}) \right\},$$

$$\text{where } \langle x^{(i)}, e^{K^i} \rangle_{\mathbb{R}^{K^i}} = 1,$$

$$\sum_{i=1}^{N} \phi^{(i)} \psi^{(i)}(x^{(i)}) = b,$$

$$\phi^{(i)} \geq 0, \; x_k^i \geq 0, \; i = 1, \ldots, N, \; k = 1, \ldots, K^i, \tag{8}$$

where the functions $H^{(i)}$, $i = 1, \ldots, N$, and $\psi^{(i)}$, $i = 1, \ldots, N$, are given by

$$H^{(i)}(x^{(i)}) := \frac{G^{(i)}(x^{(i)})}{\langle x^{(i)}, s^{(i)} \rangle} \tag{9}$$

and

$$\psi^{(i)}(x^{(i)}) := \frac{1}{\langle x^{(i)}, s^{(i)} \rangle} A_i^T x^{(i)}. \tag{10}$$

The H-problem (8) has the following properties:

- There exists a feasible point of the H-problem $\Leftrightarrow$ b is situated in the convex hull of the row vectors of A_i, $i = 1, \ldots, n$, normalized by the row sums.
- $\sum_{i=1}^{N} \phi^{(i)} = 1$.
- $\psi^{(i)}(x^{(i)}) \in \mathbb{E}^L$, $i = 1, \ldots, N$.

- b is a convex linear combination of the vectors $\psi^{(i)}(x^{(i)})$, $i = 1, \ldots, N$.
- $(\phi^{(i)}, x^{(i)})$, $i \in \{1, \ldots, N\}$, is a feasible point of the H-problem $\Leftrightarrow (f^{(i)}, x^{(i)})$ with $f^{(i)} = \frac{\phi^{(i)}}{\langle x^{(i)}, s^{(i)} \rangle}$ is a feasible point of the G-problem.
- $(\phi^{(i)}, x^{(i)})$ is a solution of the H-problem $\Leftrightarrow (f^{(i)}, x^{(i)})$ is a solution of the G-problem.

3.2 Discretization of the H-Problem

The H-problem is solved by discretization, i. e. a set of sampling points X_k^i, $k = 1, \ldots, M_i$, $M_i \in \mathbb{N}$ is chosen from the unit simplex of the space given by the phase constituents, $\mathbb{E}^{K^i}$, for each phase $i \in \{1, \ldots, N\}$. This leads to a linearized version of (8).

Let $M := \sum_{i=1}^{N} M_i$, $H := (H^m)_{m=1,\ldots,M} \in \mathbb{R}^M$ the vector containing the H-function values $H^i(X_k^i)$, $i = 1, \ldots, N$, $k = 1, \ldots, M_i$, $\Phi := (\Phi^m)_{m=1,\ldots,M} \in \mathbb{R}^M$ the vector containing ϕ_k^i and $\Psi \in \mathbb{R}^{L \times M}$ the matrix whose column vectors Ψ^m are given by $\psi^{(i)}(X_k^i)$, then the discrete H-problem is given by:

$$\min_{\phi} \left\{ \tilde{H}(\phi) = \langle \Phi, H \rangle \right\},$$

$$\text{where} \quad \Psi \cdot \Phi = b,$$

$$\Phi \geq 0 \text{ (component − wise)}. \tag{11}$$

The nonlinear phase specific Gibbs functions were replaced by point values at the sampling points. As $Y_k^i := \psi^{(i)}(X_k^i) \in \mathbb{E}^L$ and due to $\langle b, e^L \rangle_{\mathbb{R}^L} = 1$, it follows:

$$\langle \Phi, e^M \rangle_{\mathbb{R}^M} = 1, \tag{12}$$

where $e^M := (1, \ldots, 1)^T \in \mathbb{R}^M$, or, in a different formulation,

$$b \in \text{conv}\{Y_k^i | i = 1, \ldots, N, \ k = 1, \ldots, M_i\}, \tag{13}$$

where conv denotes the convex hull of the Y_k^i, $i = 1, \ldots, N$, $k = 1, \ldots, M$.

The condition (13) is necessary and sufficient for the solvability of the discrete H-problem (11).

We then consider the dual of (11):

$$\max_{\eta} \quad \langle b, \eta \rangle_{\mathbb{R}^L},$$

$$\text{where} \quad \forall_{i=1,\ldots,m} \ H^m - \langle \Psi^m, \eta \rangle_{\mathbb{R}^L} \geq 0. \tag{14}$$

The solution of the linear problem (LP) (14) can be illustrated as follows: The $L-1$-dimensional hyperplane, given by the linear function $\langle b, \eta \rangle_{\mathbb{R}^L}$, is shifted from below

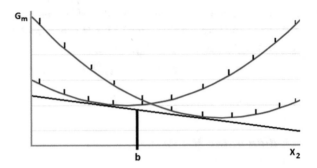

Fig. 3 Example with two phase specific Gibbs functions and a one-dimensional hyperplane. Published with kind permission of ©Fraunhofer SCAI 2016. All Rights Reserved

into the set of sampling points until an optimum in the sense of LP (14) is found. A hyperplane fixed by exactly L linear independent vectors, below which none of the points Y_k^i is located, is considered as *optimal*. Within BePhaSys, the LP is solved using the open-source linear program solver *CLP* from the software package *COIN-OR* [17]. As it tries to predict the correct phases involved, this method is called *predictor method*.

For the simplest case of two system components, i. e. $L = 2$, the situation is sketched by Fig. 3. Note that the search space degenerates to the unit interval here.

3.3 Corrector Step

Due to the discretization within the predictor method, which may be too coarse, the resulting phase selection may still be incorrect. The predictor method delivers mole fractions $x^{(i)}$ for all phases $i = 1, \ldots, N$, with $\psi^{(i)}(x^{(i)})$ being active points in $\mathbb{E}^L$. It may occur that some points $\left(\psi^{(i)}(x^{(i)}), H^{(i)}(x^{(i)})\right) \in \mathbb{R}^{L+1}$ are still located below the hyperplane for one or more phases $i \in \{1, \ldots, N\}$. If so, the corresponding phase i is chemically active with respect to the current hyperplane

$$H_\eta := \{z \in \mathbb{R}^{L+1} | \langle \hat{\eta}, z \rangle = 0\}, \ \hat{\eta} := \begin{pmatrix} \eta \\ -1 \end{pmatrix},$$

resulting from the predictor method solving LP (14), and may occur in the true chemical equilibrium. In this case, the phase selection of the predictor method is still wrong, and a subsequent corrector step is applied.

A point is located below the hyperplane if and only if the (signed) distance between the hyperplane and the corresponding function value is greater than zero, i. e.

$$\langle \psi^{(i)}(x^{(i)}), \eta \rangle - H^{(i)}(x^{(i)}) > 0.$$

In order to detect whether a point fulfills this condition, the following continuous optimization problem is solved for all phases:

$$\forall_{i=1,\ldots,N} \max_{x^{(i)} \in \mathbb{E}^i} \; d_\eta^{(i)}(x^{(i)}) := \langle \psi^{(i)}(x^{(i)}), \eta \rangle - H^{(i)}(x^{(i)}).$$

This optimization problem is solved by Newton-Lagrange using $x^{(i)}$, $i = 1, \ldots, N$, as initial guesses. Let $\bar{x}^{(i)} := \arg\max_{x^{(i)} \in \mathbb{E}^i} d_\eta^{(i)}(x^{(i)})$ be the solution for phase i. If $d_\eta^{(i)}(\bar{x}^{(i)}) \leq 0$, the corresponding function $H^{(i)}$ is completely located above or on the hyperplane. Otherwise, the corresponding phase is chemically active, which was not seen by the predictor before. In this case, the solution $\bar{x}^{(i)}$ is added to the set of sampling points, which is turned over to the predictor method again. Then, the predictor method delivers an improved solution of LP (14), i. e. a new hyperplane, and possibly a different phase selection. The corrector step corresponds to a first refinement step in usual discretization methods to solve semi-infinite problems [13].

It should be noted that the chemical activity $\alpha^{(i)}$ for a phase i is given by

$$\alpha^{(i)} = \exp\left(\frac{d_\eta^{(i)}(\bar{x}^{(i)})}{RT} \right).$$

The hyperplane coefficients $\eta_1, \ldots, \eta_L$ are the chemical potentials of the system components $j = 1, \ldots, L$.

Moreover, the chemical activity $\alpha^{(k)}$ of a phase constituent k is given by its chemical potential $\mu_k(T, P)$ and its standard chemical potential μ_k^0 subject to:

$$\alpha^{(k)} = \exp\left(\frac{\mu_k(T, P) - \mu_k^0}{RT} \right). \tag{15}$$

The process described above is repeated ℓ times (in all our test cases $\ell = 10$ was sufficient to predict the correct phase selection) in total so that, including the first solution of the predictor method, $\ell + 1$ approximate solutions with possibly different phase selections result. All $\ell + 1$ solutions are simultaneously turned over to a local minimizer, which uses them as initial guesses and solves the G-problem (4) exactly using Newton-Lagrange as well. Based on the approximate results found by the predictor-corrector approach described above, the final exact solution of the G-problem is calculated by a Newton-Lagrange method with Armijo step length control [22].

4 Automated Detection of Miscibility Gaps

If the predictor method delivers more than one active point for a single phase, it has to be checked whether a miscibility gap is present or not, cf. Fig. 4. In that situation, several composition sets in this phase have to be considered in the local

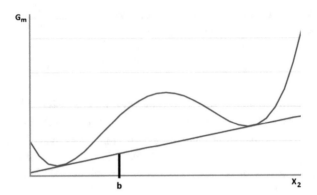

Fig. 4 Geometric illustration of the presence of a miscibility gap. Published with kind permission of ©Fraunhofer SCAI 2016. All Rights Reserved

optimization, each of which may meet the hyperplane in one of the active points. The presence of multiple active points $x^{(i,1)}, \ldots, x^{(i,Q)}$, $Q \in \mathbb{N}$, for one particular phase $i \in \{1, \ldots, N\}$, indicates that miscibility gaps may possibly be formed.

The convexity of the corresponding phase specific Gibbs energy function serves to evaluate which of the two situations is present. More precisely, it is tested whether

$$G^{(i)}\left(\zeta x^{(i,q_1)} + (1 - \zeta)x^{(i,q_2)}\right) \leq \zeta G^{(i)}\left(x^{(i,q_1)}\right) + (1 - \zeta)G^{(i)}\left(x^{(i,q_2)}\right)$$

is true for all pairwise distinct $q_1, q_2 \in \{1, \ldots, Q\}$ and $\zeta \in \{0.1, 0.2, \ldots, 0.9\}$. If yes, only one copy of the phase is given to the local optimizer. If not, $q \in \mathbb{N}$ copies of the phase are given to the local optimizer, where q is the order of the miscibility gap.

The overall algorithm of the predictor-corrector method including miscibility gap detections is visualized in Fig. 5: First of all, an initial sampling grid with a predefined grid size is determined. The predictor solves the discrete H-problem using a LP. If it finds more than one active point for one phase, it suggests that this phase has a—maybe multiple—miscibility gap. This is confirmed or disproved by the miscibility gap detector. The solution consists of a phase selection with possible multiplicities and an initial guess for the mole fractions and mole numbers. The corrector method verifies whether all suggested phases are active or not using a Newton-Lagrange algorithm, possibly finds a new phase selection and passes over new mole fractions for each phase, i. e. sampling points, to the corrector method. This procedure is repeated several times. Finally, the solution with the lowest Gibbs energy resulting from the local optimization is selected and passed over to the next step.

Stoichiometric phases are handled in a specific way: As their Gibbs energy for a given temperature and pressure is not a function of a composition but consists of one single point only, their mole fractions are fixed. If the predictor method chooses a stoichiometric phase, the corresponding point lies exactly on the calculated tangential hyperplane in the space spanned by the system components. Within the local minimization procedure, only the mole numbers of the stoichiometric phases are varied, while their mole fractions are kept constant.

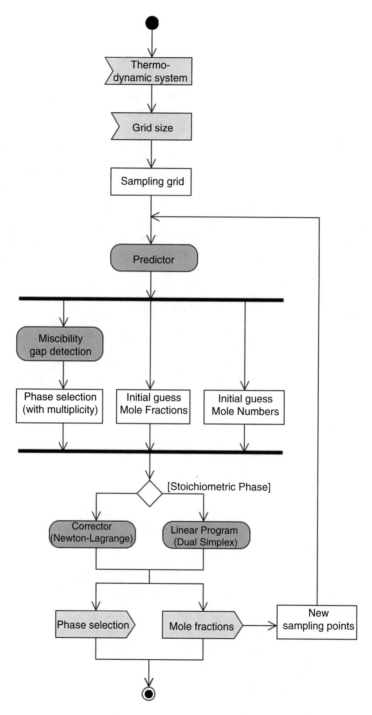

Fig. 5 Visualization of the predictor-corrector method including miscibility gap detection: After the determination of a the thermodynamic system and an initial grid of a pre-defined size, the

5 Results

5.1 Gibbs Free Energy Minimization Using BePhaSys

In this section, we briefly summarize some example results obtained by using the BePhaSys software. We consider a chemical system consisting of three components only: bismuth, lead, and tin (Bi-Pb-Sn). In principle, in this system, five mixture phases may occur, each with three phase constituents. At higher temperatures, a gas phase (G) evolves with eight phase constituents. The potential mixture phases are a liquid phase (L) as well as several solid phases with body-centered-tetragonal (BCT), face-centered-cubic (FCC), hexagonal-close-packed (HCP), and rhombohedral (RHOMBO) structure, respectively. A special feature of the system is the appearance of miscibility gaps in the chemical equilibrium. The phases BCT and HCP may be duplicated, and the RHOMBO phase may even be triplicated.

Tables 2 and 3 show the results for the Bi-Pb-Sn system for different temperatures and molar amounts of the (input) system components. Table 3 contains test cases with miscibility gaps. The pressure was always set to 1 bar. The results indicated are the Gibbs free energy of the system in chemical equilibrium and the phase selection. They are in full coincidence with the results of the reference software ChemApp.

The number of grid points M used for the discretization of the minimization problem for the phase selection (predictor method) was set to 10, which was a sufficient and reliable choice for most of the systems treated. In the case of a gas phase, even $M = 1$ was sufficient because in that case, the Gibbs function is always convex and dominated by its linear part, see the appendix for more details.

5.2 Calculation of Two-Dimensional Phase Diagrams: Interpolation and Parallelization

The calculation of phase diagrams requires phase selections for various input values of b. In this paper we confine ourselves to chemical systems consisting of three components. Phase diagrams can then be visualized as two-dimensional isothermal equilateral triangles, representing the 3D unit simplex.

In order to reduce computation time, different parallelization techniques have been applied. The phase diagram is divided into squares, which we call *quads*. In

---

Fig. 5 (continued) predictor method suggests a phase selection and miscibility gaps, which are confirmed or disproved by the miscibility gap detector. The phase selection is verified by the corrector method which possibly find new active phases. New mole fractions for each phase are passed over to the predictor method as new sampling points if the number of iterations (here: 10) has not been exceeded yet. Otherwise, the resulting eleven solutions (including the first solution resulting from the predictor) are passed over to the local optimization. Published with kind permission of ©Fraunhofer SCAI 2016. All Rights Reserved

Table 2 Results of standard test cases for the system Bi-Pb-Sn, to evaluate the accuracy of the BePhaSys algorithm: Indicated are the temperature T in K, the molar amount of substance vector b, the number of grid points M used by the predictor method, the resulting phase selection, and the Gibbs function value in equilibrium in kJ

System	Special feature	T [K]	b [mol]	M	Phase selection	Gibbs function value in equilibrium [kJ]
Bi-Pb-Sn		400	$(0.3, 0.5, 0.2)^T$	10	L, FCC	-25.8606
		373.15	$(0.5, 0.4, 0.1)^T$	10	L, HCP, RHOMBO	-23.3960
	Gaseous	3000	$(0.3, 0.5, 0.2)^T$	1	G	-432.8460
	Ternary int.	373.15	$(0.3, 0.3, 0.4)^T$	10	L, BCT, HCP	-22.3130
		373.15	$(0.6, 0.1, 0.3)^T$	10	L, BCT, RHOMBO	-21.4069
		373.15	$(0.45, 0.35, 0.2)^T$	10	L	-22.9804
	No Pb	373.15	$(0.5, 0.5)^T$	10	BCT, RHOMBO	-20.4449
		373.15	$(0.6, 0.25, 0.15)^T$	10	L, RHOMBO	-22.5318
		373.15	$(0.5, 0.4, 0.1)^T$	10	L, HCP, RHOMBO	-23.3967
		373.15	$(0.5, 0.48, 0.02)^T$	10	HCP, RHOMBO	-23.9036
	No Sn	373.15	$(0.7, 0.3)^T$	10	HCP, RHOMBO	-22.9168
		373.15	$(0.33, 0.64, 0.03)^T$	10	HCP	-24.6799
		373.15	$(0.4, 0.5, 0.1)^T$	10	L, HCP	-23.9196
		373.15	$(0.3, 0.3, 0.4)^T$	10	L, BCT, HCP	-22.2740
		373.15	$(24, 17, 59)^T$	10	L, BCT	$-2.1213 \cdot 10^3$
		373.15	$(25, 70, 5)^T$	10	FCC, HCP	$-2.4790 \cdot 10^3$
		373.15	$(0.3, 0.5, 0.2)^T$	10	BCT, HCP	-23.6794
		373.15	$(0.09, 0.11, 0.8)^T$	10	BCT, HCP	-20.4320
		373.15	$(0.07, 0.13, 0.8)^T$	10	BCT, FCC, HCP	-20.4993
		373.15	$(0.1, 0.4, 0.5)^T$	10	BCT, FCC	-22.3116
		373.15	$(0.2, 0.75, 0.05)^T$	10	FCC	-24.8987
		373.15	$(0.11, 0.8, 0.09)^T$	10	BCT, FCC	-24.7149
	No Bi	373.15	$(0.55, 0.45)^T$	10	BCT, FCC	-22.2665
	Only Pb	373.15	$(1)^T$	10	FCC	-24.4161
	No Bi	373.15	$(0.91, 0.09)^T$	10	BCT, FCC	-24.1712
	Only Sn	373.15	$(1)^T$	10	BCT	-19.3368
	Only Bi	373.15	$(1)^T$	10	RHOMBO	-21.3949
		373.15	$(0.26, 0.65, 0.09)^T$	10	BCT, FCC, HCP	-24.5264
		373.15	$(0.24, 0.67, 0.09)^T$	10	FCC, HCP	-24.5845
		373.15	$(0.23, 0.68, 0.09)^T$	10	FCC	-24.6121
		373.15	$(0.16, 0.44, 0.4)^T$	10	BCT, FCC	-22.8175
		373.15	$(0.28, 0.635, 0.085)^T$	10	BCT, FCC, HCP	-24.5003
		373.15	$(0.3, 0.63, 0.07)^T$	10	BCT, FCC, HCP	-24.5387
		373.15	$(0.37, 0.56, 0.07)^T$	10	HCP	-24.2749

The pressure was always 1 bar

Table 3 Results of standard test cases for the system Bi-Pb-Sn with miscibility gaps, to evaluate the accuracy of the BePhaSys algorithm: Indicated are the temperature T in K, the molar amount of substance vector b, the number of grid points M used by the predictor method, the resulting phase selection, and the Gibbs function value in equilibrium in kJ

System	T [K]	b [mol]	M	Phase selection	Gibbs function value in equilibrium [kJ]
Bi-Pb-Sn	373.15	$(0.01, 0.01, 0.98)^T$	10	BCT	−19.4951
(miscibility	373.15	$(0.05, 0.949, 0.001)^T$	10	BCT	−23.1661
gaps)	373.15	$(0.47, 0.27, 0.26)^T$	10	BCT double	−21.8305
	373.15	$(0.32, 0.59, 0.09)^T$	10	BCT double	−23.4033
	373.15	$(0.24, 0.01, 0.75)^T$	10	BCT double	−19.7494
	373.15	$(0.001, 0.25, 0.749)^T$	10	BCT double	−20.2170
	373.15	$(0.001, 0.06, 0.939)^T$	10	HCP	−18.8204
	373.15	$(0.01, 0.94, 0.05)^T$	10	HCP	−23.8138
	373.15	$(0.3, 0.55, 0.15)^T$	10	HCP double	−23.9574
	373.15	$(0.01, 0.215, 0.775)^T$	10	HCP double	−19.8212
	373.15	$(0.48, 0.26, 0.26)^T$	10	HCP double	−21.8464
	373.15	$(0.01, 0.03, 0.96)^T$	10	RHOMBO	−17.7189
	373.15	$(0.01, 0.5, 0.49)^T$	10	RHOMBO double	−20.7419
	373.15	$(0.45, 0.53, 0.02)^T$	10	RHOMBO double	−22.6647
	373.15	$(0.25, 0.01, 0.74)^T$	10	RHOMBO double	−18.5383
	373.15	$(0.015, 0.08, 0.905)^T$	10	RHOMBO double	−18.0837
	373.15	$(0.8, 0.1, 0.1)^T$	10	RHOMBO triple	−21.3448
	373.15	$(\frac{1}{3}, \frac{1}{3}, \frac{1}{3})^T$	10	RHOMBO triple	−20.9710
	373.15	$(0.03, 0.9, 0.07)^T$	10	RHOMBO triple	−23.3690

The pressure was always 1 bar

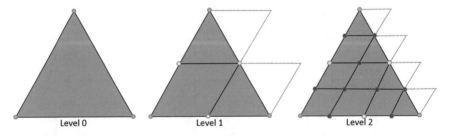

Fig. 6 Division of a diagram into quads and subquads up to level 2: At level 0, there is only one key points, marked with green dots. Corners of newly formed subquads of the next level are colored differently. Published with kind permission of ©Fraunhofer SCAI 2016. All Rights Reserved

turn, these quads are recursively divided into subquads forming the next refinement level as shown in Fig. 6. The vertices of the quads and subquads correspond to grid points of the phase diagram. The key points of subquads, i. e. their upper left vertices, are evaluated in parallel. In BePhaSys, a phase diagram is always drawn up to a certain level ℓ, which corresponds to the required resolution. In each dimension, $2^\ell + 1$ points have to be computed, while points also residing in previous quads are not re-computed, i. e. the number of quads increases exponentially. Figure 6 shows the initial quad of level 0 consisting of four points. At Level 1, this quad is divided into four subquads. In Level 2, each of these four subquads is again subdivided into four smaller subquads, resulting in 16 subquads.

Parallelization is done by concurrent computing of the newly created points of each next level. For the OpenMP approach, point computations are started on all available cores, which operate on the shared memory. In the MPI solution, a dedicated host node generates the quads of the next level, distributes the points to the compute nodes and collects their results. Each node computes the phase selection for the point of the minimal Gibbs energy. This approach needs a load balancing to ensure maximal throughput. Since the computation time of the points differ significantly, a self-adapting farming strategy is implemented: Each time when the host receives the result from a node, this node is fed with the next vacant work package.

In order to compute a phase diagrams, *Zero Phase Fraction (ZPF) Lines* [18] are essential, especially for phase diagrams which relate to three or more components. In contrast to other software, BePhaSys does not follow the ZPF lines but focuses on an efficient detection of the phase boundaries. Thereby, not every point within a phase diagram needs to be evaluated. Whenever the phase boundaries are known, all areas of coexisting phases within the diagram are defined as well.

In the first step BePhaSys scans the outer borders of the phase diagram. The border lines of Fig. 7 AB, BC, and CA are divided recursively to detect the ZPF points, where phases change. These processes are computed concurrently per line and get refined recursively until the points are determined with the required accuracy. The results are the ZPF points on the border, where phase changes occur (e. g. point X). These points are starting points for the ZPF lines to be found in further steps of the algorithm.

In the next step the interior of the phase diagram is scanned by iterative quad refinements as described before. The results can partially be used to detect areas

Fig. 7 Interpolation of the phase boundaries: The phases α, β, and γ are shown for a system with components A, B, and C. The ZPF line from point X to point Y is interpolated via the endpoints of 2-phase selections. Published with kind permission of ©Fraunhofer SCAI 2016. All Rights Reserved

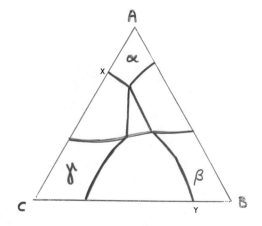

of the same phase selection. Whenever a point results in a three-phase area (e.g. in the center triangle of Fig. 7), the corner points of the area (i.e. the chemical compositions of the three phases in equilibrium) determine an area with a constant phase combination. That means, that all points within this area will result in the same phase selection. This information is used by the BePhaSys algorithm in two ways:

1. Each new point is checked for whether it lies in one of the triangles found so far. In this case, its phase selection is known, such that no further computation is required. Furthermore this point needs not be refined in a later level.
2. The three corner points of the triangle form intersection points of a ZPF line. These points are stored together with the border line points of the first step. However, in order to decide, which nodes can be connected and which course the lines follow, additional information is needed.

When a point results in a two-phase area, the points defining the chemical compositions in equilibrium of the two phases (the end points of so-called *tie-lines*) must lie on a ZPF line. In Fig. 7 such a point is found between points X and Y. To determine the shape of the ZPF line between X and Y, this phase boundary can be approximated via interpolation. In this work, the Hermite, logarithmic, and quadratic interpolation were studied. The Hermite and logarithmic interpolation provided inadequate results. In contrast, the quadratic interpolation was reliable for this task.

Finally, the algorithm checks, whether the lines found so far form a closed area. Lines can either be side lines of triangles or ZPF lines being determined by interpolation. Whenever such an area can be closed, it is handled like the three-phase triangles above. Thus, points residing in these areas need neither Gibbs minimization nor refinement.

Figure 8 juxtaposes two phase diagrams computed with BePhaSys. Figure 8a was computed without quadratic interpolation. The ZPF lines are approximated by the set of the endpoints of tie-lines. Here, the phase boundaries are not complete,

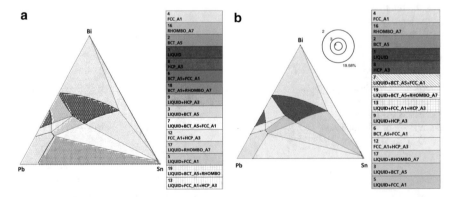

Fig. 8 Comparison of phase diagrams of the system Bi-Pb-Sn at 400 K and 1 bar, created by BePhaSys (**a**) without and (**b**) with interpolation. Published with kind permission of ©Fraunhofer SCAI 2016. All Rights Reserved

Table 4 Computation time (in sec) of OpenMP and MPI for 6 levels

Cores	OpenMP	MPI
1	106	106
2	64.4	110
4	44.5	38.6
6	38.5	25.3
8	36.6	19.4
10	35.2	16.1
12	34.2	13.5

Table 5 MPI computation time (sec) without and with quadratic interpolation for 7 levels

Cores	Without interpolation	With interpolation
2	433.4	9.0
4	143.7	4.7
6	88.0	3.9
8	63.4	3.6
10	50.0	3.6
12	41.3	3.4

but the course of the lines become visible. Figure 8b was created using quadratic interpolation, which fills the gaps between the end-points and forms closed ZPF lines. This produces small deviations to the original ZPF lines, but these are so small that they can be neglected for the visualization purpose. Most important, the gain in efficiency was significant.

Tables 4 and 5 show the wall clock time of sample calculations of Bi-Pb-Sn for 6 levels on a compute cluster with six dual-core Xenon X5650 nodes, running on 2.67 GHz.

In case of two cores, the computation time with MPI is slightly higher than with one node. This is due to the fact that the usage of two cores does not lead to parallelism because one node is the host and the other one is the worker. For OpenMP, speedup is limited for more than six nodes due to the relatively small problem size of a few thousand points to be computed. On the other hand, the speedup of MPI is nearly linear.

Table 5 compares the computation time of MPI without and with quadratic interpolation for 7 levels. Without interpolation the computational efficiency climbs over 80%, with a speedup of more than 10 on 12. With interpolation the numerical effort is drastically reduced, with the effect that parallel efficiency is slowing down. The sequential solution with quadratic interpolation required 9 s only, whereas with MPI the phase diagram can be computed in less than 4 s. This opens a way to higher dimensional cases.

Acknowledgements The authors would like to acknowledge the financial support supplied by the German Federal Ministry for Education and Research *(Bundesministerium für Bildung und Forschung, BMBF)* within the support framework *KMU-innovative grant No. 01IS11030B)*.

Appendix: The Gibbs Free Energy Function

Each phase specific Gibbs free energy function $G^{(i)}$ is the sum of a reference part $G_{\text{ref}}^{(i)}$, an ideal part $G_{\text{id}}^{(i)}$, and an excess part $G_{\text{ex}}^{(i)}$:

$$G^{(i)}(x^{(i)}) := G_{\text{ref}}^{(i)}(x^{(i)}) + G_{\text{id}}^{(i)}(x^{(i)}) + G_{\text{ex}}^{(i)}(x^{(i)}). \tag{16}$$

The reference and the ideal part are given by:

$$G_{\text{ref}}^{(i)}(x^{(i)}) := \sum_{k=1}^{K^i} G_k^{0,i} x_k^i, \tag{17}$$

$$G_{\text{id}}^{(i)}(x^{(i)}) := RT \sum_{k=1}^{K^i} x_k^i \ln x_k^i. \tag{18}$$

Here, $R = 8.31451$ J/(mol K) is the ideal gas constant and T the temperature in K. In the non-ideal part, binary interactions of phase constituents are described by the terms B^i, ternary interactions by the terms T^i and quaternary interactions by the terms Q^i (for a phase i):

$$G_{\text{ex}}^{(i)}(x^{(i)}) := \sum_{k_1 < k_2}^{K^i} B^i(\{k_1, k_2\}, x^{(i)}) + \sum_{k_1 < k_2 < k_3}^{K^i} T^i(\{k_1, k_2, k_3\}, x^{(i)})$$

$$+ \sum_{k_1 < k_2 < k_3 < k_4}^{K^i} Q^i(\{k_1, k_2, k_3, k_4\}, x^{(i)}), \tag{19}$$

with

$$B^i(\{k_1, k_2\}, x^{(i)}) := \prod_{j=1}^{2} x_{k_j}^i \sum_{\nu=1}^{m^i(k_1,k_2)} L_{k_1 k_2}^{\nu,i} \left(x_{k_1}^i - x_{k_2}^i \right)^{\nu-1}, \quad m^i(k_1, k_2) \in \mathbb{N},$$

$$\tag{20}$$

$$T^i(\{k_1, k_2, k_3\}, x^{(i)}) := \prod_{j=1}^{3} x_{k_j}^i \left(\sum_{j=1}^{3} L_{k_j}^i x_{k_j}^i + \frac{1}{3} \sum_{j=1}^{3} L_{k_j}^i \left(1 - \sum_{j=1}^{3} x_{k_j}^i \right) \right),$$

$$\tag{21}$$

$$Q^i(\{k_1, k_2, k_3, k_4\}, x^{(i)}) := \prod_{j=1}^{4} x_{k_j}^i L_{k_1 k_2 k_3 k_4}^i. \tag{22}$$

The binary excess part of the phase specific Gibbs function (20) is described here by the *Redlich–Kister–Muggianu polynomials (RKMP), cf. [15]*. Interactions of higher order could also be involved in the model but are neglected here.

If $K^i = 1$ for some $i \in \{1, \ldots, n\}$, it holds:

$$G^{(i)}(x^{(i)}) = G_1^{0,i},$$

(Gibbs energy for a stoichiometric phase).

The coefficients $G_k^{0,i}$, $i = 1, \ldots, n$, $k = 1, \ldots, K^i$ are given by:

$$G_k^{0,i}(T) = A_k^i + B_k^i T + C_k^i T \ln T + D_k^i T^2 + E_k^i T^3 + F_k^i T^{-1}$$
$$+ G_k^i T^{i_G} + H_k^i T^{i_H} + I_k^i T^{i_I} + J_k^i T^{i_J}, \tag{23}$$

where $A_k^i, \ldots, M_k^i$; $i_G, \ldots, i_M \in \mathbb{R}$, $i = 1, \ldots, n$, $k = 1, \ldots, K^i$ are coefficients taken from a material database [4]. The pressure dependence of the Gibbs energy can be involved by additional additive terms.

The RKMP coefficients $L_{(k_1, \ldots, k_I)}^i$, $i = 1, \ldots, n$, $k_1, \ldots, k_I \in \{1, \ldots, K^i\}$, $k_1 < \ldots < k_I$, where I is the interaction order, are given by:

$$L_{(k_1, \ldots, k_N)}^i(T, P) = (A_k^i)_L + (B_k^i)_L T + (C_k^i)_L T \ln T + (D_k^i)_L T^2 \tag{24}$$

$$+ (E_k^i)_L T^3 + (F_k^i)_L T^{-1} + (G_k^i)_L P + (H_k^i)_L P^2, \tag{25}$$

where the coefficients $(A_k^i)_L, \ldots, (H_k^i)_L \in \mathbb{R}$, $i = 1, \ldots, n$, $k = 1, \ldots, K^i$ were taken from a material database [4] as well. The pressure dependence can be involved by additional additive terms in this case as well.

Please note that the term $\ln P$ has to be added in the case of the presence of a gaseous phase g:

$$G^{(g)}(x^{(g)}) = G_{\text{ref}}^{(g)}(x^{(g)}) + G_{\text{id}}^{(g)}(x^{(g)}) + RT \ln P. \tag{26}$$

References

1. R.E. Balzhiser, M.R. Samuels, J.D. Eliassen, *Chemical Engineering Thermodynamics. The Study of Energy, Entropy and Equilibrium* (Prentice-Hall, Englewood Cliffs, 1972)
2. F.P. Boynton, Chemical equilibrium in multicomponent polyphase systems. J. Chem. Phys. **32**, 1880–1881 (1960)
3. S.-L. Chen, K.-C. Chou, Y.-A. Chang, On a new strategy for phase diagram calculation, CALPHAD **17**, 237–250 (1993)
4. A. Dinsdale, A. Watson, A. Kroupa, et al. *COST Action 531, Lead-Free Solders: Atlas of Phase Diagrams for Lead-Free Soldering*, COST – European Cooperation in Science and Technology (2008)
5. J.H. Dluzniewski, S.B. Adler, Calculation of complex reaction and/or phase equilibria problems. Inst. Chem. Eng. Symp. Ser. **4**, 21–26 (1972)

6. M. Emelianenko, Z.-K. Liu, Q. Du, A new algorithm for the automation of phase diagram calculation. Comput. Mater. Sci. **35**, 61–74 (2006)
7. G. Eriksson, Thermodynamic studies of high temperature equilibria. Acta Chem. Scand. **25**, 2651–2658 (1971)
8. G. Eriksson, Quantitative equilibrium calculations in multiphase systems at high temperatures, with special reference to the roasting of chalcopyrite, Ph.D. thesis, University of Umeå, Sweden, 1975
9. G. Eriksson, K. Hack, S. Petersen, Chemapp—a programmable thermodynamic calculation interface, in *Werkstoffwoche '96, Symposium B: Simulation, Modellierung, Informationssysteme*, ed. by J. Hirsch (DGM Informationsgesellschaft Verlag, 1997), pp. 47–51
10. G. Eriksson, E. Rosen, Thermodynamic studies of high temperature equilibria. Chem. Scr. **4**, 193–194 (1973)
11. G. Eriksson, W.T. Thompson, A procedure to estimate equilibrium concentrations in multi-component systems and related applications. CALPHAD **13**, 389–400 (1980)
12. P.J.F. Groenen, W.J. Heiser, The tunneling method for global optimization in multidimensional scaling. Pychometrika **61**, 529–550 (1996)
13. R. Hettich, P. Zencke, *Numerische Methoden der Approximation und Semi-Infiniten Optimierung* (Teubner, Stuttgart, 1982)
14. B.R. Kubert, S.E. Stephanou, Extension to multiphase systems of the rand method for determining equilibrium compositions, in ed. by G.H. Bahn, E.E. Zukoski, *Kinetics, Equilibria and Performance of High Temperature Systems* (Buttherworth, London, 1960) pp. 166–170
15. J.N. Lalena, D.A. Cleary, M.W. Weiser, *Principles of Inorganic Materials Design* (Wiley, Hoboken, 2010)
16. A.V. Levy, S. Gomez, The tunneling method applied to global optimization, in *Numerical Optimization*, ed. by P.T. Boggs, R.H. Byrd, R.B. Schnabel (SIAM, Philadelphia, 1985), pp. 213–244
17. R. Lougee-Heimer, The common optimization interface for operations research. IBM J. Res. Dev. **47**, 57–66 (2003)
18. J.E. Morral, Two-dimensional phase fraction charts. Scr. Metall. **18**, 407–410 (1984)
19. MPI, Message Passing Interface. http://www.mcs.anl.gov/research/projects/mpi
20. D.V. Nichita, S. Gomez, E. Luna, Multiphase equilibria calculation by direct minimization of gibbs free energy with a global optimization method. Comput. Chem. Eng. **26**, 1703–1724 (2002)
21. D.V. Nichita, S. Gomez, E. Luna, Phase stability analysis with cubic equations of state by using a global optimization method. Fluid Phase Equilib. **194–197**, 411–437 (2002)
22. J. Nocedal, S.J. Wright, *Numerical Optimization* (Springer, Berlin, 2006)
23. R.C. Oliver, S.E. Stephanou, R.W. Baier, Calculating free energy minimization. Chem. Eng. **69**, 121–128 (1962)
24. OpenMP, Portable Shared Memory Parallel Programming. http://openmp.org
25. S. Petersen, K. Hack, The thermochemistry library chemapp and its applications. Int. J. Mater. Res. **98**, 935–945 (2007)
26. W.B. White, S.M. Johnson, G.B. Danzig, Rand corporation: chemical equilibrium in complex mixtures. J. Chem. Phys. **28**, 751–755 (1957)

LC-GAP: Localized Coulomb Descriptors for the Gaussian Approximation Potential

James Barker, Johannes Bulin, Jan Hamaekers, and Sonja Mathias

1 Introduction

An important problem in computational materials science is the development of efficient and accurate *atomic potentials*. For an N-particle atomic system, an atomic potential (or simply *potential*) is a function $E^{(N)} : \mathbb{R}^{3N} \times \mathbb{Z}^N \to \mathbb{R}$ that maps from the set of three-dimensional Cartesian coordinates locating the atoms in that system, and their corresponding atomic numbers, to a potential energy.

The most accurate possible potential for any given system can be obtained from an (analytic) wavefunction solution to Schrödinger's equation [6], and all other potentials can be viewed as approximations to the same. Such solutions must be approximated numerically in all but the simplest of cases, a process which becomes computationally expensive for even relatively small systems. A variety of *first-principle* methods have been developed for this task, including most notably the density functional theory (DFT) [19]; these methods can be used to obtain potentials displaying high levels of accuracy, but are nevertheless always computationally expensive.

By contrast, empirical potentials are parametrized functions, with a fixed functional form that is usually physically motivated [12]. The parameters of an empirical potential must be fitted against the properties of a particular material, which can be

J. Barker (✉)
Fraunhofer Institute for Algorithms and Scientific Computing SCAI, Schloss Birlinghoven, 53757 Sankt Augustin, Germany

Institute for Numerical Simulation, Rheinische Friedrich-Wilhelms-Universität Bonn, Wegelerstr. 6, 53115 Bonn, Germany
e-mail: james.barker@scai.fraunhofer.de

J. Bulin • J. Hamaekers • S. Mathias
Fraunhofer Institute for Algorithms and Scientific Computing SCAI, Schloss Birlinghoven, 53757 Sankt Augustin, Germany

© Springer International Publishing AG 2017
M. Griebel et al. (eds.), *Scientific Computing and Algorithms in Industrial Simulations*, DOI 10.1007/978-3-319-62458-7_2

25

a difficult task requiring a level of intuition; once fitted, however, the evaluation of an empirical potential for an arbitrary atomic configuration is comparatively cheap. Although it is difficult to accurately reproduce material properties other than those to which the potential was fitted, limiting their use in general cases, their applicability to the problem of exploring the energy hypersurface immediately surrounding an atomic configuration makes them useful for simulations.

In recent years, machine learning (ML) approaches have also been applied to the problem of potential generation. Such approaches can capture, or "learn" information from a wider range of materials than standard empirical potentials. The resulting ML potentials are almost as fast to evaluate as empirical potentials, and retain acceptable accuracy when used to predict the energy of a larger variety of materials.

Notable ML algorithms that have already been used for generating potentials include kernel ridge regression [7, 8, 16, 17] and multilayer neural networks [10]. Another promising approach is the Gaussian approximation potential (GAP) by Bartók et al. [2], which will be explained in detail in Sect. 2. This potential requires the construction of intermediate representations of the *atomic neighborhoods* surrounding each atom in the system; the choice of this representation is crucial for the performance of the GAP. Bartók originally proposed a modified bispectrum method for producing descriptors of mono-species crystalline environments. Although this approach performs well for semiconductors with respect to accuracy, the nature of the representation as coefficients of an expansion in spherical harmonics makes it costly to evaluate. Subsequently, Bartók et al. replaced the bispectrum representation with an approach called the *SOAP kernel*, which involves expanding a density function associated with an atomic environment in spherical harmonics and to directly define the similarity between any two such density functions [1].

Here, we present an alternative approach to the encoding of atomic environments, which we refer to as the *localized Coulomb* (LC) representation. This representation is based on the Coulomb matrix, originally proposed by Rupp et al. [17], and its derivative, the sorted Coulomb matrix [10, 18]. We combine the LC representation with the Gaussian approximation potential to obtain a new method for the creation and evaluation of families of atomic potentials; we term this method *LC-GAP*.

The remaining article is organized as follows. In Sect. 2 we briefly summarize the basics for machine learning for potential energy surfaces and describe our new Coulomb matrix based descriptors in particular in Sect. 2.2. In Sect. 3 we give numerical results for several datasets and additionally discuss the distribution of individual atomic contributions in Sect. 3.3. We conclude with some remarks in Sect. 4.

2 Potential Energy Prediction Through Machine Learning

As stated above, our goal is to produce a function $E^{(N)} : \mathbb{R}^{3N} \times \mathbb{Z}^N \to \mathbb{R}$ that approximates the potential energy for an arbitrary unknown atomic system $X^* \in \mathbb{R}^{3N} \times \mathbb{Z}^N$. Given a collection of atomic systems of size N and their energies,

$\{(X_i, E_i)\}_{i=1}^M$ with $X_i \in \mathbb{R}^{3N} \times \mathbb{Z}^N$, an arbitrary ML algorithm could be used to produce, or *learn* the function $E^{(N)}$. However, such an approach will not scale well to larger systems; the dimension of $E^{(N)}$ grows linearly in the number of atoms, and the so-called "curse of dimensionality" implies that the computational cost of learning will grow exponentially in the same.

Additionally, the use of raw representations X_i as inputs for an ML process is problematic from a physical standpoint. The potential energy of any given molecule is necessarily invariant under translation and orthogonal transformation; it is also invariant under permutation of the order in which atoms are described [19]. A raw representation obeys none of these restrictions; therefore, a single unique system can be described by uncountably many raw representations. This offers a significant obstacle to the learning and prediction abilities of any ML algorithm.

The Gaussian approximation potential (GAP) framework of Bartók et al. addresses both these problems [2]. The first, by introducing an ansatz that can be used to limit the dimensionality of a representation, and the second, by using more sophisticated representations that maximize invariance. In this section, we briefly recap the GAP framework, and present some alternative, simpler, representations of atomic systems.

2.1 The GAP Framework and Gaussian Process Regression

The fundamental assumption of the GAP framework is the *atomic decomposition ansatz*: that the potential energy of an atomic system can be written as the sum of energies attributed to each of its atoms, and these atomic energies depend only on a neighborhood of the corresponding atom. For systems without long range electrostatic interactions, this assumption can be motivated by the *nearsightedness of electronic matter* [9, 13]. This allows us to write the potential energy of an N-particle system $X \in \mathbb{R}^{3N} \times \mathbb{Z}^N$ as a sum of atomic energy contributions:

$$E^{(N)}(X) = \sum_{i=1}^N E_{\text{atomic}}\left(L_i^{(N)}(X)\right). \tag{1}$$

Here, the $L_i^{(N)}(X) : \mathbb{R}^{3N} \times \mathbb{Z}^N \to \mathbb{R}^d$ are *atomic neighborhood representation functions*. Each of these maps a full atomic system to a d-dimensional representation, or *descriptor*, of the atomic neighborhood of the ith atom. $E_{\text{atomic}} : \mathbb{R}^d \to \mathbb{R}$ is then an unknown function that assigns an atomic energy contribution to an atomic neighborhood representation. Given Eq. (1) above, the problem of learning the function $E^{(N)}$ reduces to the problem of learning the function E_{atomic}. This function is necessarily dependent on the form of the functions $L_i^{(N)}$ and the chosen dimensionality of their output d, both of which will be discussed in the next section.

As originally presented, the GAP framework uses the method of Gaussian process regression (GPR) to learn the function E_{atomic}, although other suitable ML algorithms could also be used. The input to the learning process is a *training set* of M systems, each containing some number of atoms $\{N_i\}_{i=1}^{M}$, and their known energies: $X = \left\{\left(X_i \in \mathbb{R}^{3N_i} \times \mathbb{Z}^{N_i}, E_i \in \mathbb{R}\right)\right\}_{i=1}^{M}$. The ability to learn from a training set containing systems with differing numbers of atoms is a direct result of learning the function E_{atomic} rather than E_{total}, the energy of the whole molecule. Additionally, a level of Gaussian observation noise $\varepsilon_i \sim \mathcal{N}\left(0, \sigma_i^2\right)$ is associated with each training set energy E_i, such that

$$E_i = \left(\sum_{j=1}^{N} E_{\text{atomic}}\left(L_j^{(N)}(X)\right)\right) + \varepsilon_i.$$

Gaussian process regression predicts the function E_{atomic} as a linear combination of positive-definite kernel functions $\kappa : \mathbb{R}^d \times \mathbb{R}^d \to \mathbb{R}$ centered on the atomic neighborhood representations of each atom in every system in the training set. That is, for some N-particle system X^* whose energy we wish to predict,

$$E_{\text{atomic}}\left(L_k^{(N)}(X^*)\right) = \sum_{i=1}^{M} \alpha_i \sum_{j=1}^{N_i} \kappa\left(L_j^{(N_i)}(X_i), L_k^{(N)}(X^*)\right),$$

where the α_i are chosen during the learning process. Again, note that the use of E_{atomic} rather than E_{total} allows the prediction of potential energies for systems with arbitrary numbers of atoms, regardless of the contents of the training set.

2.2 Localized Coulomb Matrix Descriptors

The choice of representation functions $L_i^{(N)} : \mathbb{R}^{3N} \times \mathbb{Z}^N \to \mathbb{R}^d$ is critical to the behavior of the GAP framework, both in terms of computational performance and in accuracy. The primary contribution of this paper is the introduction of three new such functions, which are described below.

The genesis of our work is the Coulomb matrix descriptor, due to Rupp et al. [17], which allows the representation of complete molecules as square matrices. Consider an N-atom system $X \in \mathbb{R}^{3N} \times \mathbb{Z}^N$, consisting of Cartesian coordinates $\mathbf{R}_1, \ldots, \mathbf{R}_N$ describing atom locations, and the atomic numbers $Z_1, \ldots, Z_N$ associated with those atoms. The Coulomb matrix is then an $N \times N$ matrix M, whose entries are given as

$$M_{ij} = \begin{cases} 0.5 Z_i^{2.4} & i = j, \\ \dfrac{Z_i Z_j}{\|\mathbf{R}_i - \mathbf{R}_j\|_2} & i \neq j. \end{cases} \tag{2}$$

By construction, the Coulomb matrix is invariant under translations and orthogonal transformations of the set of Cartesian coordinates $\{\mathbf{R}_1, \ldots, \mathbf{R}_N\}$. It is, however, not invariant under permutation of the indexing order of the coordinates and their associated atomic numbers. To improve the handling of permuted systems, the rows and columns of the Coulomb matrix can be sorted according to their respective norms (which are identical, as the matrix is symmetric), resulting in the *sorted Coulomb matrix* [10].

The sorted Coulomb matrix is a *global descriptor* of an atomic system. As such, it can only represent molecules, and is inapplicable to infinitely-periodic crystal systems. However, it is possible to modify the Coulomb matrix into a *local descriptor*, encoding information only about the immediate neighborhood of an atom. Such a descriptor is then a candidate for unqualified use as an atomic representation function $L_i^{(N)}$ in the GAP framework, and can be applied to infinitely-periodic crystal systems as well as finite molecules.

Before we begin, it is useful to define a system of *local indices* to specify more clearly the structure of an atomic neighborhood around the ith atom in some system X. First, let p_1 indicate the ith atom itself. Then let K be the number of atoms that are located within some cutoff radius $r_{\text{cut}} > 0$ around $\mathbf{R}_i$, and let $p_2, \ldots, p_{K+1}$ specify these atoms. Finally, let m be the maximum neighborhood occupancy, which must be chosen in such a way that it is impossible to have greater than $m - 1$ atoms surrounding any atom in either the training set, or an atom in a system which will be predicted. Then let all indices $p_{K+2}, \ldots, p_m$ specify *dummy atoms*, which have arbitrary location and atomic number zero. The use of such dummy atoms is an established procedure [10], and serves to ensure that all descriptors produced in either the training or prediction process will be of the same size and therefore comparable.

With the aid of local indices, we now define the *localized Coulomb matrix* for the ith atomic neighborhood. First, similarly to Eq. (2) above, define the $m \times m$ matrix $M^{(i)}$, with entries given by

$$M_{jk}^{(i)} = \begin{cases} 0.5 Z_{p_j}^{2.4} & j = k, \\ \dfrac{Z_{p_j} Z_{p_k}}{\|\mathbf{R}_{p_j} - \mathbf{R}_{p_k}\|_2^{\alpha}} & \text{otherwise.} \end{cases} \tag{3}$$

Note that for notational simplicity, we do not explicitly indicate the dependency of the $(p_1, \ldots, p_m)$ local indices on the choice of i. Then let P be the permutation matrix such that $P_{1,1} = 1$ and

$$\left\| \left(P M^{(i)} P^T \right)_{j,*} \right\|_2 \geq \left\| \left(P M^{(i)} P^T \right)_{j+1,*} \right\|_2 \tag{4}$$

for $j = 2, \ldots, m-1$. That is, P reorders the rows and columns of $M^{(i)}$ such that those corresponding to the central atom p_1 become the upper- and leftmost, respectively, while permuting the remainder in descending order of norm. The permuted matrix $P M^{(i)} P^T$ is called $C_{\text{loc}}^{(i)}$ and is termed the *localized Coulomb matrix* for the ith neighborhood. Finally, the upper triangular entries of this matrix are packed in row-wise order into a vector $\mathbf{c}$ of size $d = \frac{m(m+1)}{2}$, and $L_i^{(N)}(X) := \mathbf{c}$ is the *localized Coulomb descriptor function*. Because the matrix is symmetric, the lower-triangular entries contain no further information.

One issue with this choice of representation function is that small changes to the state of the system that result in atoms entering or leaving the cutoff radius can cause large changes to the resulting descriptor. To reduce the significance of this effect, we modify the denominator to include distances from each atom to the central atom, creating the *decaying Coulomb matrix* $C_{\text{dec}}^{(i)} = P\hat{M}^{(i)}P^T$, with:

$$\hat{M}_{jk}^{(i)} = \begin{cases} 0.5Z_{p_j}^{2.4} & j = k = 1, \\ \dfrac{Z_{p_j}Z_{p_k}}{\left(\left\|\mathbf{R}_{p_1}-\mathbf{R}_{p_j}\right\|_2 + \left\|\mathbf{R}_{p_1}-\mathbf{R}_{p_k}\right\|_2 + \left\|\mathbf{R}_{p_j}-\mathbf{R}_{p_k}\right\|_2\right)^{\alpha}} & \text{otherwise.} \end{cases} \tag{5}$$

The permutation matrix is chosen in the same fashion as for the localized Coulomb matrix. Due to the introduction of the additional distance terms, the movement of atoms across the cutoff boundary cause significantly smaller changes, which may be lower than machine precision for some values of r_{cut} and α. The decaying Coulomb matrix can be packed into a vector of size $d = \frac{m(m+1)}{2}$ in the same manner as before.

Finally, we introduce a lower-dimensional descriptor, called the *reduced Coulomb matrix*, based in turn upon the decaying Coulomb matrix. Rather than using a permutation matrix, we require that the atoms $p_2, \ldots, p_{k+1}$ are indexed in such a way that $\|\mathbf{R}_{p_j} - \mathbf{R}_{p_1}\|_2 \leq \|\mathbf{R}_{p_{j+1}} - \mathbf{R}_{p_1}\|_2$ for $j = 1, \ldots, m-1$, similar to the approach in [18]. Then the descriptor is constructed by taking the first row and the diagonal of the matrix $\hat{M}^{(i)}$ in Eq. (5):

$$C_{\text{red}}^{(i)} = \left[\hat{M}_{11}^{(i)}, \hat{M}_{12}^{(i)}, \ldots, \hat{M}_{1m}^{(i)}, \hat{M}_{22}^{(i)}, \hat{M}_{33}^{(i)}, \ldots, \hat{M}_{mm}^{(i)}\right]. \tag{6}$$

This descriptor has dimensionality $2m - 1$, linear rather than quadratic in the maximum neighborhood occupancy, while still encoding all information about the pairwise interaction of the central particle with all its neighboring atoms.

3 Results

In order to evaluate the new atomic neighborhood functions presented in Sect. 2.2 above, we performed a series of numerical experiments over the QM7, QM7b, and GDB9 datasets. The QM7 dataset [3, 17] contains approximately 7100 biomolecules selected from the GDB13 database, each with up to seven heavy atoms,[1] and their atomization energies,[2] calculated at the PBE0 level of theory.

---

[1] All non-hydrogen atoms are considered heavy.

[2] The atomization energy is the potential energy of a molecule that has been adjusted by the combined potential energy of its isolated atoms [5].

Table 1 Decomposition of testing datasets by number of heavy atoms per molecule

Dataset	Number of heavy atoms									Total molecules
	1	2	3	4	5	6	7	8	9	
QM7	1	3	12	43	158	950	5998	–	–	7165
QM7b	1	3	12	43	158	953	6041	–	–	7211
GDB9-18K	3	5	9	31	130	618	3197	18,298	18,599	40,890

The QM7b dataset [3, 11] is a slightly expanded version of QM7, containing 7211 biomolecules with up to seven heavy atoms, including some with chlorine. As well as atomization energy, QM7b provides a total of thirteen extra properties per molecule (e.g. polarizability, HOMO and LUMO eigenvalues, excitation energies) calculated at different levels of theory (ZINDO, SCS, PBE0, GW). Finally, the GDB9 dataset [14] contains approximately 134,000 biomolecules (also selected from GDB13) containing up to nine heavy atoms, along with their atomization energies calculated at both the PM7 and B3LYP levels of theory. For computational convenience, we selected a subset of the full GDB9 database (calculated at PM7), comprised of all molecules with up to eight heavy atoms, and approximately 18,000 molecules with nine heavy atoms, chosen by stratification. We refer to this subset as the GDB9_18K dataset.

For each of the three datasets described above, we isolated subsets containing up to n heavy atoms for all $n \leq m$, where m is the maximum number of heavy atoms in any molecule in the dataset. We refer to these subsets as QM7_1, QM7_2, QM7_3, etc. A decomposition of each dataset by number of heavy atoms per molecule is given in Table 1.

Each experiment consisted of learning a potential function from one such subset, called the *training set*, and then using the resulting potential to predict the atomization energies of another (non-overlapping) subset, called the *test set*. All testing was performed using an implementation of the Gaussian approximation potential with one of the three localized Coulomb representation functions described in Sect. 2.2 above; we term this system *LC-GAP*. The underlying ML algorithm was Gaussian process regression, as described in [15].

For these experiments, we employed the Laplacian kernel

$$\kappa(x, y) = \sigma^2 \exp\left(\frac{-\|x - y\|_1}{l^2}\right),$$

which has shown good performance when applied to similar problems [7, 8]. The terms σ and l are kernel *hyperparameters*, the choice of which can have serious implications for the accuracy of the system.

To assess the quality of prediction for each test set, we used the mean absolute error (MAE), defined as

$$\text{MAE}\left(X^*\right) = \frac{1}{M} \sum_{m=1}^{M} \left| E_m^{\text{exact}} - E_m^{\text{pred}} \right|,$$

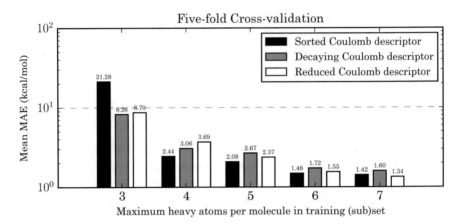

Fig. 1 Mean MAEs of the predicted atomization energies on various QM7 subsets obtained during fivefold cross-validation, employing LC-GAP potentials equipped with the different localizers described in Sect. 2.2. The precise values of α and r_{cut} used for each result can be found in Table 2. Published with kind permission of ©Fraunhofer SCAI 2016. All Rights Reserved

where M is the number of entries in the test set, the terms E_m^{exact} are the atomization energies obtained from the dataset, and E_m^{pred} the atomization energies obtained by prediction. This error metric is well-established in the statistics community for assessing the prediction accuracy of regression models [20], and has been used along with the QM7 dataset to benchmark other ML-based potentials.

3.1 Comparison of Descriptor Functions on QM7

We began by investigating the performance of the three localized Coulomb descriptors (sorted, decaying, and reduced) described in Sect. 2.2 above. To establish the ability of these descriptors to predict atomization energies of molecules with similar properties to those used for training, we performed fivefold cross-validation on each of the QM7 subsets containing at least three heavy atoms. Kernel hyperparameters (l, σ) were chosen by minimization of the negative log-likelihood on the training set. For each descriptor, the localization parameter α and the neighborhood cutoff r_{cut} were varied systematically.[3] A graph demonstrating the best results obtained on each subset can be found in Fig. 1; a table describing the values of α and r_{cut} for each can be found in Table 2.

For all but the smallest subset, the three tested descriptor types perform similarly, all producing results on the same order of magnitude and (for QM7_5 and higher) within 1 kcal/mol of MAE of each other. Although the localized Coulomb

---

[3]Here, we tested values $(\alpha, r_{cut}) \in \{3, 4, \ldots, 7\} \times \{3.0, 3.5, 4.0, 4.5, \ldots, 7.0\}$.

Table 2 Best choices of descriptor parameters (α, r_{cut}) for cross-validation over QM7, by descriptor type and data subset

Descriptor	QM7_3	QM7_4	QM7_5	QM7_6	QM7_7
Localized	(4.0, 5.0)	(3.0, 3.5)	(5.0, 3.5)	(3.0, 3.5)	(3.0, 3.5)
Decaying	(3.0, 4.0)	(3.0, 3.5)	(3.0, 3.5)	(4.0, 3.5)	(3.0, 4.5)
Reduced	(3.0, 6.0)	(5.0, 6.0)	(3.0, 5.5)	(4.0, 5.5)	(5.0, 6.5)

These choices are used to generate the results seen in Fig. 1

representation performs the best for the QM7_4, QM7_5 and QM7_6 subsets, it is interesting to note that, on the full QM7 dataset, the reduced Coulomb representation slightly outperforms both of the others. This is notable, given the lower dimensionality of that descriptor.

The cross-validation performance of each descriptor allows a direct comparison with previously published results in the literature. Hansen et al. compared several different ML potentials by using fivefold cross-validation on the complete QM7 dataset [7]. Employing non-localized kernel ridge regression coupled with randomly-sorted (global) Coulomb matrices, they report an MAE of 3.07 kcal/mol. Using multilayer neural networks, they obtained an MAE of 3.51 kcal/mol. Recently, Hansen et al. also reported an MAE of 1.5 kcal/mol, obtained through kernel ridge regression combined with the "Bag of Bonds" approach [8]. We note that both the sorted and reduced Coulomb descriptors outperform this result, with MAEs of 1.42 and 1.34 kcal/mol respectively. We conclude that all three of the tested descriptors offer competitive accuracy when compared to results in the literature (cf. Table 4).

It is also important to consider the application of an ML-based potential to datasets with potentially different properties than those used for training. To evaluate the behavior of the three LC-GAP localizer functions in such a scenario, we performed "upwards transferability" testing over the QM7 dataset. In these tests, an LC-GAP potential is trained using one of the QM7 subsets; it is then used to predict all entries in the remainder of the complete dataset, which have (by construction) at least one heavy atom more than any molecule in the training set. The results of these tests are given in Fig. 2.

The achieved MAEs are not as low as in the case of cross-validation, which is to be expected. However, potentials display consistent improvement as their training sets increase in complexity; potentials trained on the QM7_5 and QM7_6 datasets achieve MAEs below 10 kcal/mol in all cases. Unlike the cross-validation testing, the reduced Coulomb descriptor performs the best in all cases except for QM7_4. This suggests that despite its lower dimensionality, the reduced Coulomb descriptor embodies as much "useful information" as the higher-order descriptors, at a significantly lower cost with respect to both storage and computation.

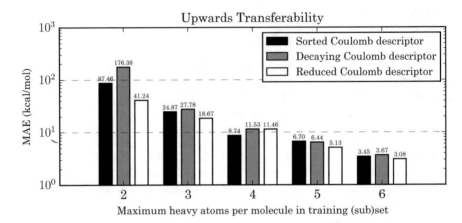

Fig. 2 MAEs of the predicted atomization energies on the remainder of the full QM7, using LC-GAP potentials trained on QM7 subsets, equipped with the different localizers described in Sect. 2.2. The precise values of α and r_{cut} used for each result can be found in Table 3. Published with kind permission of ©Fraunhofer SCAI 2016. All Rights Reserved

Table 3 Best choices of descriptor parameters (α, r_{cut}) for upwards transferability over QM7, by descriptor type and data subset

Descriptor	QM7_2	QM7_3	QM7_4	QM7_5	QM7_6
Localized	(4.0, 3.5)	(4.0, 3.5)	(4.0, 3.5)	(5.0, 3.5)	(4.0, 4.0)
Decaying	(4.0, 3.5)	(4.0, 3.5)	(4.0, 3.5)	(4.0, 4.0)	(5.0, 3.5)
Reduced	(4.0, 5.5)	(4.0, 5.5)	(4.0, 5.0)	(5.0, 5.0)	(5.0, 6.0)

These choices are used to generate the results seen in Fig. 2

Table 4 MAE for different algorithms when training and evaluating on the whole QM7 dataset

Algorithm	MAE [kcal/mol]
LC-GAP (localized Coulomb matrix)	1.42
LC-GAP (decaying Coulomb matrix)	1.60
LC-GAP (reduced Coulomb matrix)	1.34
Kernel ridge regression [7]	3.07
Multilayer neural network [7, 10]	3.51
Bag of Bonds [8]	1.5

3.2   Larger Datasets and Prediction of Multiple Properties

To further investigate the performance of LC-GAP, we repeated the cross-validation and upwards-transferability tests on the atomization energy figures provided by the remaining two datasets, QM7b and GDB9_18K, and their respective subsets. Due to the competitive performance of the reduced Coulomb descriptor, we did not perform further testing with the localized and decaying Coulomb descriptors; similarly, due to computational limitations, we did not perform cross-validation

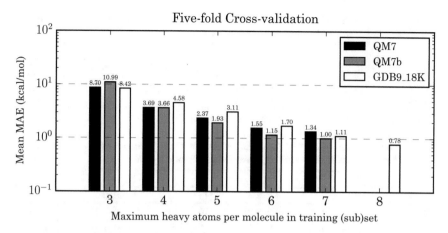

Fig. 3 Mean MAEs of the predicted atomization energies on subsets of QM7, QM7b, and GDB9_18K, obtained during fivefold cross-validation, using LC-GAP potentials equipped with the reduced Coulomb descriptor (6). The precise values of α and r_{cut} used for each result can be found in Table 5. Published with kind permission of ©Fraunhofer SCAI 2016. All Rights Reserved

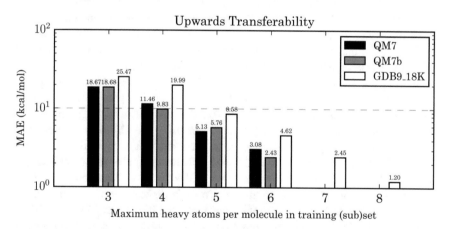

Fig. 4 MAEs of the predicted atomization energies on the remainder of the full dataset, using LC-GAP potentials trained on subsets, equipped with the reduced Coulomb descriptor (6). The precise values of α and r_{cut} used for each result can be found in Table 6. Published with kind permission of ©Fraunhofer SCAI 2016. All Rights Reserved

testing on the entire GDB9_18K dataset. As before, kernel hyperparameters were chosen by minimization of the negative log-likelihood of the training set, and a grid of descriptor parameters (α, σ) were used. Graphs displaying the results for each set of experiments can be found in Figs. 3 and 4, respectively, with the previously-obtained results for QM7 provided for comparison. The kernel hyperparameters and descriptor parameters used for each are given in Tables 5 and 6.

Table 5 Best choices of reduced Coulomb matrix descriptor parameters (α, r_{cut}) for upwards transferability, by data set and number of heavy atoms in training set

Dataset	3	4	5	6	7	8
QM7	(3.0, 6.0)	(5.0, 6.0)	(3.0, 5.5)	(4.0, 5.5)	(5.0, 6.5)	–
QM7b	(5.0, 3.5)	(5.0, 5.5)	(5.0, 5.5)	(5.0, 5.0)	(5.0, 5.5)	–
GDB9	(5.0, 4.5)	(5.0, 4.0)	(5.0, 5.0)	(6.0, 6.0)	(5.0, 4.0)	(5.0, 4.0)

These choices are used to generate the results seen in Fig. 3

Table 6 Best choices of reduced Coulomb matrix descriptor parameters (α, r_{cut}) for upwards transferability, by data set and number of heavy atoms in training set

Dataset	3	4	5	6	7
QM7	(4.0, 5.5)	(4.0, 5.0)	(5.0, 5.0)	(5.0, 6.0)	–
QM7b	(4.0, 5.5)	(4.0, 5.0)	(6.0, 5.0)	(5.0, 5.0)	–
GDB9	(5.0, 6.0)	(7.0, 4.0)	(5.0, 5.5)	(5.0, 4.0)	(6.0, 4.0)

These choices are used to generate the results seen in Fig. 4

In both cross-validation and upwards-transferability tests, the QM7b results perform similarly, although generally slightly better than the QM7 results. This is unsurprising, given that the datasets are similar, and QM7b contains only slightly more molecules than QM7. However, we note that during cross-validation over the entirety of the dataset, LC-GAP is able to predict molecules in QM7b to a mean MAE of approximately 1.00 kcal/mol, i.e., at chemical accuracy.

In contrast, LC-GAP underperforms on GDB9 when considered against both QM7 and QM7b at the same number of heavy atoms. In the upwards-transferability case, this can be explained by the fact that the vast majority of the molecules in GDB9 have eight or nine heavy atoms, increasing the difficulty of prediction by a potential trained on molecules containing fewer heavy atoms. For cross-validation, this is potentially a result of the slightly lower number of molecules in the smaller subsets (cf. Table 1). The results for GDB9 are nevertheless encouraging: a potential trained on molecules containing eight or fewer heavy atoms can predict the remainder of the dataset (i.e., all molecules with nine heavy atoms) with an MAE of only 1.20 kcal/mol, and cross-validation over the entire dataset produces a mean MAE of 0.78 kcal/mol—well above chemical accuracy.

As well as atomization energies, QM7b contains a number of different molecular properties for each molecule in the dataset, as described above. In [4], De et al. report prediction results for these properties obtained with the SOAP kernel. To investigate the applicability of the LC-GAP system to this problem, we repeatedly performed cross-validation over the entire QM7b dataset, considering each property in turn. Importantly, we did not repeat the process of hyperparameter and descriptor parameter selection for each property; rather, we used those indicated in the atomization-energy case as above. A summary of the obtained results and a comparison with those of De et al. can be found in Table 7; additionally, scatter plots indicating the distribution of both the predicted results and their absolute errors can be found in Fig. 5.

Table 7 Results obtained during cross-validation on the QM7b dataset for each of the 14 molecular properties contained therein, with comparison figures obtained from [4]

Property	Units	LC-GAP mean MAE	Reference data [4] MAE
E (PBE0)	kcal/mol	1.002 ± 0.022	0.92
E^*_{max} (ZINDO)	eV	1.717 ± 0.025	1.56
I_{max} (ZINDO)	Arbitrary	0.087 ± 0.003	0.08
HOMO (ZINDO)	eV	0.410 ± 0.009	0.13
LUMO (ZINDO)	eV	0.228 ± 0.003	0.10
E^*_{1st} (ZINDO)	eV	0.493 ± 0.007	0.18
IP (ZINDO)	eV	0.439 ± 0.010	0.19
EA (ZINDO)	eV	0.267 ± 0.004	0.13
HOMO (PBE0)	eV	0.284 ± 0.006	0.11
LUMO (PBE0)	eV	0.091 ± 0.001	0.08
HOMO (GW)	eV	0.355 ± 0.008	0.12
LUMO (GW)	eV	0.146 ± 0.003	0.12
α (PBE0)	$Å^3$	0.072 ± 0.002	0.05
α (SCS)	$Å^3$	0.086 ± 0.002	0.02

For all results, the reduced Coulomb localizer was used; kernel hyperparameters and descriptor parameters were equivalent to those used to produce the entries in graph for the full QM7b dataset

Without exception, the mean MAE results achieved by LC-GAP are on the same order of magnitude as those reported in the reference data; many agree to at least one significant figure, although in no cases are the results better than those reported; considering the slight difference between the results for atomization energy, this is unsurprising. The precise distribution of the results is harder to interpret; almost all of the lower-subrow scatter plots display similar patterns of errors to one another, regardless of the difference between the reference data accuracy and that achieved by LC-GAP. Nevertheless, these results indicate that the LC-GAP potential is, in general, quite capable of predicting molecular properties using hyperparameters and descriptor parameters that were not optimized for that specific property— potentially a significant benefit, since parameter optimization is significantly more computationally intensive than simple training and application of a potential.

3.3 Distribution of Individual Atomic Contributions

Through the atomic decomposition ansatz, the GAP approach predicts the total value of any given property for a molecule as the sum of many atomic contributions, one for each atom in that molecule. Although previous work has focused only on the total value of a property, the atomic contributions are readily available, and are worthy of interest.

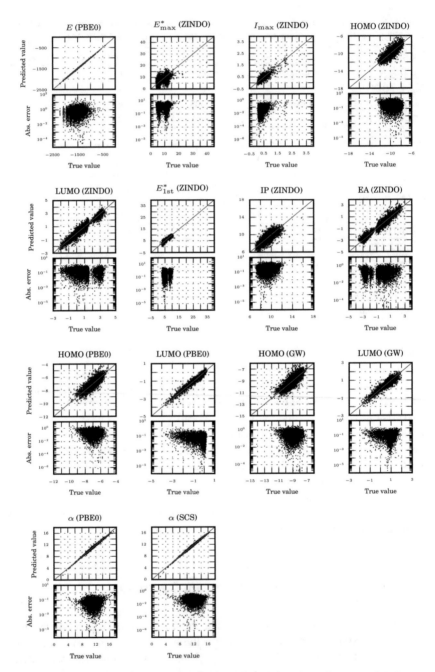

Fig. 5 Scatter plots displaying the distribution of all results obtained during cross-validation over the QM7b dataset for each of the 14 included properties, corresponding to the results in Table 7. In each main *row*, plots in the *upper subrow* display the obtained prediction values plotted against their actual values; plots in the *lower subrow* display the absolute error of the predictions, plotted against the actual value. Published with kind permission of ©Fraunhofer SCAI 2016. All Rights Reserved

To inspect the distribution of the atomic contributions for atomization energy, we repeated the computation for the cross-validation of the complete QM7 dataset with the reduced Coulomb descriptor. The full set of all atomic contributions for each cross-validation split was retained, and the values for each of the five splits were compiled; the resulting set can be considered as a collection of all atomic contributions obtained during prediction of that dataset. A histogram of the distribution of all atomic contributions for QM7 can be found in Fig. 6; additionally, a violin plot displaying the distribution of all atomic contributions (plotted by a kernel density estimator), as well as those of each of the five atom types contained in the dataset (hydrogen, carbon, nitrogen, oxygen, and sulfur) is given in the same figure.

The histogram displays a notable peak around approximately −70. This peak can be understood by inspection of the violin plots for the specific elements; a large majority of the contributions for hydrogen atoms cluster around this value. This suggests that many of the hydrogen atoms contribute similar amounts to

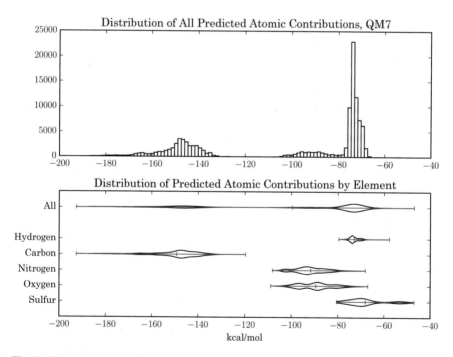

Fig. 6 Plots showing the distribution of all calculated individual atomic contributions towards the atomization energies of molecules in QM7, obtained during fivefold cross-validation. The *upper plot* is a standard histogram. The *lower plot* displays an indication of the density function of the entire collection of contributions (the *topmost object*), as well as indications of the individual density functions of all contributions by atoms of a particular element (remaining objects). *Straight lines* indicate range, with mean marked; *curves* are kernel density estimators. Published with kind permission of ©Fraunhofer SCAI 2016. All Rights Reserved

the atomization energy of their respective molecules. This is interesting, because hydrogen atoms are by far the most prevalent in the QM7 dataset. By contrast, the contributions for the other atom types are relatively spread out.

Although more work needs to be done with respect to the classification of these contributions, we can draw one particular preliminary conclusion. The tight clustering of hydrogen atoms suggests large-scale redundancies of information in the underlying dataset; in the context of the GAP framework, this will produce covariance matrices with high condition numbers and corresponding numerical inaccuracies. This is born out by our informal observations; we have noted that covariance matrices encountered when attempting to train potentials using even larger datasets than those used here (such as the full GDB9 dataset) have prohibitively high condition, particularly for suboptimal choices of kernel and descriptor parameters. We suspect that methods for reducing the redundancies contained within the training datasets (such as low-rank approximations of covariance matrices and/or dataset sparsification techniques) will yield results of significantly higher quality, while reducing computational cost.

4 Conclusions and Future Work

We have introduced three new numerical representations for atomic neighborhoods, which can be applied to the Gaussian approximation potential (GAP). Numerical experiments showed that our LC-GAP implementation is capable of predicting the atomization energies of organic molecules in the QM7 dataset at levels of accuracy competitive with similar potentials already described in the literature, for all three presented representation types. Furthermore, we have demonstrated that LC-GAP potentials trained on systems containing lower numbers of atoms can be used to predict the energy values for systems containing higher numbers of atoms, again with acceptable accuracy.

Of the three representations presented here, the reduced Coulomb descriptor has linear dimensionality in the number of particles in a local neighborhood, while the other two descriptor types have quadratic dimensionality. Despite this, the reduced Coulomb descriptor performs comparably well, if not better, in all cases. This descriptor also performs well when used to evaluate alternative molecular properties than the atomization energy, producing results comparable to the literature when applied to the QM7b many-properties dataset. Finally, when used for cross-validation on the QM7b and GDB9 datasets, LC-GAP equipped with the reduced Coulomb descriptor predicts atomization energies at and comfortably below chemical accuracy respectively, achieving MAEs of approximately 1.00 and 0.78 kcal/mol.

A number of questions remain about the utility of these representation functions when training on and predicting different kinds of atomic systems. In particular, as all systems in the QM7 dataset are equilibrium-state biomolecules, it would be interesting to investigate the behavior of the LC-GAP system over non-equilibrium systems, as well as infinitely-periodic crystal systems such as semiconductors. Additionally, an analysis of per-atom contributions produced during cross-validation suggest the existence of high levels of redundancy in the QM7 dataset; we strongly suspect that the application of techniques for lessening the impact of this redundancy will produce worthwhile results.

Acknowledgements This work was funded in part by the German Federal Ministry for Education and Research under the Eurostars project E!6935 ATOMMODEL. We would also like to thank Maharavo Randrianarivony for fruitful discussions.

References

1. A.P. Bartók, R. Kondor, G. Csányi, On representing chemical environments. Phys. Rev. B **87**, 184115 (2013)
2. A.P. Bartók, M.C. Payne, R. Kondor, G. Csányi, Gaussian approximation potentials: the accuracy of quantum mechanics, without the electrons. Phys. Rev. Lett. **04**, 136403 (2010)
3. L.C. Blum, J.-L. Reymond, 970 million druglike small molecules for virtual screening in the chemical universe database GDB-13. J. Am. Chem. Soc. **131**, 8732–8733 (2009)
4. S. De, A.P. Bartók, G. Csányi, M. Ceriotti, Comparing molecules and solids across structural and alchemical space. Phys. Chem. Chem. Phys. **18**, 13754–13769 (2016)
5. S. Fliszár, *Atoms, Chemical Bonds, and Bond Dissociation Energies*. Lecture Notes in Chemistry (Springer, Berlin, 1994)
6. M. Griebel, S. Knapek, G. Zumbusch, *Numerical Simulation in Molecular Dynamics: Numerics, Algorithms, Parallelization, Applications*. Texts in Computational Science and Engineering, vol. 5 (Springer Science & Business Media, Heidelberg, 2007)
7. K. Hansen, G. Montavon, F. Biegler et al., Assessment and validation of machine learning methods for predicting molecular atomization energies. J. Chem. Theory Comput. **9**, 3404–3419 (2013)
8. K. Hansen, F. Biegler, R. Ramakrishnan et al., Machine learning predictions of molecular properties: accurate many-body potentials and nonlocality in chemical space. J. Phys. Chem. Lett. **6**, 2326–2331 (2015)
9. W. Kohn, Density functional and density matrix method scaling linearly with the number of atoms. Phys. Rev. Lett. **76**, 3168–3171 (1996)
10. G. Montavon, K. Hansen, S. Fazli et al., Learning invariant representations of molecules for atomization energy prediction, in *Advances in Neural Information Processing Systems* (Springer, Berlin, 2012), pp. 440–448
11. G. Montavon, M. Rupp, V. Gobre et al., Machine learning of molecular electronic properties in chemical compound space. New J. Phys. **15**, 095003 (2013)
12. S.J. Plimpton, A.P. Thompson, Computational aspects of many-body potentials. MRS Bull. **37**, 513–521 (2012)
13. E. Prodan, W. Kohn, Nearsightedness of electronic matter. Proc. Natl. Acad. Sci. U. S. A. **102**, 11635–11638 (2005)
14. R. Ramakrishnan, P.O. Dral, M. Rupp, O.A. von Lilienfeld, Quantum chemistry structures and properties of 134 kilo molecules. Sci. Data **1** (2014). doi:10.1038/sdata.2014.22
15. C.E. Rasmussen, C.K.I. Williams, *Gaussian Processes for Machine Learning* (The MIT Press, Cambridge, 2006)
16. M. Rupp, Machine learning for quantum mechanics in a nutshell. Int. J. Quantum Chem. **115**, 1058–1073 (2015)

17. M. Rupp, A. Tkatchenko, K.-R. Müller, O.A. von Lilienfeld, Fast and accurate modeling of molecular atomization energies with machine learning. Phys. Rev. Lett. **108**, 058301 (2012)
18. M. Rupp, R. Ramakrishnan, O.A. von Lilienfeld, Machine learning for quantum mechanical properties of atoms in molecules. J. Phys. Chem. Lett. **6**, 3309–3313 (2015)
19. E. Tadmor, R. Miller, *Modeling Materials: Continuum, Atomistic and Multiscale Techniques* (Cambridge University Press, Cambridge, 2011)
20. C.J. Willmott, K. Matsuura, Advantages of the mean absolute error (MAE) over the root mean square error (RMSE) in assessing average model performance. Clim. Res. **30**, 79–82 (2005)

River Bed Morphodynamics: Metamodeling, Reliability Analysis, and Visualization in a Virtual Environment

Tanja Clees, Igor Nikitin, Lialia Nikitina, Sabine Pott, and Stanislav Klimenko

1 Introduction

Numerical simulations of river bed evolution can be used for evaluating river engineering concepts and the efficient operation of natural and man-made waterways. An example is illustrated in Figs. 1 and 2. The requirements for the precision of such simulations as well as their ability to represent the real behaviour of the river bed morphodynamics are very high due to the large impact of civil water engineering to nature and society. In morphodynamical models, parameters and initial conditions possess uncertainties related to natural variability, imperfect description of physical processes and imprecision of model parameters. Reliability analysis is used to detect the origin of uncertainties, to track their propagation in the model and to evaluate their contribution to the result of the modeling.

Generally, a numerical simulation process produces a result for a given set of simulation parameters. Usually the result is a field value distributed in space and time, i.e. in every point of space-time the result is a function of the simulation parameters. The parameters possess uncertainty, i.e. they are considered as random numbers with a given input probability distribution. Therefore, the simulation result is also uncertain. In every point of space-time the result is a random number with a certain output probability distribution. The purpose of reliability analysis is to define a relationship between input and output distributions. More practically, one needs to determine certain probabilistic characteristics of an output distribution such as quantiles in terms of given input distributions. The success of this intention depends

T. Clees (✉) • I. Nikitin • L. Nikitina • S. Pott
Fraunhofer Institute for Algorithms and Scientific Computing SCAI, Schloss Birlinghoven, 53757 Sankt Augustin, Germany
e-mail: tanja.clees@scai.fraunhofer.de

S. Klimenko
Institute of Computing for Physics and Technology, Protvino, Moscow Region, Russia

© Springer International Publishing AG 2017
M. Griebel et al. (eds.), *Scientific Computing and Algorithms in Industrial Simulations*, DOI 10.1007/978-3-319-62458-7_3

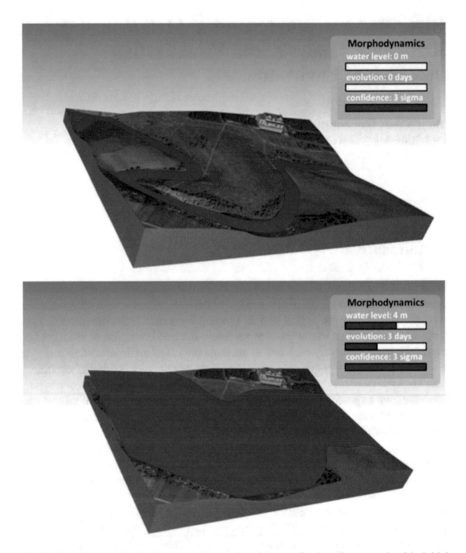

Fig. 1 Visualization of a flood event on Protva river, Moscow Region. *Top*: water level in initial state. *Bottom*: maximal water level. Published with kind permission of ©Fraunhofer SCAI 2016. All Rights Reserved

on the properties of the functional dependence between input parameters and output values, as well as on the form of the distribution of input parameters.

In case of linear functional dependencies and normally distributed input parameters the output distribution will also be normal. In this case, the first-order reliability method (FORM [3]) is applicable. It allows defining sensitivities, i.e. constant coefficients of the linear dependence, either by a finite difference scheme [10] or by automatic differentiation [12]. This estimation can be performed with minimal

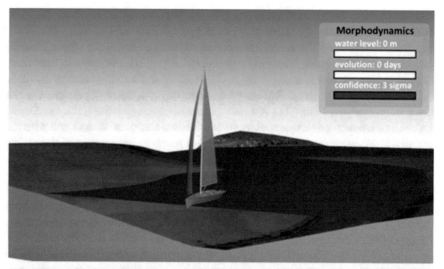

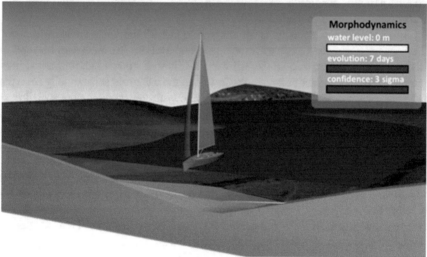

Fig. 2 Visualization of a flood event on Protva river, Moscow Region. *Top*: river bed profile in initial state. *Bottom*: river bed profiles in the final state for three quantiles—median (*green surface*), lower 3-sigma quantile (*yellow surface*), upper 3-sigma quantile (*textured surface*). The quantiles represent an interval containing the result with 99.7% confidence. Published with kind permission of ©Fraunhofer SCAI 2016. All Rights Reserved

computational cost. If sensitivities are known, one can determine the quantiles by an analytical formula.

In case of a functional dependence other than a linear one and/or a distribution of input parameters other than a normal one, FORM is not applicable. Indeed, the output distribution is not normal anymore; it is distorted, e.g. its mean and median do not coincide, upper and lower quantiles are not symmetric, standard deviation does not characterize quantiles etc. For a special case of quadratic functional dependence and normal input distribution, analytical methods are still applicable such as the second-order reliability method (SORM [2]). For generic functional dependences and distributions only a direct computation of quantiles by their definition [20] is possible. Furthermore, when the functional dependence is not given in an analytical form, but is a result of numerical simulations, only statistical approximation methods are available, such as Monte Carlo. In comparison with the sensitivity-based methods, Monte Carlo needs a much larger number of simulation results as input. However, it provides direct estimation of quantiles which, for distorted distributions, can differ a lot from the first-order approximation.

A considerable speedup of the Monte Carlo method can be achieved if the functional dependence is modeled by interpolation of simulation results, e.g. with the aid of an RBF metamodel [1]. A small set of simulations is used as a basis for modeling; a much larger number of sample experiments is used for estimating quantiles. Since the estimation is applied to the model function, the computational cost of which is negligible in comparison with a simulation run, such a method features a significant acceleration.

Another source of speedup is the usage of quasi-Monte Carlo techniques. Low-discrepancy sequences, such as Sobol [17] and Halton [8] ones, can be used to improve the precision of quantile estimation. It is well known that numerical integration of smooth functions by Monte Carlo method can profit substantially from the usage of low-discrepancy sequences. Namely, low-discrepancy sequences in dimension n and sample length N provide an approximation error of $O(N^{-1}(\log N)^n)$, while direct Monte Carlo gives only $O(N^{-1/2})$. A technical difficulty is that quantile estimation requires integration of non-smooth functions, similar to the Heaviside step function $\theta(t)$. In this paper we will show how to overcome this problem.

Below, in Sect. 2 we present our metamodeling technique, in Sect. 3 we overview various methods of quantile estimation, and in Sect. 4 we apply these methods for a reliability analysis of coupled morphodynamic-hydrodynamic simulations of a river-bed evolution scenario. Interactive 3D visualization in a virtual environment is described in Sect. 5. Finally, Sect. 6 summarizes the results.

2   RBF Metamodel

Numerical simulations define a mapping $y = f(x) : \mathbb{R}^n \rightarrow \mathbb{R}^m$ from an n-dimensional space of simulation parameters to an m-dimensional space of simulation results. In morphodynamic simulation, the dimensionality of the simulation parameters x is moderate ($n \sim 10$–30), while the simulation results y are dynamical fields sampled on a large grid, typically containing $\sim 10^5$ nodes and ~ 10 time steps, resulting in values of $m \sim 10^6$. High computational complexity of morphodynamic models restricts the number of simulations available for analysis (typically $N_{\exp} < 10^3$), and this number shall be as small as possible. Metamodeling is an approximation technique allowing efficient representation of these large datasets for the purpose of data analysis, robust optimization and real time visualization. Metamodeling naturally involves in the analysis the uncertainties in optimization variables and other control parameters influencing the simulation. Metamodeling with radial basis functions (RBF) is a representation of the form

$$f(x) = \sum_{i=1\ldots N_{\text{exp}}} c_i \Phi(|x - x_i|), \tag{1}$$

where x_i are the points with known function values $y_i = f(x_i)$. A suitable choice for the RBF is the multi-quadric function $\Phi(r) = \sqrt{b^2 + r^2}$, which provides non-degeneracy of the interpolation matrix $\Phi = (\Phi_{ij}) = (\Phi(|x_i - x_j|))$ for all finite datasets of distinct points and all dimensions n, cf. [1]. The result can be written in a form of a weighted sum $f(x) = \sum_i w_i(x) y_i$, with the weights

$$w_i(x) = \sum_j \Phi_{ij}^{-1} \Phi(|x - xj|). \tag{2}$$

RBF interpolation can be extended by adding polynomial terms to (1). This allows for an exact reconstruction of polynomial (including linear) dependencies and generally improves the precision of interpolation. Adaptive sampling and hierarchy of metamodels with appropriate transition rules are used for a further improvement of precision [19].

RBF metamodeling is directly applicable to interpolation of high dimensional bulky data. This means that complete simulation results can be interpolated at a rate linear in the size of data, and even faster in combination with dimensional reduction techniques based on principal component analysis (PCA) [6]. The precision can be controlled by means of cross-validation: one data point is removed, the remaining data are interpolated to this point, and the result is compared with the actual value at this point. In case of an RBF metamodel, this can be expressed explicitly [4]:

$$\text{err}_i = f_{\text{interpol}}(x_i) - f_{\text{actual}}(x_i) = -c_i / (\Phi^{-1})_{ii}. \tag{3}$$

Metamodeling performed with controlled precision can replace simulation results in computationally intensive procedures, such as quantile estimation. Various quantile estimators are known [9, 11, 16], they all can profit from a combination with meta-models. Another approach has been used in [18], where a Monte Carlo method was combined with Kriging metamodeling for setting up additional simulations, used for corrections of the result. All the above methods show a theoretical convergence rate of the order $O(N^{-1/2})$ with increasing number of metamodel evaluations N, making them hardly applicable for analysing bulky data. The convergence can be improved by means of quasi-Monte Carlo integration schemes [14].

3   Quantile Estimation

In the case considered here, quantile estimation means the determination of quantiles Q_p for simulation results, i.e.

$$P(y < Q_p) = p, \tag{4}$$

where P is the probability measure and p is a user-specific threshold value. For example, the median med $= Q_{0.5}$ corresponds to 50% of the distribution, i.e. $P(y < \text{med}) = 0.5$; 68% of the distribution is located in the interval $[Q_{0.16}, Q_{0.84}]$, etc. Several approaches for quantile estimation have been developed, for details see [2, 3, 8, 9, 11, 13, 14, 16–18] and the references therein.

3.1   Sensitivity-Based Approach

3.1.1   First-Order Approximation

A first-order approximation is applicable for a linear mapping $y = f(x)$ and a normal distribution of simulation parameters, with $x_0 = \, <x>$ being the mean value of x and $((\text{cov}_x)_{ij}) = (<dx_i dx_j>)$ being the covariance matrix of x and $dx = x - x_0$. In this case, y is also normally distributed with mean value $y_0 = \, <y> = \text{med}(y) = f(x_0)$ and covariance matrix $\text{cov}_y = J\text{cov}_x J^T$, where $J = (J_{ij}) = \partial f_i/\partial x_j$ is the Jacobian matrix of $f(x)$. The diagonal part of cov_y gives the standard deviations σ_y^2, directly defining $Q_{p(y)}$, e.g.

$$Q_{68\%} = \, <y> \pm \sigma_y, \quad Q_{99.7\%} = \, <y> \pm 3\sigma_y. \tag{5}$$

A finite difference scheme for computing the Jacobian matrix of $f(x)$ requires $N_{\text{exp}} = O(n)$ simulations, e.g. $2n$ simulations for the central difference scheme plus one simulation at x_0, summing up to $N_{\text{exp}} = 2n + 1$ simulations. The algorithm has a computational complexity of $O(nm)$ and can be implemented efficiently since data from N_{exp} open data streams can be read and processed simultaneously for writing quantiles to a single output data stream. In this way, memory requirements can be minimized and parallelization can be done in a straightforward manner.

3.1.2   Second-Order Approximation

A second-order approximation applies a correction to the above first-order approximation associated with the Hessian matrix

$$(H_{jk}^i) = (\partial^2 f_i/(\partial x_j \partial x_k)). \tag{6}$$

In particular, the average of this correction

$$< H_{jk}dx_jdx_k > = \mathrm{Tr}(H\mathrm{cov}_x) \tag{7}$$

can be directly used e.g. as robustness measure in optimization, distinguishing a sensitive maximum (large H) from a wide one (small H) being viewed in comparison with the size of the scatter measured by cov_x.

Also higher order moments such as $< (y_{max}-y)^2 >$ can be directly computed, as well as the cumulative distribution function $\mathrm{CDF}(f) = P(y < f)$ and all quantiles.

However, this method requires the full Hessian matrix, which needs $N_{exp} = O(n^2)$ simulations. Practically, the usability of this second-order method is limited anyway, because strongly nonlinear functions can involve higher order terms and distributions of simulation parameters can strongly deviate from normal ones.

3.2   Monte Carlo

Monte Carlo methods can be applied to a generic nonlinear mapping $f(x)$ and arbitrary distribution $\rho(x)$ to estimate the probability

$$P_N(y < Q_p) = |\{y_n < Q_p\}|/N \tag{8}$$

for a finite sample $\{y_1, \ldots, y_N\}$. By the central limit theorem, the error of this estimation $\mathrm{err}_N = F_N - F$ for large N is distributed normally with zero mean and standard deviation $\sigma \sim \sqrt{F(1-F)/N}$.

The algorithm for quantile estimation performs sorting of m samples $\{y_1, \ldots, y_N\}$ in ascending order and selecting the k-th item in every sample with $k = (N-1)p+1$ as a representative for Q_p. The algorithm possesses a computational complexity of $O(mN \log N)$ and can be efficiently implemented using data stream operations similar to the sensitivity based methods.

The standard lower and upper quantiles are defined from a condition that a given portion α of the distribution is located in the interval $[Q_{min}, Q_{max})$:

$$P(y < Q_{min}) = (1 - \alpha)/2,$$
$$P(y \geq Q_{max}) = (1 - \alpha)/2, \tag{9}$$
$$P(Q_{min} \leq y < Q_{max}) = \alpha.$$

For an interpretation of the results it is convenient to define deviations subtracting the median value from the upper/lower quantiles:

$$dQ_{min} = \mathrm{med} - Q_{min}, \quad dQ_{max} = Q_{max} - \mathrm{med}. \tag{10}$$

The choice of the median as a central value has the advantage that both deviations are positive then: $dQ_{min} > 0$, $dQ_{max} > 0$. In other definitions, e.g. by subtracting the mean value instead, one deviation can become negative for strongly distorted distributions.

An RBF metamodel can represent the mapping $f(x)$ in this method, reducing the number of required simulations. While the metamodel can be constructed with a moderate number of simulations, e.g. $N_{exp} \sim 100$, the determination of Q_p can be done with $N \gg N_{exp}$. The respective algorithm performs a precomputation of the RBF weight matrix $(w_{ik}) = (w_i(x_{sk}))$ and its multiplication with the data matrix y_{di}, comprising $O(mNN_{exp})$ operations. Effort for the latter usually prevails over $O(mN \log N)$ operations needed for sorting the interpolated samples. Here, $x_{sk}, k = 1, \ldots, N$ are sample points, $x_j, j = 1, \ldots, N_{exp}$ are simulation points, and the data matrix y_{di} stores the whole simulation result of size m for each experiment $i = 1, \ldots, N_{exp}, d = 1, \ldots, m$.

3.3 Weighted Monte Carlo

Harrell and Davis [9] propose to use a weighted sum of order statistics to improve the precision of the quantile estimation:

$$Q_p = \sum_{i=1\ldots N} \omega_i y_{(i)}, \tag{11}$$

where

$$\omega_i = I_{i/N}(p(N+1), (1-p)(N+1)) - I_{(i-1)/N}(p(N+1), (1-p)(N+1)). \tag{12}$$

Here, $y_{(i)}$ denotes the i-th point in the sample sorted in ascending order, and the weights ω_i are defined in terms of the incomplete beta function $I_x(a, b)$.

3.4 Quasi-Monte Carlo (QMC)

Low-discrepancy sequences, such as Sobol [17] and Halton [8] ones, cf. Fig. 3, can be used to improve the precision of quantile estimation. It is well known [13] that numerical integration of smooth functions by Monte Carlo method can profit substantially from the usage of low-discrepancy sequences, providing an integration error of $O(N^{-1}(\log N)^n)$. Here, N is the number of points in the sample and n the dimension of the parameter space. The problem of quantile determination is related to the integration of nonsmooth functions. A quantile can be found by inverting the cumulative distribution function. This can be expressed as

$$\text{CDF}(y) = \int d^n x \rho(x) \theta(y - f(x)). \tag{13}$$

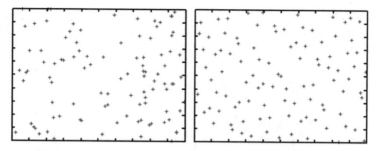

Fig. 3 Examples of random sequences in dimension $n = 2$, standard pseudo-random on the *left*, Sobol quasi-random on the *right*. Quasi-random sequences provide a more regular, quite gap-free sampling of the design space. Published with kind permission of ©Fraunhofer SCAI 2016. All Rights Reserved

Here, $\theta(t) = \{1, t \geq 0; 0, t < 0\}$ is the Heaviside step function (theta-function). For discontinuous functions, quasi-Monte Carlo has a weaker performance [13]. In particular, the error of the integration of the theta-function can be estimated as $O(N^{-1/2-1/(2n)})$.

3.5 Quasi-Random Splines (QRS)

In our previous paper [7] we have developed a version of a QMC integration method, where $n - 1$ variables are generated as quasi-random points, while the dependence of the function $f(x)$ on the remaining variable is represented by a cubic spline. It is designed for the practically important special case, when the dependence on the remaining variable is monotonous and can be unambiguously inverted.

In this case, an iso-surface $f(x_1, x_2 \ldots x_n) = y$ can be represented as a graph of a function $x_1 = f^{-1}(y, x_2 \ldots x_n)$, and one integration in (13) can be performed analytically.

In practically relevant cases, the function f^{-1} turns out to be sufficiently smooth, and integration over $x_2 \ldots x_n$ variables by the QMC method provides the desired convergence in $O(N^{-1}(\log N)^{n-1})$.

To be more specific, the following construction is used. At first, a transformation $x \rightarrow u$ to new variables possessing a uniform probability distribution $\rho(u) = 1$ is performed. In the case of independent random variables, i.e. when the original distribution is composed as a product $\rho(x) = \rho_1(x_1) \cdots \rho_n(x_n)$, the cumulative distribution functions of the input variables give the desired transformation: $u_i = \text{CDF}_i(x_i)$. The integral is now transformed to

$$\text{CDF}(y) = \int d^n u \, \theta(y - f(u)). \tag{14}$$

In the case when the function $f(u)$ is monotonous w.r.t. one variable, say u_1, a surface $f(u) = y$ can be re-expressed as a graph of a function $u_1 = f^{-1}(y, u_2, \ldots, u_n)$, and integration over u_1 leads to

$$\text{CDF}(y) = \int du_2 \ldots du_n f^{-1}(y, u_2, \ldots, u_n). \tag{15}$$

Here, the integrand is the function $f(u_1)$, inverted at fixed $u_2, \ldots, u_n$ and extended as follows: $u_1 = f^{-1}(y)$ for $f(0) \leq y \leq f(1)$, 0 for $y < f(0)$, and 1 for $y > f(1)$.

Note that, for definiteness, we consider the case of a monotonously increasing function $f(u_1)$, while for a monotonously decreasing function similar formulae can be derived.

The function $y = f(u_1)$ can be represented by a cubic spline to speed up the inversion, avoiding too many computationally intensive function evaluations. For this purpose u_1 is sampled regularly in [0,1] with K points, the $f_k = f(u_{1k})$ are evaluated by calling the metamodel, and a sequence (u_{1k}, f_k) is interpolated with a cubic spline. For monotonous functions one can revert the pairs and use the sequence (f_k, u_{1k}) directly for a spline representation of the inverse function $u_1 = f^{-1}(y)$. Finally, one introduces a regular sampling y_p, $p = 1, \ldots, K$ in $[y_{min}, y_{max}]$, uses the above constructed splines to find $f^{-1}(y_p)$ and QMC-averaging to obtain $\text{CDF}(y_p)$.

The QRS algorithm requires NK evaluations of the metamodel, where N is the number of quasi-random samples. This effort $O(NKN_{exp})$ prevails over the own computational complexity of the algorithm $O(NK)$. Processing bulky data with this algorithm can be done straightforwardly by the data stream operations described above. The precision of the algorithm is balanced between the quality of spline interpolation $O(K^{-4})$ and the quality of QMC integration, where for a sufficiently smooth integrand, a precision of $O(N^{-1}(\log N)^{n-1})$ is expected.

4   Numerical Tests

The methods above have been applied for the analysis of morphodynamic simulations of river bed evolution [5]. The simulations have been performed by our colleagues from German Federal Waterways Engineering and Research Institute (BAW, Karlsruhe). A 10 km long stretch of the river Danube including a 270° bend (see Fig. 4) is modeled with the open source software suite TELEMAC-MASCARET (www.opentelemac.org) including the morphodynamic module SISYPHE [21] coupled with the hydrodynamic module TELEMAC-2D [10].

Nearly 10^5 grid elements were used for the discretization with a mean node distances of about 6 m in the river channel and up to 30 m at the flood planes. A synthetic hydrograph of 9 days was simulated including two high flood events. Using 32 processors, one simulation run needed about 45 min. For the analysis 13 simulation parameters have been considered to be uncertain, from which five most influencing parameters have been detected. A probability distribution of distorted

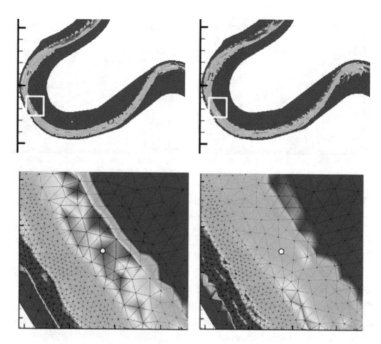

Fig. 4 Quantiles of the resulting distribution for the "river bed evolution criterion", ranging from 0 m (*blue*) to 0.5 m (*red*), at the last time step. *Left*—dQmin(99.7%), *right*—dQmax(99.7%), *bottom*—corresponding closeups, the *white circle* indicates the location of the probe point. Published with kind permission of ©Fraunhofer SCAI 2016. All Rights Reserved

normal type has been used for modeling:

$$\rho(x) \sim \exp(-(x-x_0)^2/2\sigma-), \quad \text{for} \quad x < x_0,$$
$$\rho(x) \sim \exp(-(x-x_0)^2/2\sigma+), \quad \text{for} \quad x \geq x_0. \tag{16}$$

A detailed description of simulation parameters can be found in [5, 12, 15].

For the application of quantile estimators, an RBF metamodel has been constructed based on 151 experiments for a variation of the five most influencing parameters ($n = 5$). The typical result of determined quantiles is shown in Fig. 4. The asymmetry of the distribution is clearly visible from the closeups. In particular, for a probe point indicated by a white circle in the closeups, we get $dQ_{min} = 0.48$ m, $dQ_{max} = 0.25$ m.

In Fig. 5 (left) different methods for quantile determination are compared:

- MC—Monte Carlo based on a standard pseudo-random sequence,
- WMC—Monte Carlo with Harrell-Davis weights,
- QMC—quasi-Monte Carlo based on Sobol sequence,
- WQMC—quasi-Monte Carlo with Harrell-Davis weights.

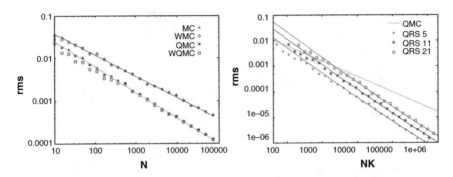

Fig. 5 Convergence of different methods for quantile estimation. Published with kind permission of ©Fraunhofer SCAI 2016. All Rights Reserved

For this comparison the quantiles are estimated for the "river bed evolution criterion" at the last time step in the probe point. The estimation visualized in Fig. 5 is done for the median ($p = 0.5$), while the other quantiles show similar behavior. The root mean square error (rms) of the determined quantile is computed on 100 trials. The dependence on the sample length N is tracked in the range $N = 10, \ldots, 105$. Harrell-Davis weighting gives an improvement at small N, particularly near $N \sim 100$ the improvement is 10% for MC and 20% for QMC. For comparison, [9] report improvements of 5–10% for MC, obtained in the tests on model functions at $N = 60$. [14] shows improvements of 10–40% for MC achieved at $N = 100$ for other model functions. Furthermore, with increasing N, the effect of Harrell-Davis weighting becomes smaller. The observed asymptotic performance of the methods, namely

$$\text{rms(MC)} \sim N^{-0.5}, \quad \text{rms(QMC)} \sim N^{-0.6}, \tag{17}$$

coincide with the estimations above:

$$\text{rms(MC)} \sim N^{-0.5}, \quad \text{rms(QMC)} \sim N^{-1/2-1/(2n)} \quad \text{at} \quad n = 5. \tag{18}$$

For the application of the QRS method we have tested that the "river bed evolution criterion" is a monotonously decreasing function of parameter x_2 ("coefficient for slope effect"). The test has been done by numerical evaluation of the derivative $\partial f / \partial x_2$ in 107 randomly selected points. Figure 5 (right) compares the convergence of QMC and QRS methods. As expected, the QRS method provides an asymptotic performance

$$\text{rms(QRS)} \sim N^{-1}(\log N)^{n-1}. \tag{19}$$

In particular, rms(QRS) can be fitted well by the dependence $c_1 N^{-1}(\log N)^4 + c_2 N^{-1}$.

Fig. 6 Bias of median for
QRS method. Published with
kind permission of
©Fraunhofer SCAI 2016. All
Rights Reserved

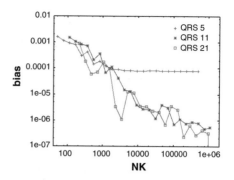

Tests of QRS method with three different $K = 5, 11, 21$ have been performed. Smaller K require less metamodel evaluations, larger K correspond to a better precision of the spline interpolation. Figure 6 shows the bias between the median computed by the QMC and the best method (QRS with $K = 21$ and $N = 10^5$). Saturation of the dependence corresponds to the quality balance between spline interpolation and QMC integration. Comparing the methods, we see that QRS provides the best performance, in particular, the number of function evaluations necessary to achieve a rms $\sim 10^{-4}$ precision is reduced from 10^6 for the direct Monte Carlo technique to 6×10^3 for quasi-Monte Carlo (QMC).

5 Visualization in Virtual Environment

We developed a novel way to visualize and explore not only simulation results, but also statistical quantities, based on a virtual environment. The principal scheme is sketched in Fig. 7 and its components are described in the following. Since input data and simulation results for the realistic application test case considered in the previous section are not available in 3D (yet), we demonstrate the newly developed scheme for another morphodynamic application case, see Figs. 1 and 2.

Avango (www.avango.org) is an object-oriented programming framework for building interactive applications in a virtual environment. Its applicability extends from simple displays to fully immersive large-scale projection systems such as CAVE and iCONE and can be adapted to arbitrary hardware configurations by means of flexible abstraction schemes. The framework provides a developer with a high-level interface for representing and graphical processing of complex geometrical scenes. The developer is shielded from interaction with low-level graphical and system interfaces and can focus on the development of the virtual environment itself.

Avango is an open source software package, based on OpenSceneGraph, see www.openscenegraph.org, which in turn is built on the top of OpenGL. The purpose of OpenGL is to control the graphics board operations, such as rendering of primitives (triangles, lines, points), usage of various graphical modes (enlightening, texturing, transparency), projection of 3D objects on the screen plane, depth

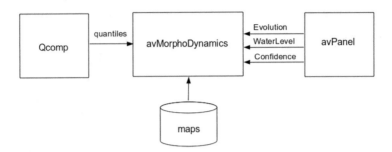

Fig. 7 Principal scheme of the Avango application for visualization of river bed morphodynamics in a virtual environment. Published with kind permission of ©Fraunhofer SCAI 2016. All Rights Reserved

buffering etc. The resulting image can be directed to a monitor, a beamer, or a head-mounted display. The objects of OpenGL are geometrical primitives, textures mapped to them, projection matrices etc.

The list of OpenGL commands necessary to render a scene is formed by OpenSceneGraph. Its purpose is an optimization of graphical processes using a data hierarchy (representation of the scene in form of a tree-like structure, a scene graph), special techniques for rendering acceleration (view frustum culling, adaptive level of details), and simple algorithms for collision detection. The objects of OpenSceneGraph are scene graph nodes, geometrical models (geosets) specified as lists of primitives, materials describing the graphical modes (geostates) including various enlightening and texturing models as well as screen configurations (mono, parallel or sequential stereo).

In principle, these capabilities are sufficient for developing virtual environment applications. However, in OpenSceneGraph this means developing a new C++ object which encapsulates a part of the scene graph and defines material properties, dynamical behavior and interaction with other objects and with the user. Every change like positioning or scaling of a scene component requires recompiling the object. Connections to external devices are also the responsibility of the developer.

The purpose of Avango is to provide a convenient interface for application development, where the application can be assembled from building blocks directly, without recompiling. These blocks are elementary C++ objects, which are compiled to shared libraries and can be loaded to memory at runtime. Connections between the objects and external devices are implemented in Avango in form of a data-flow graph, which can also be changed at runtime. To provide this functionality, the public variables are equipped by an interface enabling an access to their values from the Python interpreter (www.python.org). Using the interpreter commands, the user can set/get the values of such variables and connect them together. These variables are called fields, following the terminology of the earlier virtual environment frameworks OpenInventor and VRML.

In our application, the main module `avMorphoDynamics` (see also Fig. 7) possesses the following fields:

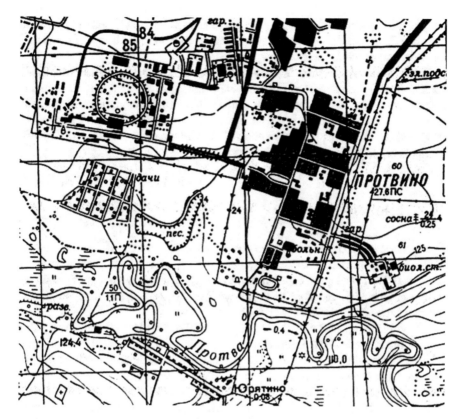

Fig. 8 Topographic map of Protvino used for modeling. Published with kind permission of ©Fraunhofer SCAI 2016. All Rights Reserved

- SFFloat Evolution: model time (in days)
- SFFloat WaterLevel: current water level (in meters)
- SFFloat Confidence: confidence level (0–99.7%, for a normal distribution corresponding to 0–3σ variation)

which allow interactive exploration of river bed morphodynamics (see Figs. 1 and 2).

The module uses an elevation map and corresponding textures for representation of the terrain. Our experience shows that commonly available digital elevation maps (DEM) based on satellite radar measurements (SRTM) have insufficient precision. Some of them even show that the river goes up the hill at the location in Fig. 8. This artefact is probably related to nearby trees, which distort the elevation measurement, making the map inappropriate for modeling. Local topographic maps after their digitalization give better precision.

The avMorphoDynamics module also receives data from the computational Qcomp module for visualization of the current water level, river bed evolution and its quantiles. The quantiles are displayed in a form of stratified river bed profile, as shown in Figs. 1 and 2: a green surface represents a median, a yellow surface a lower quantile, and a textured surface an upper quantile.

For stereoscopic visualization we use a single 3D-capable beamer with DLP-Link technology. Such commonly available beamers do not require special projection screens and can turn any regular office into a virtual laboratory providing full immersion into the model space. This opens wide opportunities for the application of virtual environments for the solution of real problems in a variety of scientific and technological disciplines.

6 Conclusion

Four generic methods for quantile estimation have been tested: Monte Carlo (MC), Monte Carlo with Harrell-Davis weighting (WMC), quasi-Monte Carlo with Sobol sequence (QMC) and quasi-random splines (QRS). The methods are combined with RBF metamodeling and applied to the analysis of coupled morphodynamic and hydrodynamic simulations of the river bed evolution.

The comparison of the methods shows the following: Harrell-Davis weighting gives a moderate 10-20% improvement of precision at small number of samples $N \sim 100$. Quasi-Monte Carlo methods provide significant improvement of quantile precision, e.g. the number of function evaluations necessary to achieve rms $\sim 10^{-4}$ precision is reduced from 10^6 for MC to 10^5 for QMC and to 6×10^3 for QRS. On the other hand, RBF metamodeling of bulky data allows to speed up the computation of one complete result for the considered problem from 45 min (on 32CPU) to 20 s (on 1CPU), providing rapid quantile estimation for the whole set of bulky data.

An interactive 3D virtual environment has been used in a novel way to visualize and explore also statistical results. The developed virtual environment application allows for a detailed inspection of the numerical model, including the evolution of the water level, river bed profiles and their quantiles. In this way, one achieves full immersion into the model space and an intuitive understanding of morphohydrodynamical processes.

References

1. M.D. Buhmann, *Radial Basis Functions: Theory and Implementations* (Cambridge University Press, Cambridge, 2003)
2. L. Cizelj, B. Mavko, H. Riesch-Oppermann, Application of first and second order reliability methods in the safety assessment of cracked steam generator tubing. Nucl. Eng. Des. **147**, 359–368 (1994)
3. T. Clees, I. Nikitin, L. Nikitina, C.-A. Thole, Nonlinear metamodeling and robust optimization in automotive design, in *Proceedings of 1st International Conference on Simulation and Modeling Methodologies, Technologies and Applications SIMULTECH 2011*, Noordwijkerhout, July 29–31 (SciTePress, Setúbal, 2011), pp. 483–491

4. T. Clees, I. Nikitin, L. Nikitina, Nonlinear metamodeling of bulky data and applications in automotive design, in *Progress in industrial mathematics at ECMI 2010*, ed. by M. Günther et al. Mathematics in Industry, vol. 17 (Springer, Berlin, 2012), pp. 295–301
5. T. Clees, I. Nikitin, L. Nikitina, R. Kopmann, Reliability analysis of river bed simulation models, in *CDROM Proceedings of EngOpt 2012, 3rd International Conference on Engineering Optimization*, ed. by J.Herskovits. vol. 267. Rio de Janeiro, Brazil, July 1–5 (2012)
6. T. Clees, I. Nikitin, L. Nikitina, C.-A. Thole, *Analysis of bulky crash simulation results: deterministic and stochastic aspects*, in *Simulation and Modeling Methodologies, Technologies and Applications: International Conference*, ed. by N. Pina, J. Kacprzyk, J. Filipe. SIMULTECH 2011 Noordwijkerhout, The Netherlands, July 29–31, 2011 Revised Selected Papers (Springer, Berlin, 2013), pp. 225–237
7. T. Clees, I. Nikitin, L. Nikitina, S. Pott, Quasi-Monte Carlo and RBF metamodeling for quantile estimation in river bed morphodynamics, in *Simulation and Modeling Methodologies, Technologies and Applications*, ed. by M.S. Obaidat, S. Koziel, J. Kacprzyk, L. Leifsson, T. Ören. Advances in Intelligent Systems and Computing, vol. 319 (Springer International Publishing, Berlin, 2015), pp. 211–222
8. J.H. Halton, *Algorithm 247: Radical-inverse quasi-random point sequence*. Commun. ACM **7**, 701–702 (1964)
9. F.E. Harrell, C.E. Davis, A new distribution-free quantile estimator. Biometrika **69**, 635–640 (1982)
10. J.-M. Hervouet, *Hydrodynamics of Free Surface Flows: Modelling with the Finite Element Method* (Wiley, New York, 2007)
11. M.C. Jones, The performance of kernel density functions in kernel distribution function estimation. Stat. Probab. Lett. **9**, 129–132 (1990)
12. R. Kopmann, A. Schmidt, Comparison of different reliability analysis methods for a 2D morphodynamic numerical model of River Danube, in *Proceedings of River Flow 2010 - International Conference on Fluvial Hydraulics*, Braunschweig, 8–10. Sept. 2010, ed. by A. Dittrich, K. Koll, J. Aberle, P. Geisenhainer (2010), pp. 1615–1620
13. W.H. Press, S. Teukolsky, W. T. Vetterling, B.P. Flannery, *Numerical Recipes in C*, Chap. 7.7 (Cambridge University Press, Cambridge, 1992)
14. B. Rhein, T. Clees, M. Ruschitzka, Robustness measures and numerical approximation of the cumulative density function of response surfaces. Commun. Stat. B-Simul. **43**, 1–17 (2014)
15. J. Riehme, R. Kopmann, and U. Naumann, Uncertainty quantification based on forward sensitivity analysis in SISYPHE, in *Proceedings of V European Conference on Computational Fluid Dynamics: ECCOMAS CFD 2010*, Lisbon, Portugal,14–17 June 2010, ed. by J.C.F. Pereira, A. Sequeira, ECCOMAS (2010)
16. M.E. Sfakianakis, D.G. Verginis, A new family of nonparametric quantile estimators. Commun. Stat. Simul. Comput. **37**, 337–345 (2008)
17. I.M. Sobol, On the distribution of points in a cube and the approximate evaluation of integrals. USSR Comput. Math. Math. Phys. **7**, 86–112 (1967)
18. B. Sudret, Meta-models for structural reliability and uncertainty quantification, in *Proceedings of 5th Asian-Pacific Symposium on Structural Reliability APSSRA 2012*, Singapore, ed. by K. Phoon et al., (2012), pp. 53–76
19. G. van Bühren, T. Clees, N. Hornung, L. Nikitina, Aspects of adaptive hierarchical RBF metamodels for optimization. J. Comput. Methods Sci. Eng. **12**, 5–23 (2012)
20. A.W. van der Vaart, *Asymptotic Statistics* (Cambridge University Press, Cambridge, 1998)
21. C. Villaret, *Sisyphe 6.0 User Manual H-P73-2010-01219-FR*, Department National Hydraulics and Environ-ment Laboratory, Electricité de France (2010)

Cooling Circuit Simulation I: Modeling

Tanja Clees, Nils Hornung, Éric Lluch Alvarez, Igor Nikitin, Lialia Nikitina, and Inna Torgovitskaia

1 Introduction

This is the first of two associated articles on temperature distributions in cooling circuits. Our focus is to simulate and optimize heat flow along the cooling circuit in order to reduce overall energy consumption for relevant operating scenarios, and this first contribution gives a general introduction to the underlying simulation model. The second article [3], also contained in this book, then provides a detailed case study of a real system within a supercomputing center for which measurements and technical data are available.

For the simulation method to be general, we consider here the case of a typical cooling or heating circuit present in most private, public, or corporate facilities. Cooling or heating circuits are usually controlled either manually or automatically. For example, pumps and valves can be used to locally obtain a given temperature or pressure. In particular, pump *speed* or valve *aperture* are chosen in such a way that the temperature or pressure goals are fulfilled.

Control situations almost invariantly raise the question of the consequences of certain operating decisions. Typically, we like to know in which way a variation in controlled or uncontrolled parameters might affect the cooling or heating power of the system. A related question is how a given operation mode can be improved with an optimization goal in mind. In most cases, energy resources are narrow, which implies that their consumption needs to be minimized. Thus, a typical optimization goal is to reduce the electric power needed to run a system. Regarding a combination of both questions, we are, eventually, interested in a numerical model that is apt to facilitate heat flow prediction and energy management *together.*

T. Clees (✉) • N. Hornung • É.L. Alvarez • I. Nikitin • L. Nikitina • I. Torgovitskaia
Fraunhofer Institute for Algorithms and Scientific Computing SCAI, Schloss Birlinghoven, 53757 Sankt Augustin, Germany
e-mail: tanja.clees@scai.fraunhofer.de

© Springer International Publishing AG 2017
M. Griebel et al. (eds.), *Scientific Computing and Algorithms in Industrial Simulations*, DOI 10.1007/978-3-319-62458-7_4

The goal of this article is, therefore, to provide an up-to-date introduction to a reader with a background in scientific computation, yet unexperienced in the field of the specific application. Always referencing literature that enables further exploration of the subject, the introduction should address all relevant issues needed for a first implementation. In order to achieve this goal, we provide models for the most typical network elements of a cooling or heating system, such as pipes, pumps, valves, and energy exchangers. Numerical issues are discussed with the availability of common software for the solution and optimization of systems of equations in mind. Actually, we assume that *some* interior point method for continuous optimization problems is at hand, as well as a Newton-type algorithm for nonlinear root finding.

Above all, we address the question how to combine the mentioned elements into networks that may include closed loops and input and output flows. In fact, the spatial domain of computation of energy circuits has the character of a network, the notation of which is established in Sect. 2. As a basis for subsequent sections, Sect. 3 elaborates on the description of fluid flow through a pipe, making all assumptions of the model transparent. This results in a system of partial differential equations, for which Sect. 3.3 conveys a discretization specific to the particular graph-like domain. Then we build on this proposal and introduce more involved network elements in Sect. 4. Finally, in Sect. 5, we discuss how to locally target or enforce pressure, flow, and temperature.

2   Network

For a network of pipes, we denote the domain of computation as a directed graph

$$\mathcal{G} = (\mathcal{N}, \mathcal{E})$$

with nodes $\mathcal{N}$ and edges

$$\mathcal{E} \subset \{(n_1, n_2) : n_1, n_2 \in \mathcal{N}\}.$$

Each edge constitutes a pipe segment that is parametrized along its length by $x \in [0, L] \subset \mathbb{R}_0^+$. The beginning ($x = 0$) and the end ($x = L$) of an edge $e = (n_1, n_2)$ coincide with nodes n_1 and n_2, respectively. This identification of edge boundaries via nodes helps to define a topology on the entire graph. If some network consists of more involved elements, the edges need to be labeled, accordingly, so that an edge can always be related to its element type. Moreover, the model might need to encompass additional element-typical links between nodes or edges. For example, a heat exchanger connects two independent circuits in order to transfer heat power between them. In our graph description, this link is reflected by a reference between the corresponding two edges. Similarly, a mixing valve is modeled as a couple of independent edges, i.e., as two simple valves whose apertures are linked.

Selected nodes $N_0 \subset N$ incorporate sources and sinks (water inflow and outflow) with corresponding boundary conditions for volume flow or pressure—and temperature. The sources or sinks themselves are modeled as special elements with either fixed pressure or fixed volume flow. They are connected to only one node, i.e., to the location where the flow is introduced into or removed from the network. Such a model may seem complicated at first sight. It helps to automatically enforce necessary topological assumptions, though, while the user can (seemingly) be given almost arbitrary freedom to define sources and sinks. Notice that at least one source with a pressure boundary condition is needed per connected component of the graph, even for closed circuits. This condition corresponds to the fact that the pressure in real systems is maintained by so-called extension tanks. For each source, the temperature of the inflowing water must also be defined, of course.

Within the network, the only way to fix pressure, temperature, or volume flow is via controllable elements, see Sect. 5. This seemingly severe restriction again corresponds to the way in which such quantities are adjusted in real systems.

3 Water Pipes

All further elements can be seen as variations of one basic case, an ideal fluid flow through a single pipe. We derive equations for this case from the basic conservation laws of continuum mechanics in Sect. 3.1 and state assumptions in Sect. 3.2 that lead us to further simplifications. One outstanding way to simplify the equations is to consider the flow being constant over the cross-sectional area of the pipe. Finally, Sect. 3.3 proposes a discretization of the continuum equations, which keeps in mind that typical Newton-type solvers show difficulties dealing with zero derivatives. Therefore, physically motivated regularization is described for some of the terms in the equations.

3.1 Continuum Mechanics

For our purpose, it is sufficient to first consider water as an incompressible fluid flowing through a pipe with constant circular cross-section. In continuum mechanics, partial differential equations that describe the motion of an incompressible fluid across a domain are derived from the basic conservation laws for momentum and energy. Incompressibility, i.e., preservation of fluid volume for any open and connected subdomain, completes this set of equations. For a more detailed introduction to continuum mechanics with application to fluid flow, refer to [7].

Let (t, x) denote a point in space-time occupied by the fluid at time t. By $v(t, x)$ we denote the spatial velocity field at (t, x). A motion that preserves volume must satisfy the relation

$$\operatorname{div} v = 0. \tag{C1}$$

Since mass must also be preserved—which is stated by the relation $\partial_t \rho + \mathrm{div}(\rho v) = 0$, known as the continuity equation—incompressibility can be seen to amount to $\dot{\rho} := \partial_t \rho + v \cdot \nabla \rho = 0$.

A fluid flow is called *Eulerian* if the stress within the fluid can be described by pressure p as $(-pI)$, where I denotes the identity. Gravity, a conservative body force, can be written as the gradient $(-\rho g \nabla x_3)$ of the height potential x_3. The Eulerian flow, with gravity as body force, conserves linear and angular momentum if

$$\partial_t v + (v \cdot \nabla)v + \frac{\nabla p}{\rho} + g\nabla x_3 = 0. \tag{M1}$$

Incompressible Eulerian fluids are called *ideal*.

Let the density of internal energy be denoted by $\rho \epsilon$, the heat flux per area by q, and the volumetric heat supply by r. If we take implications of the continuity and motion equations into account, the energy equation is usually given in the simplified form

$$\rho \left(\partial_t \epsilon + v \cdot \nabla \epsilon \right) = r - \mathrm{div}\, q - pI \cdot \nabla v. \tag{E1}$$

3.2 Simplifying Assumptions

Following [11, 13], we describe possible simplifications of an ideal flow. For pipes, one usually considers averaged values over the (fixed) cross-sectional area, such that only one spatial dimension remains, and the incompressibility equation reads

$$\partial_x v = 0. \tag{C2}$$

Friction is often incorporated into momentum conservation by an additional force term $\left(-\frac{\rho \lambda v |v|}{2D} \right)$, where λ denotes the friction factor. In addition, the convective term $v \partial_x v$ vanishes, which follows directly from the incompressibility equation of a volume preserving Eulerian flow. Thus we arrive at the equation of motion of an ideal fluid in a pipe

$$\partial_t v + \frac{\partial_x p}{\rho} + g\nabla x_3 + \frac{\lambda v |v|}{2D} = 0. \tag{M2}$$

The energy equation takes the form

$$\partial_t \epsilon + v \partial_x \epsilon = \frac{r - \partial_x q}{\rho}, \tag{E2}$$

since the term involving pressure vanishes due to the incompressibility assumption. In terms of specific enthalpy $h = \epsilon + \frac{p}{\rho}$ we get

$$\partial_t h + v \partial_x h = \frac{1}{\rho} (\partial_t p + v \partial_x p) + \frac{r - \partial_x q}{\rho}, \tag{E3}$$

since $\left(\frac{p}{\rho}\right)^{\cdot} = \frac{\dot{p}}{\rho} - \frac{p}{\rho^2}\dot{\rho} = \frac{\dot{p}}{\rho}$.

Internal heat flow q can simply be omitted, as is often done, see [13]. By preference, other possibly simplified laws may be applied to q. In Sect. 3.3 we are going to use a regularization of energy balance, for instance, which acts similarly to heat *transfer*.

The friction factor λ is a function of the Reynolds number and may also depend on the roughness k in turbulent flow regimes. One can mainly distinguish between laminar flow [12], where Re < 2320 and

$$\lambda = \frac{64}{\text{Re}},$$

and turbulent flow, for which, among others, an approximate implicit formula exists [5] that determines λ as

$$\frac{1}{\sqrt{\lambda}} = -2 \log \left(\frac{k}{3.7D} + \frac{2.51}{\text{Re}\sqrt{\lambda}} \right).$$

If, as we suppose, a Newton-type algorithm is available, the implicit formula can be incorporated into the system of equations and handed over to the solver without major adaption. Nevertheless, several explicit approximations exist, such as the one by Nikuradse for higher Reynolds numbers, see also [5]. For more details, we refer to survey texts such as [1, 10]. Notice that if the flow within a simulation changes between different regimes, our assumption of a Newton-type solver requires the corresponding formulae to be combined *smoothly* in order for derivatives to exist. Examples of regularizations for nonsmooth functions are given in the context of Sect. 3.3.

A possible further simplification stems from the assumption that pressure effects contribute but are negligible. Then we finally arrive at

$$\partial_t h + v \partial_x h = \frac{r}{\rho}. \tag{E4}$$

To relate specific enthalpy to temperature T, one of the simplest ways is to introduce a constant $c_p = \partial_T h$ or, within small temperature ranges,

$$c_p T = h. \tag{1}$$

A simple approximation of heat supply r from the environment assumes a linear relationship between r and T

$$r = \frac{4c_{ht}}{D}(T - T_s),\tag{2}$$

where c_{ht} denotes the heat transfer coefficient and T_s the surrounding temperature.

Making use of all described simplifications, we arrive at a system of partial differential equations (C2), (M2), and (E4) with enthalpy and heat supply given by (1) and (2), respectively. If we introduce mass flow $m = \int_{\Omega_A} \rho v \, d\mathcal{H}^2(x)$ through cross section Ω_A with area A, the resulting system takes the form

$$\partial_x m = 0,$$

$$\frac{1}{A}\partial_t m + \partial_x p + \rho g \nabla x_3 + \frac{\lambda m |m|}{2DA^2\rho} = 0,$$

$$c_p\left(\rho A \partial_t T + m \partial_x T\right) = \frac{4Ac_{ht}}{D}(T - T_s).$$

A rigorous treatment would need to address the averages of nonlinear quantities. Instead we assume almost constant temperature across Ω_A. We also remark that $d\mathcal{H}^2(x)$ denotes the Hausdorff measure on Ω of dimension 2.

The additional assumptions of quasi-stationarity and perfect pipe insulation finally lead to a system of equations, which forms the basis of everything that is going to follow,

$$\partial_x m = 0,$$

$$\partial_x p + \rho g \nabla x_3 + \frac{\lambda m |m|}{2DA^2\rho} = 0,$$

$$m \partial_x T = 0.$$

3.3 Discretization and Regularization

In order to define a discretization along a single edge, we almost entirely follow an implementation of *gas* transport in networks [2]: The well-known staggered (non-collocated) grid [8, Fig. 2] in one spatial dimension will be used together with central differencing. Consider a pipe segment $e = (n_1, n_2)$ again. Let p_i denote the pressure value at node n_i, $i = 1, 2$, m the flow at the center of the edge, i.e., at $x = \frac{L}{2}$. Thus the equation of motion becomes

$$\frac{p_2 - p_1}{L} + \rho g \nabla x_3 + \frac{\lambda m |m|}{2DA^2\rho} = 0.\tag{M3}$$

The discretized form of the incompressibility equation must now be defined across pipe junctions. It is instructive to turn to the integral form of the equation in order to derive an appropriate law,

$$\int_{\Omega} \partial_x m \, dx. \tag{C3}$$

Here, Ω denotes all half pipe segments connected to a node n, averaged across A. By Gauss's theorem we have

$$\int_{\Omega} \partial_x m \, dx = \int_{\partial\Omega} m \cdot \eta \, d\mathcal{H}^0(x),$$

where η, not to be confused with velocity v, denotes the outward normal to $\partial\Omega$. Since mass flow m_i is defined for each edge e_i at the center in the direction of the pipe segment, we can write

$$\sum_i \eta_i m_i = 0 \tag{C4}$$

over all edges $e_i = (n_1, n_2)$ for which $n_1 = n$ or $n_2 = n$. Here, $\eta_i = 1$ if e_i ends in n, and $\eta_i = -1$ if e_i starts in n.

Notice that (C4) corresponds to the well-known Kirchhoff law of electric circuits. Sources or sinks given by a mass flow m at n can be considered within the sum, too,

$$\sum_i \eta_i m_i + m = 0,$$

where the sign of m determines the flow direction as before. As explained in Sect. 2, m acts as an additional inflow or outflow edge with fixed direction that does not contribute equations for its opposite node.

Similar to the mass flow, the temperature T is considered to be constant along pipes. At nodes, the total energy is preserved, i.e., the energy entering a node equals the energy leaving the same node. When water flows of different temperatures unite at a node, the resulting mixed temperature is also determined by the energy balance equation. Due to the second law of thermodynamics, the temperature at the node cannot split up again into separate temperatures for different outgoing pipes, though.

A nodal discretization of temperature would be sufficient for networks of pipes only. In order to allow for other components, we introduce a discretization of temperature with values at all nodes *and* at all edge centers, so that the different effects may be well separated. For simplicity, we again start from the integral formulation of energy (recall $m = \int_{\Omega_A} \rho v \, d\mathcal{H}^2(x)$ and $\partial_x m = 0$, T almost constant across A)

$$0 = c_p \rho \int_\Omega \int_{\Omega_A} v \partial_x T \, d\mathcal{H}^2(x) \, dx = c_p \int_\Omega m \partial_x T \, dx$$

$$= c_p \int_{\partial\Omega} mT \cdot \eta \, d\mathcal{H}^0(x). \tag{E5}$$

First consider the domain Ω of all half pipes that start or end at a distinguished node n. Then partition the domain $\Omega = \Omega_1 \cup \Omega_2$ into ingoing and outgoing flows. The boundary $\partial\Omega_1$ consists of the centers of the pipes of ingoing flows and of their common junction point. Let us denote the temperature and flow at the junction by T_0 and m_0, respectively. Then, we have

$$0 = c_p \int_{\partial\Omega_1} mT \cdot \eta \, d\mathcal{H}^2(x) = \sum_{i:\, \eta_i m_i > 0} \eta_i m_i T_i - m_0 T_0$$

$$= \sum_{i:\, \eta_i m_i > 0} \eta_i m_i T_i - \sum_{i:\, \eta_i m_i > 0} \eta_i m_i T_0 \tag{E6}$$

over all edges $e_i = (n_1, n_2)$ such that $(n_1 = n) \wedge (m < 0)$ or $(n_2 = n) \wedge (m > 0)$. We can rewrite the law as

$$0 = \sum_{i:\, \eta_i m_i > 0} \eta_i m_i (T_i - T_0) = \sum_i \max(\eta_i m_i, 0) (T_i - T_0). \tag{E7}$$

The term $\max(\eta m, 0)$ has a discontinuous derivative at $m = 0$: Due to our simplifying assumptions, the temperature is not defined if the velocity of the fluid becomes identically zero in some part of the system. Depending on the solver, $\max(\eta m, 0)$ thus needs to be regularized as

$$\max[\epsilon](\eta m, 0) = \frac{1}{2} \left(\sqrt{(\eta m - \epsilon)^2 + \epsilon^2} + \eta m + \epsilon \right) \tag{3}$$

with small regularization values $\epsilon > 0$, for example. This regularization introduces slight thermal transfer against the direction of flow between connected pipes, similar to a heat exchange (2) with the environment. Consider first the simpler regularization $\max[\epsilon](\eta m, 0) = \frac{1}{2} \left(\sqrt{(\eta m)^2 + \epsilon^2} + \eta m \right)$ [12, (15)]. Then (3) is derived by addition of ϵ and subsequent translation along m by $\eta \epsilon$. As a consequence, $\max[\epsilon](\eta m, 0) \to \epsilon$ for $\eta m \to -\infty$ and $\max[\epsilon](\eta m, 0) \to \eta m$ for $\eta m \to \infty$. In the latter case, the regularization acts as $\max(\eta m, 0)$. For $\eta m_i \to -\infty$, we obtain a summand $\epsilon(T_i - T_0)$ in (E7), which can be understood as the previously postulated thermal transfer with small heat transfer coefficient ϵ. This transfer is not physically justified. Albeit, such a behavior agrees with the effect of slight diffusion up to a certain extent.

We have, hence, determined a *node* law (E7) for incoming flows. The outgoing flows need to be considered *per edge* due to the second law of thermodynamics. We write the edge law as

$$\max(m, 0)(T - T_1) + \max(-m, 0)(T_2 - T) = 0 \tag{E8}$$

over all edges $e_i = (n_1, n_2)$, where we denote values at $n_i, i = 1, 2$ by their respective indices and values at the pipe center without index. Regularization may again be applied as before.

Eventually, the discretized system consists of $|\mathcal{E}|$ equations of the form (M3) and $|\mathcal{N}|$ equations of the form (C4) as well as of $|\mathcal{N}| + |\mathcal{E}|$ equations, possibly regularized, of the form (E7) or (E8), respectively.

4 Further Devices

Although pipe flow represents the basis of any other element in the network, it is often not straightforward to derive numerically stable equations for other devices. We are going to discuss several simpler and more challenging cases, starting with a quite universally applicable way to model resistance within controls and instruments.

4.1 Resistors and Valves

In general, many complex elements can be expected to act as resistors

$$p_2 - p_1 = \zeta \frac{m\,|m|}{2A^2\rho} \tag{M4}$$

if not noted otherwise [9, 3.2.1.2.2]. The coefficient $\zeta \geq 0$ must be determined experimentally or from characteristic data. Examples of such elements are system coolers, heat exchangers, adsorbers, etc.

Also, the above formula is often written as

$$k_v^2(p_2 - p_1) = \frac{m\,|m|}{\rho^2}, \tag{M5}$$

k_v given with unit $\left[\frac{m^3}{h}\right]$, p_1, p_2 with unit [bar]. If a control valve can be gradually closed, $k_v(h)$, as a function of aperture h, may range between 0 (closed) and k_{vs}

(fully open), where k_{vs} denotes resistance at maximum aperture $h_{\max}$. The dependence $k_v(h)$ is determined by the shape of the control valve. Typical characteristics are linear, i.e.,

$$\frac{k_v(h)}{k_{vs}} = \frac{h}{h_{\max}},$$

or "equal percentage", i.e.,

$$\frac{k_v(h)}{k_{vs}} = \frac{k_{vo}}{k_{vs}} \left(\frac{k_{vs}}{k_{vo}}\right)^{\frac{h}{h_{\max}}} = \left(\frac{k_{vs}}{k_{vo}}\right)^{\frac{h}{h_{\max}}-1},$$

where k_{vo} corresponds to the k_v-value at $h = 0$, see [14]. Often, k_{vo} is indirectly given via a theoretical rangeability $\frac{k_{vs}}{k_{vo}}$. In other cases, it can be approximated from the smallest controllable flow k_{vr} (sometimes conveyed via rangeability $\frac{k_{vs}}{k_{vr}}$). Similar to "equal percentage", the linear characteristics can also incorporate rangeability, which behaves like a regularization for $h = 0$ in both cases, see the initial "jump" in Figs. 1 and 2. A three-way mixing valve acts as two valves, where apertures h_1 and h_2 are related by

$$\frac{h_1 + h_2}{h_{\max}} = 1.$$

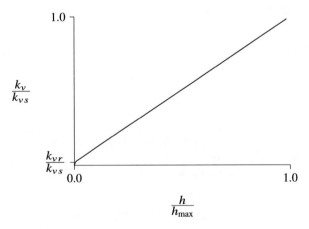

Fig. 1 Valve—linear characteristics. Published with kind permission of ©Fraunhofer SCAI 2016. All Rights Reserved

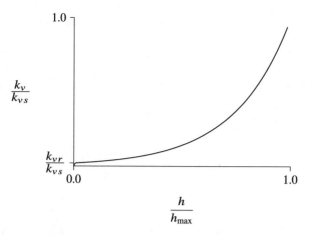

Fig. 2 Valve—equal percentage characteristics. Published with kind permission of ©Fraunhofer SCAI 2016. All Rights Reserved

4.2 Pumps

Unregulated pumps are inserted to a network in order for the outgoing pressure to be increased in an edge. The increase of pressure of a pump is usually given as a function H_ω of volumetric flow Q, where $\rho Q = m$ and $H_\omega(Q)$ is scaled to have unit [m],

$$\frac{\partial_x p}{\rho} = g H_\omega \left(\frac{m}{\rho} \right). \tag{M6}$$

$H_\omega(Q)$ depends on the (rotational) frequency ω, which can be coupled to a goal such as pressure or temperature for regulated pumps. In this case, ω becomes a free parameter determined by an additional equation such as $p = p_{\text{goal}}$ or $T = T_{\text{goal}}$. If the goal cannot be reached, optimization methods need to be applied. A more detailed discussion of goals, control, and optimization can be found in Sect. 5.

The dependence of H_ω on ω obeys the following similarity condition: If (H, Q) is an element of the graph of $H_{\omega_0}(Q)$, then $(\tilde{H}, \tilde{Q})$ is an element of the graph of $H_\omega(Q)$, where

$$\tilde{H} = \left(\frac{\omega}{\omega_0} \right)^2 H, \quad \tilde{Q} = \frac{\omega}{\omega_0} Q.$$

Thus

$$H_\omega(\tilde{Q}) = \tilde{H} = \left(\frac{\omega}{\omega_0} \right)^2 H_{\omega_0}(Q) = \left(\frac{\omega}{\omega_0} \right)^2 H_{\omega_0} \left(\left(\frac{\omega}{\omega_0} \right)^{-1} \tilde{Q} \right).$$

Usually, $H(Q)$ is available for a maximum frequency ω_0 and the working frequency is given as $\frac{\omega}{\omega_0} \cdot 100\,\%$.

Apart from the pressure increase, it is important to calculate the electric power consumption of a pump. Power consumption P_ω depends again on the volumetric flow through the pump and is usually scaled to have unit [kW]. The dependence of P_ω on ω again obeys a similarity law: If (P, Q) is an element of the graph of $P_{\omega_0}(Q)$, then $(\tilde{P}, \tilde{Q})$ is an element of the graph of $P_\omega(Q)$, where

$$\tilde{P} = \left(\frac{\omega}{\omega_0}\right)^3 P, \quad \tilde{Q} = \frac{\omega}{\omega_0} Q.$$

Hence,

$$P_\omega(\tilde{Q}) = \tilde{P} = \left(\frac{\omega}{\omega_0}\right)^3 P_{\omega_0}(Q) = \left(\frac{\omega}{\omega_0}\right)^3 P_{\omega_0}\left(\left(\frac{\omega}{\omega_0}\right)^{-1} \tilde{Q}\right).$$

It is interesting to notice that power consumption does not enter other computations except as a possible optimization goal. In many *simulation* cases, we can, therefore, decouple its calculations from the rest of the system.

The above exposition closely follows [9], where more information on characteristic data can be found. Figures 3 and 4 show examples for pump height and power characteristics from practice.

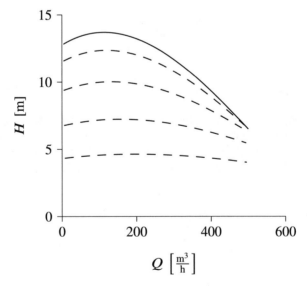

Fig. 3 Pump—height characteristics Q vs. $H_{1474\,\mathrm{min}^{-1}}$ (*solid line*) and vs. $H_{\omega \cdot 1474\,\mathrm{min}^{-1}}$ with $\omega = 95\,\%, 90\,\%, 85\,\%, 80\,\%$ (*dashed lines*) for the pump Grundfos NKE 150-200/224 [6]. Published with kind permission of ©Fraunhofer SCAI 2016. All Rights Reserved

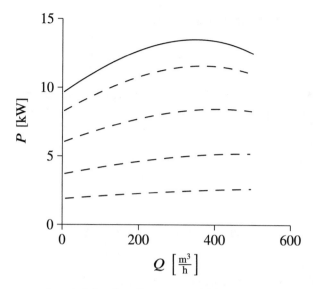

Fig. 4 Pump—power characteristics Q vs. $P_{1474\,\mathrm{min}^{-1}}$ (*solid line*) and vs. $P_{\omega\cdot1474\,\mathrm{min}^{-1}}$ with $\omega = 95\,\%, 90\,\%, 85\,\%, 80\,\%$ (*dashed lines*) for the pump Grundfos NKE 150–200/224 [6]. Published with kind permission of ©Fraunhofer SCAI 2016. All Rights Reserved

4.3 Heat Exchangers

A countercurrent heat exchanger transfers energy between two separate water circuits: The cool circuit is heated and the hot circuit cooled. A heat exchanger can be modeled by two parallel elements with directions opposite to each other. Let us denote the elements by $e_1 = (n_1, n_3)$ and $e_2 = (n_2, n_4)$ and first assume mass flows $q_{e_i} \geq 0$ for $i = 1, 2$. The energy equation for each element must now encompass the heat transfer r between them, i.e.

$$c_p m_{e_1}(T_{e_1} - T_{n_1}) = r,$$
$$c_p m_{e_2}(T_{e_2} - T_{n_2}) = -r. \tag{E9}$$

Here, T_{n_1} and T_{n_2} denote the inflow temperatures. The transferred amount of energy r is determined by

$$r = \frac{\frac{4c_{\mathrm{ht}}L}{D}\left((T_{e_1} - T_{n_2}) - (T_{n_1} - T_{e_2})\right)}{\ln\left|\dfrac{T_{e_1} - T_{n_2}}{T_{n_1} - T_{e_2}}\right|}, \tag{4}$$

where L and D, respectively, refer to the length and distance of the pipes along which heat transfer occurs. The coefficient $\frac{c_{\mathrm{ht}}L}{D}$ must be determined experimentally or from characteristic data.

In order to derive (4) following [4], consider

$$c_p m_{e_1} \partial_x T_1(x) = \frac{4c_{ht}}{D} \left(T_1(x) - T_2(x) \right), \tag{E10}$$

which describes the heat exchange at each contact point x of the pipes. Here, $T_i(x)$, $i = 1, 2$, is the temperature at x of edge e_i averaged over its cross section. Notice that (E9) is concluded from (E10) by integration. For further calculations also recall that by (E9)

$$c_p m_{e_1} (T_{e_1} - T_{n_1}) = c_p m_{e_2} (T_{n_2} - T_{e_2}),$$

of which

$$\frac{m_{e_1}}{m_{e_2}} = \frac{T_{n_2} - T_{e_2}}{T_{e_1} - T_{n_1}} = \frac{T_2(x) - T_{e_2}}{T_1(x) - T_{n_1}}$$

is an immediate consequence. Now we can rewrite (E10) as

$$\frac{1}{\frac{4c_{ht}}{D} (T_1(x) - T_2(x))} \partial_x T_1(x) = \frac{1}{c_p m_{e_1}}$$

and integrate

$$\int_0^L \frac{1}{\frac{4c_{ht}}{D} (T_1(x) - T_2(x))} \partial_x T_1(x) dx = \int_0^L \frac{1}{c_p m_{e_1}} dx$$

$$\iff \int_{T_{n_1}}^{T_{e_1}} \frac{1}{\frac{4c_{ht}}{D} (T_1 - T_2)} dT_1 = \frac{L}{c_p m_{e_1}}. \tag{E11}$$

The integral on the left hand side can be solved, too,

$$\frac{1}{\frac{4c_{ht}}{D} \left(1 - \frac{m_{e_1}}{m_{e_2}}\right)} \int_{T_{n_1}}^{T_{e_1}} \frac{\left(1 - \frac{m_{e_1}}{m_{e_2}}\right)}{\left(T_1 - \frac{m_{e_1}}{m_{e_2}} T_1 + \frac{m_{e_1}}{m_{e_2}} T_{n_1} - T_{e_2}\right)} dT_1$$

$$= \frac{1}{\frac{4c_{ht}}{D} \left(1 - \frac{m_{e_1}}{m_{e_2}}\right)} \int_{T_{n_1}}^{T_{e_1}} \frac{1}{\left(T_1 + \frac{m_{e_1}}{m_{e_2} - m_{e_1}} T_{n_1} - \frac{m_{e_2}}{m_{e_2} - m_{e_1}} T_{e_2}\right)} dT_1$$

$$= \frac{1}{\frac{4c_{ht}}{D} \left(1 - \frac{m_{e_1}}{m_{e_2}}\right)} \ln \left| \frac{T_{e_1} + \frac{m_{e_1}}{m_{e_2} - m_{e_1}} T_{n_1} - \frac{m_{e_2}}{m_{e_2} - m_{e_1}} T_{e_2}}{T_{n_1} + \frac{m_{e_1}}{m_{e_2} - m_{e_1}} T_{n_1} - \frac{m_{e_2}}{m_{e_2} - m_{e_1}} T_{e_2}} \right|$$

$$= \frac{1}{\frac{4c_{ht}}{D} \frac{(T_{e_1} - T_{n_2}) - (T_{n_1} - T_{e_2})}{T_{e_1} - T_{n_1}}} \ln \left| \frac{m_{e_2} (T_{e_1} - T_{n_2})}{m_{e_2} (T_{n_1} - T_{e_2})} \right|.$$

Substituting the result back into (E11) leads to

$$c_p m_{e_1} (T_{e_1} - T_{n_1}) = \frac{\frac{4c_{ht}L}{D} ((T_{e_1} - T_{n_2}) - (T_{n_1} - T_{e_2}))}{\ln \left| \dfrac{T_{e_1} - T_{n_2}}{T_{n_1} - T_{e_2}} \right|},$$

from which (4) can eventually be derived. The derivation is based on the more general case of [4] and takes into account that we deal with a countercurrent flow.

If the mass flows can change direction, (E9) and (4) need to be adapted in the spirit of (E8). When the right hand side of (4) is to be employed as a term within an equation that shall be solved by a Newton-type algorithm, a series of obvious problems may occur. For $T_{e_1} \to T_{n_2}$ or $T_{n_1} - T_{e_2} \to \pm\infty$, all terms $\partial_{T_{n_i}} r$ and $\partial_{T_{e_i}} r$, with $i = 1, 2$, may have a limit that is not defined. For $T_{e_1} - T_{n_2} \to \infty$ or $T_{n_1} \to T_{e_2}$, the limits of both $\partial_{T_{n_i}} r$ and $\partial_{T_{e_i}} r$, with $i = 1, 2$, might vanish. Similar situations occur for $(T_{e_1} \to T_{n_1}) \wedge (T_{e_2} \to T_{n_2})$, for example. We, therefore, have to expect any Newton-type solver to exhibit a great variety of numerical problems.

In the context that we assume, it is necessary to find a smooth reformulation with non-zero derivatives. As to our knowledge, there does not seem to exist an appropriate proposal in the literature. However, several possible solutions come to mind. One might be able to find a suitable approximation of the function $r(T_{e_1}, T_{n_1}, T_{e_2}, T_{n_2})$, for instance. Yet, a thorough investigation of novel methods is beyond the scope of this article. Hence, we just give a brief hint how some regularization similar to that of Sect. 3.3 may be applied. We have observed such a regularization to still exhibit numerical issues, however.

If we consider (4), (E9) takes the form

$$c_p m_{e_1} (T_{e_1} - T_{n_1}) = \frac{\frac{4c_{ht}L}{D} ((T_{e_1} - T_{n_2}) - (T_{n_1} - T_{e_2}))}{\ln \left| \dfrac{T_{e_1} - T_{n_2}}{T_{n_1} - T_{e_2}} \right|},$$

$$c_p m_{e_2} (T_{e_2} - T_{n_2}) = -\frac{\frac{4c_{ht}L}{D} ((T_{e_1} - T_{n_2}) - (T_{n_1} - T_{e_2}))}{\ln \left| \dfrac{T_{e_1} - T_{n_2}}{T_{n_1} - T_{e_2}} \right|}.$$

In order to avoid the above mentioned problems, we suggest to rewrite the above formulae as

$$\frac{\frac{c_p D}{4 c_{ht} L} (\ln |T_{e_1} - T_{n_2}| - \ln |T_{n_1} - T_{e_2}|)}{(T_{e_1} - T_{n_2}) - (T_{n_1} - T_{e_2})} (m_{e_1} (T_{e_1} - T_{n_1})) - 1 = 0,$$

$$\frac{\frac{c_p D}{4 c_{ht} L} (\ln |T_{e_1} - T_{n_2}| - \ln |T_{n_1} - T_{e_2}|)}{(T_{e_1} - T_{n_2}) - (T_{n_1} - T_{e_2})} (m_{e_2} (T_{e_2} - T_{n_2})) + 1 = 0.$$

Applying regularization, one might pursue different goals. The denominator $(T_{e_1} - T_{n_2}) - (T_{n_1} - T_{e_2})$ should be prevented to become identically zero and, similarly, the arguments of $|\cdot|$ ought not to vanish. Of course, it makes sense for m_{e_i}, $i = 1, 2$, to be regularized as in Sect. 3.3, which may avoid undefined temperatures for zero flow velocities. All put together, a possible regularization choice reads

$$\frac{\frac{c_p D}{4 c_{ht} L} \left(\ln \sqrt{(T_{e_1} - T_{n_2})^2 + \epsilon^2} - \ln \sqrt{(T_{n_1} - T_{e_2})^2 + \epsilon^2} \right)}{\sqrt{((T_{e_1} - T_{n_2}) - (T_{n_1} - T_{e_2}))^2 + \epsilon^2}} \cdot \alpha_1 = 0,$$

$$\frac{\frac{c_p D}{4 c_{ht} L} \left(\ln \sqrt{(T_{e_1} - T_{n_2})^2 + \epsilon^2} - \ln \sqrt{(T_{n_1} - T_{e_2})^2 + \epsilon^2} \right)}{\sqrt{((T_{e_1} - T_{n_2}) - (T_{n_1} - T_{e_2}))^2 + \epsilon^2}} \cdot \alpha_2 = 0,$$

where

$$\alpha_1 = \left(\max[\epsilon] \, (m_{e_1}, 0) \, (T_{e_1} - T_{n_1}) \right) - 1,$$

$$\alpha_2 = \left(\max[\epsilon] \, (m_{e_2}, 0) \, (T_{e_2} - T_{n_2}) \right) - 1.$$

We have been able to successfully solve some practical issues with this choice. It might still be necessary to handle the case $T_{e_1} - T_{n_2} \to T_{n_1} - T_{e_2}$, though. On the other hand, regularizations, applied too generously, can contradict each other or lead to unacceptable errors. In total, we believe that a more detailed analysis and discussion is necessary, which is beyond the ambit of this text. However, we hope to have been able to point towards several promising directions that might be of help in practical cases.

5   Element Control

We have already introduced several devices that allow for certain control of the network elements. Boundary conditions such as inflow and outflow can be seen as controllable variables in some cases, too. Pressure can, for instance, be controlled by an inflow node that behaves similar to an extension tank.

Another network control example, already mentioned, are regulated pumps or valves. They both incorporate an additional parameter that can be used to achieve certain goals. For example, a pump increases pressure. Therefore, pump speed, if variable and controllable, can be used to attain a fixed pressure value at a specified point of the network. Valve aperture influences the flow ratio at a junction node and can thus be used to control the flow through a specified element.

Letting pump speed or valve aperture be an independent variable of the system of equations, we can, hence, attach an additional control equation to the system, setting flow or pressure to given values m_{goal} or p_{goal}, i.e.,

$$m = m_{goal}, \quad p = p_{goal},$$

or, by preference, enforcing a specified temperature at the mixing point or a specified flow and pressure difference. Within real systems, such regulating tasks are assumed by algorithms, seldom unveiled to the users. It can, therefore, not be expected that goals implemented by the above approach mimic the behavior of the system exactly.

There might also be situations where the equation implementing our goal cannot be fulfilled. In fact, the regularization of (E7) and (E8) can already pose a problem. Consider the case where two different flows m_1 and m_2 with corresponding temperatures T_1 and T_2 are mixed via a three-way valve in order to fulfill a temperature goal T_{goal} for the combined flow $m = m_1 + m_2$ with resulting temperature T. Energy balance then has the form

$$Tm = m_1 T_1 + m_2 T_2.$$

Now let $\tilde{m}_i$, for $i = 1, 2$, denote the flow regularized by (3). The mixed temperature of the regularized system reads

$$T = \frac{\tilde{m}_1 T_1 + \tilde{m}_2 T_2}{m_1 + m_2}.$$

For illustrative purposes, we set $m_1 + m_2 = 1$, $T_1 = 10$, $T_2 = 30$, see Fig. 5, where units are completely omitted without loss of validity. In this case, $\epsilon = 0.1$ is chosen larger than usual in order to emphasize the visual impression. This simple example already shows that the full range $[T_1, T_2]$ is not always covered. We can even get spurious temperatures $T > T_2$. Moreover, it can be observed that $\frac{\partial T}{\partial m_1}$ vanishes within the interval $[0, 1]$ of values of m_1. An interrelation with valve characteristics and their rangeability complicates matters even more and we have, indeed, noticed numerical issues in combination with Newton-type equation solvers. In several cases, the problems could be remedied partly by the use of an interior point optimization algorithm, which is why such an algorithm is assumed to be available. In general, the reader is advised to carefully consider the close interplay between controlled variable, goal, equation formulation, and numerical solver.

The use of an optimization method does not only allow for a treatment of numerical issues. Its main use can be found in the reduction of quantities such as consumed electric power. If the electric power consumption of a pump is incorporated into the system of equations, for example, a deep coupling of all effects, including optimization and simulation, is within eyeshot: Power minimization and, if necessary, control goals are dealt with by an interior point method, whereas the entire system of equations is considered as an equality constraint to the optimization problem.

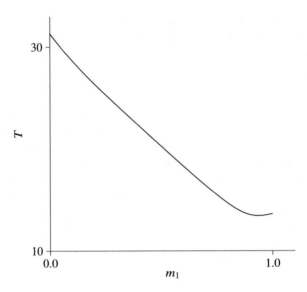

Fig. 5 Node—mixed temperature for a regularized system with $\epsilon = 0.1$. Published with kind permission of ©Fraunhofer SCAI 2016. All Rights Reserved

6   Conclusion

Simplified continuum mechanics for the pipe flow in heating or cooling circuits have been described. A graph represents the circuit domain itself, for which the flow equations are discretized. Further typical elements of such circuits, like resistors, pumps, valves, and energy exchangers, have been introduced and discussed— always in the light of their numerical treatment. These elements available, it is possible to assemble typical cooling or heating situations. Numerical considerations have been presented with a Newton-type solver in mind and, where necessary, accessibility of an interior point optimization algorithm. Element control by a wide range of control goals has been approached as well as optimization of certain computable quantities such as of consumed power.

Numerical considerations have been examined on a very basic and pragmatic level only. While the simulation of liquid flow through pipes has been thoroughly addressed in the literature, we believe numerical methods for some of the other elements might benefit from further work. Although all simplifying assumptions have been chosen deliberately, it will be of interest to relax these simplifications in many practical cases. Another possible direction of thought is sparked by the question how different optimization and control goals should best be combined. To ensure a wide applicability to real situations, equations for variations of the presented elements (such as more advanced types of heat exchangers or simple radiators) and for elements not considered yet (such as hydraulic compensators, cooling or heating based on adsorption, absorption, or evaporation) need to be developed.

In [3], an exemplary use case is executed, which opens the discussion of more practical issues: A typical live system might suffer from a considerate lack of technical documentation. Necessary measurements to bridge the lack of technical data might not be available, are difficult to access, or spurious, while automatic control behavior may not be disclosed, etc. Thus [3] is going to report on simple possibilities to determine system properties approximately from insufficient data and compare the simulation of a real case to previously recorded measurements.

References

1. D. Brkić, Review of explicit approximations to the Colebrook relation for flow friction. J. Pet. Sci. Eng. **77** 34–48, (2011)
2. K. Cassirer, T. Clees, B. Klaassen, I. Nikitin, L. Nikitina, MYNTS User's Manual, Release 3.7, Fraunhofer SCAI, Sankt Augustin, Germany, http://www.scai.fraunhofer.de/mynts, Dec. 2015
3. T. Clees, N. Hornung, D. Labrenz, et al., Cooling circuit simulation II: A numerical example, in *Scientific Computing and Algorithms in Industrial Simulations*, ed. by M. Griebel, A. Schüller, M.A. Schweitzer (Springer, Cham, 2017). doi:10.1007/978-3-319-62458-7
4. A.P. Colburn, Mean temperature difference and heat transfer coefficient in liquid heat exchangers. Ind. Eng. Chem. **25**, 873–877 (1933)
5. C.F. Colebrook, Turbulent flow in pipes, with particular reference to the transition region between the smooth and rough pipe laws. J. Inst. Civ. Eng. **11**, 133–156 (1939)
6. Grundfos Product Center, http://product-selection.grundfos.com/. Accessed 13 Nov 2015
7. M.E. Gurtin, *An Introduction to Continuum Mechanics*. Mathematics in Science and Engineering, vol. 158 (Elsevier, Amsterdam, 1981)
8. F.H. Harlow, J.E. Welsh, Numerical calculation of time-dependent viscous incompressible flow of fluid with free surface. Phys. Fluids **8**, 2182–2189 (1965)
9. KSB Aktiengesellschaft, *Auslegung von Kreiselpumpen*, 5th edn. (KSB Aktiengesellschaft, Frankenthal, 2005)
10. J. Mischner, Zur Ermittlung der Rohrreibungszahl, in *gas2energy.net: Systemplanerische Grundlagen der Gasversorgung* (Oldenbourg Industrieverlag, München, 2011), pp. 402–420
11. B.-U. Rogalla, A. Wolters, Slow transients in closed conduit flow–Part I numerical methods, in *Computer Modeling of Free-Surface and Pressurized Flows*, ed. by M.H. Chaudhry, L.W. Mays, NATO ASI Series, vol. 274 (Springer, Dordrecht, 1994), pp. 613–642
12. M. Schmidt, M.C. Steinbach, B.M. Willert, High detail stationary optimization models for gas networks. Optim. Eng. **16**, 131–164 (2015)
13. V.D. Stevanovic, B. Zivkovic, S. Prica, et al., Prediction of thermal transients in district heating systems. Energy Convers. Manag. **50**, 2167–2173 (2009)
14. VDI, Strömungstechnische Kenngrößen von Stellventilen und deren Bestimmung, VDI/VDE 2173 (Verein deutscher Ingenieure, Verband deutscher Elektrotechniker, Berlin, 1962)

Part II
Products

Algebraic Multigrid: From Academia to Industry

Klaus Stüben, John W. Ruge, Tanja Clees, and Sebastian Gries

1 Introduction

Looking back in time, multigrid has been—and still is—one of the most important research topics in numerical solvers for over 40 years now. While the multigrid idea for Poisson-like partial differential equations (PDEs) in 2D had been described already in the early works of Fedorenko and Bakhvalov [4, 26], Brandt [9, 10] and Hackbusch [31] realized in the early 1970s the enormous practical potential as well as the generality of the multigrid methodology.

Since the early papers, a vast number of publications demonstrated the extraordinary efficiency of multigrid (cf. references in [54]). Although multigrid became accepted as the most efficient approach for the solution of wide classes of PDEs, to our knowledge there is still no industrial/commercial technical simulation package available which relies on 'classical' *geometric multigrid* (GMG) in a strict sense. Two major drawbacks have hindered industrial exploitation: On the one hand, there was no satisfactory way for existing simulation codes to exploit the potential benefits of GMG without a complete code re-writing. On the other hand, for the design of new complex simulation codes, GMG was not an obvious option either, a major reason being technical limitations in applying GMG in industrially very important situations such as *unstructured grids* and/or non-smooth, in particular, *discontinuous coefficients*.

K. Stüben (✉) • T. Clees • S. Gries
Fraunhofer Institute for Algorithms and Scientific Computing SCAI, Schloss Birlinghoven, 53757 Sankt Augustin, Germany
e-mail: klaus.stueben@scai.fraunhofer.de

J.W. Ruge
University of Colorado at Boulder, Boulder, CO, USA
e-mail: john.ruge@colorado.edu

© Springer International Publishing AG 2017
M. Griebel et al. (eds.), *Scientific Computing and Algorithms in Industrial Simulations*, DOI 10.1007/978-3-319-62458-7_5

Instead of geometric multigrid, *algebraic multigrid* (AMG) finally brought a breakthrough regarding the industrial use of multigrid ideas. This is remarkable because at its beginning in the early 1980s [13, 14, 44, 45, 48], AMG was regarded an exotic academic approach demonstrating that, to a certain extent, the exploitation of multigrid ideas does not necessarily require grids.[1] However, mainly due to its relatively high overhead (i.e., the setup cost), AMG did not appear to be competitive with either GMG or with much simpler classical iterative methods such as conjugate gradient preconditioned by standard one-level methods. Actually, this assessment was not too surprising because problem sizes of those days were still so small that AMG indeed seemed to be (and actually was) overkill.

Since the early 1990s, however, the solution of discretized PDEs had become a serious bottleneck for the practicability of industrial simulation software, resulting in an increasing demand for more efficient linear solvers. Although AMG was still in its infancy, its principle advantages (compared to GMG) made this approach highly interesting from a practical point of view. In particular, its formal integration into existing codes promised to be as easy as for any standard solver. Moreover, AMG was applicable to structured as well as unstructured grids, in 2D as well as in 3D, with smooth and non-smooth coefficients.

Finally, around 1995, this new situation gave a boost to the further research and development (R&D) on AMG-based approaches (see, for example, [54, Appendix A], [48] and the references given therein). Different AMG variants and related approaches as well as other hierarchically based algebraic solvers were developed. Soon afterwards, AMG became a major topic of research which still today is continuously represented at the international conferences on multigrid, namely, the 'European Multigrid Conference' and the 'Copper Mountain Multigrid Conference'. In a sense, regarding the improvement of robustness and the extension of generality, AMG has undergone a development similar to GMG. While the early AMG investigations essentially referred to linear systems with M-matrices, later on increasingly complex linear systems—mostly from discretized coupled systems of PDEs—moved into the foreground. For some specific references, see [36, 45, 55], and for a general survey of references, we refer to the proceedings of the above-mentioned conferences, in particular [2, 21].

When SCAI became a Fraunhofer institute in 2001, the development of the software package SAMG (System Algebraic MultiGrid) was started, and continues to today, with strict focus on its use in industrial simulation programs. As part of its development, the application of AMG to special highly complex industrial applications has been investigated in detail, see [16, 29, 51].

This paper is neither an introduction to GMG nor to AMG. Rather, we will review the path of AMG from an interesting academic toy to one of the most important new solution methods of today's industrial simulation tools. We outline

---

[1]Since AMG by default does not exploit any information about underlying geometric grids, one should actually use the term multi*level* rather than multi*grid*. Nevertheless, multi*grid* is still used for historical reasons.

the reasons for this success and describe the scientific and non-scientific efforts needed to 'bridge the gap' between academia and industry. Although this paper is written mainly from the point of view of Fraunhofer SCAI based on its experience with the software package SAMG, we try to embed SCAI's activities into the context of other developments.

In Sect. 2 we recall some GMG-typical difficulties which have turned out to be historically relevant for the introduction of AMG. Regarding the development of AMG, we then distinguish three phases: the 'basic phase' (1982–1987, Sect. 3), the 'resumption phase' (1995–2000, Sect. 4) and the 'main phase' which is characterized by an increasing work on software and software-relevant aspects as well as research on advanced applications of AMG (2001–today, Sect. 5). While Sect. 5 also summarizes the basics of SAMG, Sect. 6 describes three advanced industrial applications in some more detail which, among others, have driven the development of SAMG. Note that the technical challenges imposed by the parallelization of AMG are discussed in [42] in this volume.

2   From Geometric to Algebraic Multigrid

In geometric multigrid, discretization and solution were no longer considered separate processes but rather were interwoven. Assuming a given hierarchy of grids, information from finer grids was used only when absolutely necessary, with far most information taken from coarser grids. The optimality (scalability with respect to problem size) of concrete multigrid methods relies on the proper interplay between two basic principles:

- **Smoothing:** Local iterative processes (e.g., relaxation) are used to efficiently reduce *high frequency error components* of a fine-grid approximation, that is, those components which really need the fine-grid resolution for their representation.
- **Coarse-grid correction:** Once *low frequency error components*—that is, those components which are also seen on a coarser grid—dominate the error of the fine-grid approximation, a correction can be computed via the coarse-grid projection of the fine-grid residual equation.

In the early days of multigrid, traditional coarse-grid correction approaches were based on simple coarsening strategies (typically by $h \rightarrow 2h$ coarsening, that is, doubling the mesh size in each spatial direction), straightforward geometric grid transfer operators (linear interpolation and restriction by weighted averaging) and coarse-grid operators being natural analogs of the finest-grid one. A lot of R&D went into optimizing this approach for model problems such as Poisson-like equations on uniform grids. Optimized smoothing processes and transfer operators as well as techniques like *Full Multigrid* have been used to squeeze the last out of the performance for Poisson-like problems.

Unfortunately, in more complex applications, the traditional multigrid components were no longer able to ensure a good interplay between smoothing and
coarse-grid correction. Both the smoother and the coarse-grid correction process
(consisting of its components interpolation, restriction and coarse-grid operator)
needed careful, problem-dependent construction. As a consequence, in the late
1970s, there was a shift of interest from 'highly efficient' towards 'robust' multigrid
approaches. The following two model situations fostered this trend:

- **MS1: Anisotropic problems.** In GMG, smoothness on any grid is defined relative to the next coarser grid. For example, assuming uniform $h \to 2h$ coarsening,
  traditional pointwise relaxation is very efficient for essentially isotropic problems. For anisotropic problems, however, it exhibits good smoothing properties
  locally only 'in the direction of strong couplings'. More 'robust' smoothers, such
  as alternating line-relaxation or ILU-type smoothers, are required in order to
  maintain fast multigrid convergence in combination with uniform coarsening.
  While the implementation of robust smoothers was not difficult in 2D model
  situations, for 3D applications on complex meshes their realization tended to
  become rather cumbersome. For instance, the robust 3D analog of alternating line
  relaxation is alternating *plane* relaxation (realized by robust 2D multigrid within
  each plane) which, in complex geometric situations, becomes very complicated,
  if possible at all. ILU smoothers, on the other hand, loose much of their
  smoothing property in general 3D situations.
- **MS2: Diffusion equations with discontinuous coefficients.** Independent of the
  employed smoother, traditional coarse-grid correction processes are no longer
  appropriate. First, errors typically exhibit the same discontinuous behavior as
  the solution itself and geometric (linear) interpolation does not correctly transfer
  corrections to finer levels. Since the discretization stencils themselves *do* reflect
  the discontinuities correctly, so-called *operator-dependent interpolation* was
  introduced to replace geometric (linear) interpolation. Second, the use of coarse-
  grid analogs of the fine-grid operator gives rise to problems in properly capturing
  fine-grid discontinuities on coarser levels. The so-called *Galerkin operator* [32]
  was introduced as an alternative to the geometrically 'natural' coarse-grid operators. Galerkin-based coarse-grid correction processes with operator-dependent
  interpolation became increasingly popular since then (see, for example, [1, 22]).

Driven by **MS1** and **MS2**, the development of AMG started in the early 1980s (cf.
next section). A key point was the observation that reasonable operator-dependent
interpolation and the Galerkin operator can be derived directly from the underlying
matrices, without any reference to the grids. To some extent, this fact had already
been exploited in the first 'black-box' multigrid code [23]. However, regarding
the selection of coarser levels, this code was still geometrically based. In a purely
algebraic setting, the coarsening process itself had to be automated as well, based
solely on algebraic information contained in the given matrix. As a consequence,
the applicability of AMG would neither depend on the dimension of a given
problem nor on the type of grid, and AMG would even allow the solution of certain
linear systems which were not derived from PDEs at all. We will see later that

the automatic AMG coarsening also allows circumventing the GMG difficulties described in **MS1**.

Remark 2.1 In GMG, quite some research was invested to circumvent **MS1** by simplifying the smoother without sacrificing convergence. However, in order to maintain an efficient interplay between smoothing and coarse-grid correction, more effort had to be put into the coarse-grid correction process. Sophisticated semi-coarsening techniques were developed, in the most general case by *multiple semi-coarsening* (semi-coarsening in multiple directions on each level of the multigrid hierarchy) [24, 37, 38, 60]. Compared to such very complicated approaches, AMG can be viewed as the most radical approach to simplify the smoother and still maintain good convergence.

3 The Early Phase of Algebraic Multigrid (1982–1987)

Before outlining the beginning of AMG, let us review the very first documented non-PDE application [13] which was investigated when AMG was in its very early stage, with the choice of coarse grids and interpolation somewhat 'quaint', compared to what was developed just a year or so later (Fig. 1). In fact, this application probably contributed more to advances in AMG than AMG contributed to its solution. A number of interpolation schemes were tested, some actually involving geometric information that would probably not be needed by AMG today. On the other hand, this problem was closely related to later applications in elasticity, particularly those involving 'truss structures' that were nearly identical in form and proved a challenge to AMG over the next several years, and in some cases, continues to this day.

Fig. 1 In the early 1980s, the first AMG prototype has been developed by J. Ruge on an Apple II computer (8 bit, O(1) KFlops peak, 48 KByte RAM). Published with kind permission of ©Fraunhofer SCAI 2016. All Rights Reserved

3.1 The First Documented AMG Application

The main motivation for the development of AMG was that it could be useful in situations where the application of geometric multigrid was difficult or impossible (cf. **MS1, MS2**). However, it was envisioned that there was an even wider range of (non-PDE) problems for which the basic multigrid 'divide and conquer' principles of smoothing (to reduce high-frequency error) plus a coarse grid correction (to reduce low-frequency error) could apply. Of course, it was also seen that there were some restrictions. While AMG was supposed not to rely on (or even know about) any underlying geometry, some sense of locality of the underlying problem was necessary in order to both provide local smoothing and to avoid runaway complexity (where the coarse grid matrices loose sparsity).

The so-called *North American geodetic survey problem* seemed to be ideally suited for demonstrating the capabilities of the early AMG ideas. Over many years (over a century or more), in an effort to map out the country, markers (or stations) were placed across the US, and measurements were made between them, incorporating distances between markers and angle measurements from the magnetic north pole. Roughly 250,000 stations and 1.6 million measurements were involved. After the advent of computers, a natural question was whether the locations of all markers could be determined from this data. This problem was probably overdetermined, so a direct solution was not possible. Since measurement methods were known to vary widely over time, some measurements were assumed to be more accurate than others. A natural problem was to determine all marker locations to minimize the error of all measurements, in a least-squares sense, with weights assigned according to the estimated accuracy of the measurements.

Here, it is worth going into a little more detail. For simplicity, ignoring the orientation measurements and restricting the problem to a plane, the idea was to solve for location vectors $x = (x_1, \ldots, x_N)$ and $y = (y_1, \ldots, y_N)$, with (x_k, y_k) being the location of marker k that minimize the least-squares functional:

$$F(x, y) = \sum_i \sum_{j \in N(i)} w_{ij}(m_{ij} - d_{ij}(x, y))^2, \tag{1}$$

where $N(i)$ is the set of markers connected to i by a measurement, w_{ij} is the weight assigned to the corresponding measurement, m_{ij} is the measured distance between them, and d_{ij} is the actual distance between the two markers using the computed solution:

$$d_{ij}(x, y) = \sqrt{(x_i - x_j)^2 + (y_i - y_j)^2}. \tag{2}$$

One way to approach this problem is simply to linearize d_{ij}, and then deal with the modified functional. Expanding the solution (x_k, y_k) as $(x_k^0 + \delta x_k, y_k^0 + \delta y_k)$, where the superscript indicates the current approximate location (either given a priori or from a previous Newton iteration), this gives:

$$d_{ij} \approx d_{ij}^0 - c_{ij}^x(\delta x_i - \delta x_j) - c_{ij}^y(\delta y_i - \delta y_j), \tag{3}$$

where

$$c_{ij}^x = \frac{x_i^0 - x_j^0}{d_{ij}^0} \quad \text{and} \quad c_{ij}^y = \frac{y_i^0 - y_j^0}{d_{ij}^0}. \tag{4}$$

Setting the appropriate derivatives of the resulting functional to zero and writing the equations in a 2×2 block form results in

$$\sum_{j \in N(i)} A_{ij} \left(\begin{bmatrix} \delta x_i \\ \delta y_i \end{bmatrix} - \begin{bmatrix} \delta x_j \\ \delta y_j \end{bmatrix} \right) = \begin{bmatrix} f_i \\ g_i \end{bmatrix} \quad \text{with} \quad A_{ij} = w_{ij} \begin{bmatrix} c_{ij}^x c_{ij}^x & c_{ij}^x c_{ij}^y \\ c_{ij}^y c_{ij}^x & c_{ij}^y c_{ij}^y \end{bmatrix}. \tag{5}$$

Note that $(c_{ij}^x)^2 + (c_{ij}^y)^2 = 1$, so that each A_{ij} is positive semi-definite. Reordering this as a linear system for the updates of marker i, the off-diagonal matrix coefficients $-A_{ij}$ are negative semi-definite, and the diagonal coefficients $\sum_j A_{ij}$ should be positive definite (unless we have a degenerate case where it is actually singular, such as i and all $j \in N(i)$ being collinear).

In an intuitive way, this fits nicely into a multigrid setting. Most measurements should be between a marker and its geographically nearest neighbors. Here, with a relatively small average number of measurements per marker, a natural relaxation method would be to proceed through the markers one at a time, and determine the location of each in a way that minimizes the least-squares functional (obviously only involving terms associated with that marker). It is easy to imagine that, locally, each marker should end up well-situated with respect to its neighbors, although large smooth global errors could persist. Perhaps an area of North Dakota could be placed miles north of its actual location, although local residuals were small. Some sort of more global correction would be needed, and this is where AMG came in.

In testing, both manufactured problems (using markers placed on a structured grid with nearest-neighbor connections) and a subset of the real problem were tested. Promising results were obtained, although the full problem was never solved by AMG. Actually, the full problem went away with more accurate satellite measurements, although it could easily be solved today by modern computers, even with direct methods.

3.2 The Basics of 'Classical' AMG

As mentioned before, the introduction of Galerkin coarse-grid operators and operator-dependent interpolation, formally enabled the 'algebraization' of multigrid. But only their combination with automatic coarsening processes (that is, selecting subsets of variables to serve as the coarse grid) led to true algebraic multigrid approaches. This automated coarse grid selection is the basis of the most important conceptual difference between geometric and algebraic multigrid.

Geometric multigrid employs fixed, pre-defined grid hierarchies and, therefore, an efficient interplay between smoothing and coarse-grid correction requires the selection of appropriate smoothing processes (cf. **MS1** in Sect. 2). In contrast to this, AMG fixes the smoother as some simple relaxation scheme (typically plain point Gauss-Seidel) and enforces an efficient interplay with the coarse-grid correction by choosing the coarser levels and interpolation appropriately. Geometrically speaking, AMG attempts to coarsen only in directions in which relaxation really smoothes the error for the problem at hand, producing coarser levels which are *locally* adapted to the smoothing properties of the given smoother. In standard (close to M-matrix) cases, the relevant information can directly be extracted from the matrix (in terms of size and sign of coefficients) allowing the coarsening process to be performed based only on matrix information. The guiding principle in constructing the operator-dependent interpolation is to *force* its range to approximately contain those error components which are unaffected by relaxation (that is, the 'algebraically smooth' components). It turns out that this is the crucial condition to obtain efficient coarse-grid correction processes (for details, we refer to [45]).

The first fairly general AMG program to solve linear systems,

$$Au = f \quad \text{or} \quad \sum_{j=1}^{n} a_{ij} u_j = f_i \quad (i = 1, 2, \ldots, n), \tag{6}$$

is described and investigated in [44, 45, 48]. AMG's coarsening process is directly based on the connectivity pattern reflected by A, and interpolation is constructed based on the matrix entries following the ideas outlined above. The Galerkin coarse-level matrix for level $n + 1$ is recursively defined as

$$A_{n+1} := I_n^{n+1} A_n I_{n+1}^n \quad \text{with} \quad I_n^{n+1} := \left(I_{n+1}^n \right)^T$$

and I_{n+1}^n denoting the interpolation from level $n + 1$ to n. Level 1 represents the original matrix equation (6).

The early phase of AMG development was characterized by systematic studies regarding the robustness and efficiency of AMG if applied to problems (6) with A being 'close' to an M-matrix. Typical applications were scalar elliptic model PDEs exhibiting various types of inherent difficulties considered in multigrid testing those days, but also non-PDE model VLSI design applications as well as some problems with randomly created matrices. Results were promising, especially for problems difficult for GMG. It turned out that the 'nicest' problems (such as Poisson's equation discretized on a uniform mesh) actually posed more of a challenge for early AMG algorithms.

In addition, in [11, 45], theoretical 2-level convergence proofs were presented for symmetric positive definite M-matrices or 'essentially positive type' matrices.

The possibility of applying AMG to coupled systems was demonstrated as well, for instance to linear elasticity problems. Also, various alternative coarsening strategies as well as possible modifications/improvements of interpolation were discussed. For an extensive list of references regarding this early phase of *classical* AMG development, see [45, 48], [54, Appendix A].

Around 1985, a simplified code version for scalar problems, amg1r5, was made publicly available. After that though, there was no substantial further research and development on AMG for many years. AMG was generally regarded as a nice but complicated academic toy, useful to demonstrate that the idea behind multigrid does not necessarily require grids. For the relatively small problems considered those days, however, there was no obvious benefit compared to much simpler standard one-level methods such as preconditioned conjugate gradient. This changed only in the 1990s, both in academia and industry.

4   The Renaissance of AMG (1995–2000)

Since the early nineties, and even more since the mid nineties, numerical simulation got increasingly sophisticated and costly. The growing geometrical complexity of applications technically limited the immediate use of geometric multigrid causing a continuously growing demand for efficient 'plug-in' solvers. For industrial codes, this demand was driven by increasing problem sizes which clearly exhibited the limits of classical one-level solvers used in those packages. Up to millions of degrees of freedom in the underlying numerical models required *hierarchical* approaches for an efficient solution, and AMG provided a promising methodology without requiring complete restructuring of existing software packages.

4.1   Resumption of Major Research on AMG

As a consequence of this development, there was a remarkable increase of research and development on 'classical' AMG, several variants and new hierarchical approaches came up, some of which differed substantially from the original AMG ideas as outlined above. In the following, we mention only a few major research activities. For an extensive list of references, see [54, Appendix A].

A lot of papers were focusing on improving the *quality of coarsening* and *interpolation*. These included new strategies for exploiting additional information (besides mere matrix information)—for instance the location of grid points and/or user-provided 'smooth' test vectors—to improve AMG's interpolation by means of some fitting process. Since eigenvectors (of A) corresponding to the smallest eigenvalues are required to be interpolated particularly well, several approaches tried to find improved ways to achieve this. AMGe, for instance, has been developed for finite element methods, assuming the element stiffness matrices to be known [15].

New approaches have been investigated for the solution of certain *coupled PDE systems*, in particular, from *linear elasticity*. As an important alternative to 'classical' coarsening by selecting subsets of variables [45], *aggregation-based* coarsening was introduced. Due to the rigorous way it ensures exact interpolation of the rigid body motions, the so-called *smoothed aggregation AMG* [55, 56] has turned out to be highly suited for linear elasticity applications, including singular and near-singular ones.

The parallelization of AMG started to become a research topic, although major steps forward have been achieved only later in the 2000s, see [42]. Several papers investigated multigrid methods for solving finite-element problems on unstructured grids. Although some of them were also based on algorithmical components which were, more or less, algebraically defined, most of them were not meant to be algebraic multigrid solvers in the sense understood here. Finally, further hierarchical approaches were developed which were not directly based on multigrid principles, such as approximate block Gauss elimination ('Schur-complement' methods), the introduction of multilevel structures into ILU-type preconditioners, and some hybrid methods using ideas both from ILU and from multigrid.

For clarity, we point out that here and in the following we use the term AMG for all algebraic 'multigrid-like' approaches, that is, all approaches for which smoothing and coarse-grid correction play a role similar to the one they play in standard multigrid. This includes approaches based on 'classical' as well as aggregation-based coarsening.

4.2   Towards Industry

Due to the increasing complexity of industrial simulation software, run times of hours and even days were not unusual. By far most computing time was spent in solving ever larger linear systems of equations, especially since solution times for most solvers grew more than linearly with respect to problem size. However, almost no industrial providers of simulation software were aware of the rapid numerical solver development which took place in academia. Typical solvers in industrial codes were Krylov methods preconditioned by some standard method such as Jacobi or ILU. Knowledge about AMG was very limited and, if at all, restricted to the most simple (M-matrix) cases like those covered by the publicly available amg1r5.

Instead, industrial software developers were relying on new parallel hardware ('high-performance computing', or HPC) to reduce their performance bottleneck. This was fostered by Europort, a large European Initiative—funded by the European Commission—which migrated 35 industrial codes to parallel platforms, covering almost the whole spectrum of industrial applications in Europe for which HPC was essential at that time. Europort took place 1994–1996 and was managed by SCAI (together with Smith System Engineering in the UK). Although it was regarded a great success by all partners as well as the European Commission (cf. [47]), it became clear that the exploitation of HPC systems—although being a very important step forward—did not really remove the performance bottleneck. A real breakthrough would require the combination of HPC and scalable (optimal) numerical solvers, see [42].

Hence, after SCAI's strong involvement in the industrial Europort project, it was very natural to invest major efforts in helping to transfer know-how regarding scalable AMG-based solvers to industry. Right after the end of Europort in 1997, SCAI was offered a project to develop an MPI-parallel AMG code specifically to accelerate the commercial general purpose CFD code STAR-CD of Computational Dynamics Ltd.[2] While the general status of AMG development was still far from being mature enough for industrial applications and parallelization experience was still very limited, the use of amg1r5 as a starting point to solve the pressure equations in the underlying pressure correction approach seemed fairly straightforward. Although SCAI was still a GMD institute, this pioneering project was paving the way for the future AMG work at Fraunhofer.

4.2.1 Computational Fluid Dynamics

Segregation (or *pressure correction*) still today belongs to the most established approaches to solve the Navier-Stokes equations in general-purpose CFD codes,

$$\mathbf{u}_t - \frac{1}{Re}\Delta\mathbf{u} + \mathbf{u}\cdot\nabla\mathbf{u} + \nabla p = f, \quad \nabla\cdot\mathbf{u} = 0.$$

The advantage of the segregation approach is that, at each time step, the approximate solution of just a series of *scalar* equations is required (rather than the solution of the coupled Navier-Stokes system, cf. [36]) with the Poisson-like pressure-correction system being by far the most expensive one (for details, see [54, Appendix A]).

Since the numerical solution of the (Poisson-like) pressure equations by AMG was fairly mature (e.g., via amg1r5) (Fig. 2), an installation of AMG was expected to be fairly straightforward. Nevertheless, this project was an interesting exercise towards a direct cooperation with industrial software developers. Some of the lessons learned—although partly obvious in hindsight—have been completely underestimated:

- In contrast to our former expectations, memory was a critical issue. For instance, customers from the car industry wanted most of the memory for themselves to refine and store their models. The AMG solver was expected to be fast but its complexity[3] was requested to not exceed a factor of two, say. Unfortunately, amg1r5's typical complexity values were much higher than that: In unstructured

---

[2]Computational Dynamics Ltd. was one of the partners in the Europort project.

[3]That is, the ratio of the total memory required for all matrices on all levels and the memory required to store the finest-level matrix.

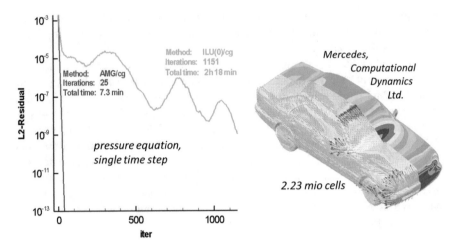

Fig. 2 Flow around a full Mercedes E-Class (picture courtesy Computational Dynamics and Daimler Benz). The plots show the convergence histories of (serial) AMG vs. a standard one-level solver for a single pressure equation (computed in the late 1990s). This application was among the largest ones solved at that time. Published with kind permission of ©Fraunhofer SCAI 2016. All Rights Reserved

3D applications it could easily go up to 8 or even more, completely unacceptable! This has led to a complete redesign of the original amg1r5 algorithm:

- Coarsening in amg1r5 was a two-step process (see [45]). While the second step tried to ensure good interpolation by a posteriori introducing additional coarse-level variables, it had the tendency to turn fairly regular coarse levels (produced by the first step) into quite disturbed ones. The recursive coarsening process then led to increasingly irregular and complex Galerkin operators, often causing unacceptably high complexity values. In order to avoid this, coarsening was replaced by a one-step process by just skipping the second step.
- To compensate for the latter, interpolation had to be improved *without* adding additional coarse-level variables (cf. 'standard interpolation' in [54, Appendix A]).
- 'Aggressive coarsening' was introduced to reduce memory requirement further. This was, in particular, necessary for non-compact discretization stencils typically used in finite volume based codes. Otherwise, just the second AMG level alone tended to require much more memory than the first level (for details, see [54, Appendix A].
- Finally, to compensate for some loss of convergence due to all these simplifications, AMG had to be used as a preconditioner rather than a stand-alone solver.

- The pressure-correction equation is just one component within an outer iteration and there is no need to solve each equation accurately, in particular not in steady-state simulations. For low accuracy computations, however, AMG's setup cost dominated the total solution cost by far. As a consequence, the possibility to (fully or partially) reuse setups had to be introduced.
- Although parallelization details are not discussed in this paper, we want to mention that the parallelization of AMG is not at all trivial, compare [42]. Looking back in time, it was rather demanding to do the very first parallelization of a classical AMG code in a commercial environment.
- User-friendliness is a very important issue. The average user does not want to be involved in any details and/or parameter settings of a linear solver, in particular not, if parameter settings are required for purely numerical rather than physical reasons. Unfortunately, a true black-box integration of AMG, to some extent sacrificed optimal performance which was essential as well.

We realized that, under the industrial/commercial conditions, our developments took (much!) longer than expected. This was for various reasons. For instance, the integration of AMG into the whole simulation process and its optimal tuning was far more work than originally anticipated. Also, while efficiency was important, robustness was even more important, including 100% clean and informative error exits. Because AMG was not a fixed final solver but still under further development and optimization, it was unexpectedly demanding to achieve this. The linear systems—although in principle perfectly suited for a program like amg1r5—turned out to be more complicated than expected because matrices obtained in industrial cases rarely satisfy all requirements from nice model situations. Finally we had to realize that our visions as numerical analysts and those of simulation engineers were diametrically opposed. In particular, the development of a fancy linear solver was certainly not part of the engineer's vision, it just has to be fast enough. This was not too surprising but it caused many intense discussions.

In any case, the project was finished successfully and the resulting AMG variant provided the basis for the new code, SAMG, whose systematic development started only a few years later.

4.2.2 Streamline Approach in Oil Reservoir Simulation

Today, oil reservoir simulation is the most important application area for SAMG (see Sect. 6.3). The beginning of this, however, dates back to the late 1990s. At that time, *IMPES*-like reservoir simulators have frequently been considered.

In classical *IMPES* (IMplicit Pressure, Explicit Saturation) [20], the 'total' pressure p is computed in each time step. Since all other quantities are taken from previous time steps, we just have to deal with a series of Poisson-type *IMPES* pressure equations of the form,

$$- \nabla \cdot (\mathbf{T} \, \nabla p) = Q. \tag{7}$$

In contrast to the pressure-correction equation from the previous section, the tensor **T** is *strongly discontinous* (directly corresponding to the permeability tensor **K**, see Sect. 6.3).

As with any other explicit time stepping method, classical *IMPES* suffers too much from limited time step sizes (CFL condition) to be practicable for large problems and is rarely used anymore. A way to overcome the sharp time step restriction is given in [6, 50]. Rather than marching in time directly on the discretization grid, a *streamline method* is used instead. By transporting fluids along periodically changing streamlines, the streamline approach is actually equivalent to a dynamically adapting grid that is decoupled from the underlying *static* grid used to describe the reservoir geology (and to compute the pressure).

Although this approach cannot be applied to all relevant situations occuring in oil-reservoir simulation, it is well suited for large heterogeneous multi-well problems that are convection-dominated. Since the computational time for the linear solver accounts for the largest part of the total simulation time by far, an efficient solver for the pressure equation becomes highly important.

During the development phase of the streamline-based reservoir simulator 3DSL, amg1r5 had been efficiently used for solving (7). A newly founded company, Streamsim (San Francisco), wanted to commercialize this simulator and continue to use amg1r5 as a solver. Since amg1r5 was a freely available research code, there would have been no problem to give the permission. However, according to US law, every agreement requires 'consideration', meaning something of value, however small, in return for the subject of the agreement, here amg1r5. Since amg1r5 was a research code without any maintenance or support, a formal licensing contract was set up and a license fee of one dollar was agreed (see Fig. 3).

Fig. 3 Licensing fee for amg1r5: According to US law, every agreement requires 'consideration', meaning something of value, however small, in return for the subject of the agreement (here amg1r5). Note that, although the cheque is issued on fool's day, this is not a joke. Published with kind permission of ©Fraunhofer SCAI 2016. All Rights Reserved

Although this fee was really marginal, the related effort has more than payed off. Nobody expected at that time that this 'licensing story' would pave the way for SCAI's work in reservoir simulations many years later. In fact, once the development of SAMG had started several years later, Streamsim became the first 'real' customer.

5 The Main AMG Development Phase (2000–Today)

By 2000, AMG had become the most discussed approach in linear solvers. The development of AMG software and research on software-related aspects such as robustness and generality as well as concrete applications more and more moved into the focus of interest. Looking back in time, one might say that the overall development of AMG has undergone a similar phase as GMG many years earlier. Before we summarize industry-related research and development on AMG at SCAI, we want to briefly recall the general trend in AMG development at the time.

5.1 The General Trend

One of the major research topics referred to the question of how AMG could be used to automatically improve itself, more precisely, its smoothing and/or the interpolation operator. New ideas like 'compatible relaxation', 'bootstrap AMG' and 'adaptive AMG' came up and initiated an extensive research on the improvement of robustness and generality of AMG (for a selection of major papers in this direction, see [12] and the references given therein). Besides classical AMG, many developments referred to aggregation-based AMG approaches, both smoothed and unsmoothed. A large variety of advanced applications have been considered, and, finally, all aspects regarding parallelization of AMG became a major topic of investigation, compare [42]. Altogether, there have been too many publications on algebraic multigrid to list here. Instead, we only make a general reference to the proceedings of the regular conferences on multigrid, namely, the 'European Multigrid Conference' and the 'Copper Mountain Multigrid Conference'. In particular, see [21] where all Copper conference presentations are listed in detail. The yearly 'AMG Summit' is focusing specifically on AMG-related advances and questions. Although not a conference in the usual sense, the topics of the AMG Summit [2] give a good overview on the development of AMG.

Lots of software, mostly MPI-parallel, have been developed. Among the most prominent free AMG programs are *Hypre BoomerAMG* from Lawrence Livermore Labs [25] and *Trilinos ML* from Sandia Labs [33]. Whereas BoomerAMG relies on classical AMG, ML is based on smoothed aggregation AMG. In this context we also mention *UG* from the University of Heidelberg [5]. This package, a quite general platform for the numerical solution of PDEs supporting unstructured grids and adaptive grid refinement, can be regarded as a multigrid-based toolbox for academic purposes. Several universities are contributing to the open source project *DUNE* [8]. Besides these large, well-known software packages, *PEBBLES* and *AMuSe* from the University of Linz [43, 57] should be mentioned as well as the somewhat special AMG codes *PyAMG* from the University of Illinois [7] and *LeanAMG* from the University of Chicago [35].

Commercially available AMG programs are *AGMG* from the Université libre de Bruxelles (which is known to be particularly fast for Poisson-like problems) [39], *AmgX*, a solver for Poisson-like problems (which is tuned for GPUs) [40], the *SuperMatrixSolver SMS-AMG* from VINAS Ltd. [49][4] and, finally, SCAI's software framework SAMG.

The remainder of this paper refers to SAMG. Its development started at SCAI in 2001 and is the only AMG-based software package which has been and is still being further developed with the main focus on its use in industrial simulation codes. By today, SAMG has been commercially licensed to various software developers worldwide. We will review the history of SAMG, its major components as well as some major industrial applications.

5.2   Bridging the Gap

When SCAI became a Fraunhofer institute in 2001, a new requirement was that work on AMG had to be strictly for the benefit of industry.[5] Since new mathematical ideas are published and freely accessible for everybody, it was clear that, to be successful in this new Fraunhofer setting, work would have to go substantially beyond simple implementation of these ideas. However, there was quite some uncertainty about how to proceed, in particular, how to maximize the impact on industrial software developers and through this ensure enough return on investment to allow a basic development of AMG to continue. In any case, although SCAI already had some limited industrial experience (see Sects. 4.2.1 and 4.2.2), there was neither a final piece of robust and general AMG software available which could be licensed, nor did there exist 'serious' experience for the application of AMG (beyond scalar problems) which could serve as the basis for industrial projects. A 'business plan' (see Fig. 4) had to be set up as a first step to clarify the market potential as well the general strategy to achieve the goal.

From SCAI's involvement in the European initiative Europort (see Sect. 4.2) it was clear that software developers increasingly suffered from tremendous comput-ing times, mainly due to inefficient linear solvers. It had also been realized that parallel hardware alone was not enough to remove the performance bottleneck.

---

[4]To our knowledge, the SMS library is only used in Japan.

[5]Which is a result of the requirement of over a 70% return on investment, most of it coming directly through cooperation with industry.

Hence, there was an enormous demand for scalable linear solvers, ready to be plugged into industrial simulation software. But no such software was available off-the-shelf, and, since the research on AMG was still ongoing, this situation was not likely to change in the near future.

Academic R&D was widely done on the basis of selected model applications and under well-defined model situations in order to study both practical and theoretical aspects of AMG. What was missing was an attempt to continuously consolidate all relevant scientific experiences with focus on the requirements of an industrial user. In particular, there was no serious attempt to drive AMG-based software development to the point where it could be used safely and robustly by an inexperienced industrial user for solving his 'real-world' applications, let alone to incorporate it as a fast and robust solver in commercial simulation software. On the other hand, due to a lack of insight, the average industrial user was overwhelmed by the attempt to apply this methodology himself. In light of the rapid development of AMG variants and new ideas, it became virtually impossible for a non-specialist to easily separate aspects which might be relevant for him from those which were very interesting from an academic point of view, but without obvious/immediate impact on the development of industrial simulation software.

Hence, there was a clear 'market' for scalable solvers and, at the same time, there was no indication that industrial simulation software developers would be able to easily benefit from the exiting AMG development. The gap between the rapid academic research and concrete industrial needs was simply too big. There was also no indication that this situation would change in the near future, because *the* AMG goal—a *final solver* that would be efficiently and robustly applicable in all types of industrial simulation codes, once and for all—was not in sight. To bridge the gap, a lot of work would be required focusing on industrial needs rather than scientifc interests, most of it service and support rather than scientific R&D.

Serious and persistent attempts to change this situation became the driving force behind the AMG 'business' of SCAI. Based on the experience outlined in

Sects. 4.2.1 and 4.2.2, SCAI started to develop the AMG-based solver package SAMG, specifically for industrial use. That is, it had to be *simple enough* to be integrable into existing software, *robust enough* to be safely usable, *fast enough* to provide sufficient benefit, and, finally, *flexible enough* to be applicable under complex industrial 'real-world' situations. The (still today ongoing) development of SAMG was a long-lasting dynamic process which needed permanent cooperations with various industrial partners, strictly focusing on those questions which were relevant for them. In other words, R&D work at SCAI had started to become request- and product- rather than research-driven, that is, a combination of focused research, service & support, and software development.

Over many years, the further development and algorithmic generalization of SAMG has been driven by industrial applications from various areas such as computational fluid dynamics (Euler and Navier-Stokes equations), linear elasticity, electro-chemistry, casting and molding, reservoir simulation (oil and groundwater), circuit simulation, and semi-conductor process and device simulation. To maximize the impact, most of the R&D work had to be done in close cooperation with our clients (often based on direct projects), including continuous service, support (such as focused testing, adapting/generalizing, optimizing and benchmarking) and maintenance. In order to be acceptable for our partners, basic software requirements such as portability, extendability, backwards compatibility and the like had to be satisfied.

Besides the actual software development, a substantial part of the work had to go into 'pre-sales activities', that is, all kinds of efforts needed to convince industrial clients of their benefits in using SAMG. In this context, one should also point out that a serious industry-funded cooperation requires binding contracts. In particular, if external software such as SAMG is to be used in commercial products, a binding contract has to define responsibilities and liabilities in case something goes wrong. Clearly, in a complex numerical environment based on algebraic multigrid there is always the risk of failure, but without taking a contractual responsibility, nobody would ever use the software in a commercial context. Finally, a lot of other work had to be scheduled for many things such as setting up contracts, licensing software, providing service, support and maintenance, documentation, acquisition of projects and new clients, consulting for software houses, and the like.

Free software packages are not generally an alternative to SAMG because it is not so much the software that will make it interesting in a commercial context, it is the service, support and maintenance which comes along with it as well as a hotline for immediate help in case something does not work as expected. This kind of work usually dominates the total cooperative work by far, and realizing it in a commercial context—that is, based on contractual obligations—is often very demanding and time consuming.

Major user requests concerned the applicability of SAMG to various types of coupled PDE systems. Since corresponding research with AMG was very limited (essentially to linear elasticity) and no well-settled approach existed yet, SAMG was not designed to become a fixed solver but rather a framework of individual modules, flexible enough to allow SAMG's adaptation to a variety of PDE systems whose numerical properties may differ substantially. This framework will be outlined in the following while all issues related to parallelization of SAMG are discussed in [42] of this volume.

5.3 SAMG for Coupled PDE Systems

Typical industrial simulation packages consist of an implicit discretization in time and space, a Newton-type method to treat the nonlinearities and direct and/or iterative one-level methods to solve the arising systems of linear equations. The corresponding matrices are large, sparse, frequently ill-conditioned, often not symmetric positive definite, and usually far from being M-matrices. Hence, these matrices do not exhibit the properties that typical 'scalar' AMG approaches—such as the original amg1r5—principally rely on.

Relevant systems often consist of diffusion equations with additional convection, drift (migration) or reaction terms. The individual PDEs are often of first order in time (if time-dependent) and of second order in space. They can be nonlinear and/or strongly coupled, the latter normally enforcing a 'fully-coupled' solution approach, that is, a simultaneous solution for all physical functions involved.

Several ways to generalize SAMG to industrially relevant applications have been investigated, and there is still an ongoing development of new AMG and AMG-like approaches. For a detailed review, we refer to [16]. Basic versions of the two approaches, the *unknown-based* and the *point-based approach*, have already been proposed in the early paper [45].

5.3.1 Unknown-Based Approach

The so-called *unknown-based* approach in SAMG is very similar to scalar AMG except that all 'unknowns', i.e. *scalar* physical functions of the PDE system to be solved, are treated separately. To be more specific, coarsening and interpolating the variables of the n-th unknown is strictly based on the submatrix of A reflecting the couplings of the (variables of the) n-th unknown to itself. In particular, interpolation to any variable i involves only coarse-level variables corresponding to the same unknown i belongs to. The Galerkin matrices, however, are usually computed w.r.t. all unknowns, so coupling is actually preserved on all levels.

The essential conditions for the unknown-based approach to work are that, for each unknown, the submatrix of A reflecting the couplings of this unknown to itself is close to being an M-matrix and that smoothing results in an error which is smooth separately for each unknown. Advantages of unknown-based AMG are that it can easily cope with anisotropies which are different between the different unknowns and that unknowns can be distributed arbitrarily across mesh points. This simple approach works quite efficiently for some important applications. However, it may become inefficient if the cross-function couplings are too strong.

5.3.2 Point-Based Approach

In [16] a flexible framework for constructing so-called *point-based* approaches to
solve various types of strongly coupled PDE systems was systematically developed.
Point-based AMG operates on the level of points rather than variables as do scalar
and unknown-based AMG. Typically, points are physical grid nodes (in space).
However, it is only relevant whether there are (disjoint!) blocks of variables (corresponding to different unknowns) which may be coarsened, maybe also interpolated,
simultaneously.

In order to coarsen A, a so-called primary matrix P is to be selected or
constructed, with the number of points being its dimension. Its entries can be
seen to result from a 'condensation' of the point-coupling matrices (each of which
representing the couplings of the variables of a point to the variables of another
point) to scalar values of P so that the resulting P reflects the couplings between
the points reasonably well. A 'scalar' coarsening process is applied to P, and the
resulting is transferred to all unknowns. Note that this is different from unknown-
based AMG where each function is associated with its own hierarchy. Many
different variants for P and the interpolation are possible. Note, however, that a
reasonable choice of components (if possible) strongly depends on the application
at hand.

5.3.3 Status of the Solver Framework SAMG

SAMG is a general framework for solving coupled systems of PDEs. SAMG real-
izes all coarsening strategies mentioned above, namely, the scalar AMG approach,
as well as the unknown- and point-based AMG approach for solving coupled
systems. In point-based AMG, coarsening can be based on various types of
primary matrices (defined by norms of coupling matrices, coordinates, submatrices
of A or simply user-defined). Several types of interpolation are integrated for
point-based AMG, classified as so-called single-unknown, multiple-unknown and
block-interpolation. Depending on the application, the coarsening and interpolation
strategies can be combined in any reasonable way and be adapted to the requirements of the given application.

Originally designed for 'classical' AMG approaches [45], [54, Appendix A],
SAMG now contains also all kinds of aggregation-based approaches, smoothed
[55] as well as unsmoothed ones [39]. The unsmoothed variants are accelerated by
various means to approximately ensure scalability, in particular the highly robust
k-cycles [41]. As a unique feature, classical and aggregative strategies are fully
integrated in SAMG and can even be combined in a hybrid way. For instance, on
fine grids one can apply some cheap aggregative process, on coarser grids one can
switch to a more expensive (but also more robust) classical approach. This feature
is of particular interest on parallel computers.

Finally, SAMG features different smoothers, including several ILU-type and Uzawa-variants, as well as standard smoothers adapted for parallel implementation. SAMG can be used as preconditioner for several Krylov methods, namely CG/BiCGstab, GCR, flexible CG and restarted GMRes (cf. [46]). Developments of automatic, adaptive solver and parameter switching strategies, in particular for ILU-type smoothing, are in progress; for first very promising results, see [17, 19].

SAMG has been in development for over 15 years now. The employed algorithms are state-of-the-art with particular focus on its industrial use. There has been the continuous attempt to include all those algorithmic approaches which appeared to be most relevant and promising for that purpose. Since linear systems from industrial applications are often not exactly as expected (for instance, due to constraints or singularities), there are various fallback approaches to decouple critical parts of the matrix from uncritical ones (for instance, based on overlapping Schwarz methods). AMG is then only applied to the uncritical (but largest) part.

Finally, although not described in this paper, SAMG is fully parallel, MPI and OpenMP can be used in a hybrid way. It has been tested for many industrial 'real-life' cases and under industrial conditions (up to 3000 cores), compare [42].

6 Industry-Driven Applications

This section summarizes three industrial applications—*semiconductor simulation, multi-ion transport and reaction*, and *oil reservoir simulation*—as a demonstration for the complexity of problems being solved by SAMG. These applications have been investigated in detail in [16, 51] and [29], respectively. All related work has been initiated by industrial partners. If not explicitly stated otherwise, all SAMG approaches outlined below are based on classical coarsening modules.

6.1 Semiconductor Applications

Due to the complexity of the models and grids used, industrial semiconductor *process* and *device simulation* is recognized as an important and challenging area for numerical simulation. A well-known example is the simulation of the manufacturing process and electrodynamic properties of a transistor, e.g., a metal-oxide semiconductor field-effect transistor (MOSFET), see Fig. 5.

Among the different PDE systems to be solved here are the *Lamé equations* (linear elasticity) as well as *reaction-diffusion(-convection)* and *drift-diffusion(-convection-reaction)* systems. The standard linear solvers used for all three application classes are ILU-preconditioned BiCGstab, CGS or GMRes. In the following, we briefly outline the PDE systems to be solved and which type of SAMG configuration is suitable to outperform the standard solvers. Regarding details, we refer to [16, 18, 27, 28].

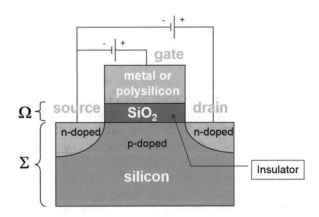

Fig. 5 Structure of a metal-oxide semiconductor field-effect transistor (MOSFET). Published with kind permission of ©Fraunhofer SCAI 2016. All Rights Reserved

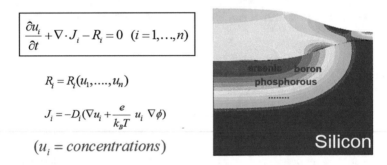

$$\frac{\partial u_i}{\partial t} + \nabla \cdot J_i - R_i = 0 \quad (i = 1, \ldots, n)$$

$$R_i = R_i(u_1, \ldots, u_n)$$

$$J_i = -D_i(\nabla u_i + \frac{e}{k_B T} u_i \nabla \phi)$$

$$(u_i = concentrations)$$

Fig. 6 *Left*: Reaction-diffusion system for process simulation. *Right*: Simulation result obtained by DIOS from ISE AG. Published with kind permission of ©Fraunhofer SCAI 2016. All Rights Reserved

While unknown-based SAMG is suitable for speeding up stress simulations based on the Lamé equations, the situation is considerably more complicated for reaction-diffusion and drift-diffusion models. Where standard iterative solvers often converge only slowly (or even fail), suitable point-based approaches, accelerated by BiCGstab or GMRes, can achieve remarkable speedups as outlined next.

Process simulation based on reaction-diffusion systems is important, for instance, in the simulation of annealing steps after ion implantation into a wafer. Of particular interest are the concentration profiles in and near the reaction front, a narrow region, moving from the 'implantation surface' of the wafer towards the interior, where fast reactions occur due to large concentration gradients (cf. Fig. 6). In this region, the reaction terms cause very large positive or negative off-diagonal entries in the corresponding rows of the matrices, leading to serious problems for the standard iterative solvers mentioned above. Also straightforward unknown-based approaches fail because of these large reaction terms.

Reaction-diffusion problems can efficiently be solved by point-based AMG used as preconditioner for BiCGstab. However, the proper choice of the primary matrix $P = (P_{kl})$ is crucial, in particular, it must not involve any reaction terms. It was found that a reasonable choice is based on geometric distances,

$$p_{kl} = -1/\delta_{kl}^2 \; (k \neq l) \quad \text{and} \quad p_{kk} = -\sum_l p_{kl},$$

and an interpolation which is separate for each unknown with weights being also defined based on distances. Note that the shape of the mesh is not important, just the point coordinates need to be known (which is generally the case anyway). Block-Gauss-Seidel is an efficient smoother here. Based on the resulting system approach, reaction-diffusion problems have extensively been tested for the commercial simulation code DIOS from ISE AG.

In device simulation, the behavior of transistors is simulated based on drift-diffusion systems. More precisely, the simulation domain usually consists of two parts, Σ and Ω (see Fig. 5). The subdomain Σ usually represents the semiconductor region(s) (doped silicon, the wafer), in which typically three coupled equations, a Poisson-like equation and the electron and hole continuity equations, are solved for the electrostatic potential and the electron and hole carrier concentrations, respectively (see Fig. 7). The second subdomain, Ω, consisting of at least one region (usually an oxide), is treated as an insulator, so that (nearly) no charge carrier currents can occur. Hence, the PDE system here degenerates to Laplace's equation.

Drift-diffusion problems are also solved efficiently by point-based AMG used as preconditioner for BiCGstab. Here, however, a reasonable primary matrix $P = (P_{kl})$

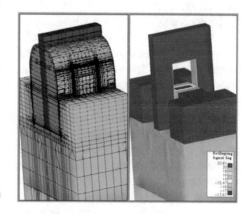

$$-\nabla \cdot (\varepsilon_s \nabla \phi) + q(n - p - N) = 0$$
$$\nabla \cdot (\mu_n n \nabla \phi - D_n \nabla n) + R_n = 0$$
$$\nabla \cdot (-\mu_p p \nabla \phi - D_p \nabla p) + R_p = 0$$

$$J_n = -q\mu_n n \nabla \phi + q D_n \nabla n$$

$$J_p = -q\mu_p p \nabla \phi - q D_p \nabla p$$

(ϕ = *electrostatic potential,*
n, p = electron/ hole density)

Fig. 7 *Left*: Drift-diffusion system for device simulation. *Right*: Exemplary mesh for a device simulation with TAURUS from Synopsys Inc.: a FinFET transistor (double-gate MOSFET). Published with kind permission of ©Fraunhofer SCAI 2016. All Rights Reserved

should be based on norms of the point coupling matrices, that is,

$$p_{kl} = -||A_{kl}|| \quad (k \neq l) \quad \text{and} \quad p_{kk} = -\sum_l p_{kl}.$$

Interpolation can be chosen to be the same for each unknown with weights being also based on the entries of the primary matrix. For smoothing, Gauss-Seidel methods can no longer be used, but ILU(0) has turned out to be an efficient smoother.

Based on the resulting system approach, drift-diffusion problems have extensively been tested for the commercial simulation code TAURUS of Synopsys Inc. An involved example is shown in Fig. 7 (on the right), namely, a FinFET transistor (double-gate MOSFET). The corresponding mesh has approx. 70,000 points, the systems to be solved nearly 100,000 variables, and the arising matrices contain approximately 1.5 million non-zero entries. Although this problem is relatively small, the SAMG configuration outlined above is able to provide a speedup of a factor of 1.33 (compared to the standard TAURUS solver ILU-CGS). Note that this refers to the overall runtime of the simulator, including preprocessing, setups and solver steps for nonlinear systems, setups and solution of linear systems as well as postprocessing.

6.2   Multi-Ion Transport and Reaction

Electrochemical processes play an increasingly important role in industry, for instance, for plating a large variety of parts, or for controlled etching by means of electrochemical corrosion when mechanical stress should not be applied to a part. In general, a complex interplay of the following phenomena takes place: electrochemical electrode kinetics, electrolyte hydrodynamics (movement of the electrolyte due to forced convection), ionic mass transport, gas evolution at the electrodes, and heat generation in the bulk of the electrolyte and at the electrode–electrolyte interfaces.

The modeling approach that takes these phenomena into account is commonly denoted as the *multi-ion transport and reaction model* (MITReM), which consists of the *transport equation* for each ion, the *electro-neutrality constraint* or the *Poisson equation*, and the nonlinear *Butler–Volmer boundary conditions* describing the electrochemical processes at the electrodes. In addition to Navier-Stokes systems, solved to obtain initial flow and pressure conditions, PDE systems consisting of terms describing migration (drift), convection, reactions and diffusion have to be solved.

On the one hand, such a migration-diffusion-convection-reaction system strongly resembles the PDE system to be solved in device simulation and, hence, a point-based SAMG configuration can indeed also be applied successfully here (see [53]). On the other hand, there are several important differences from the drift-diffusion

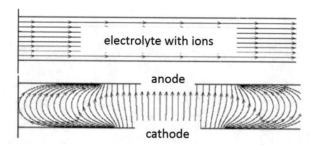

Fig. 8 Simple, yet practically relevant system of an anode, cathode, and the electrolyte to be simulated. *Top*: Convection field. *Bottom*: Migration field (electrons moving from cathode to anode through moving electrolyte). Published with kind permission of ©Fraunhofer SCAI 2016. All Rights Reserved

systems described in the previous section. Typically, the total number of equations of the resulting PDE system is between 5 and 10 (since typically more than two types of charged particles exist). More importantly, usually there are three main directions of 'flow' inside the simulation domains, namely, the movement of the electrolyte (due to convection) and the movement of charged particles to either the anode or cathode (see Fig. 8 for an illustration). The resulting fields are partly perpendicular to each other. As a consequence, designing an appropriate smoother or solver which treats both convection and migration more or less simultaneously well, is difficult. In our experience, (blocked versions of) standard ILU-based smoothers, alternatively acting in the direction of convection and migration, can yield a good compromise, see also [51–53]. In addition, one often has to deal with high aspect ratios arising from an adaptive meshing of relevant, geometrically complex domains. For an example, see Fig. 9.

In order to obtain an efficient yet robust solver for many industrially relevant applications, a point-based SAMG configuration with physics-aware smoothing and coarse-grid correction techniques has been developed in [51]. In particular, a reordering framework was developed allowing for physics-oriented matrix-based measures for strength of connectivity to derive application-specific point orderings for smoothing. In addition, a heuristic Peclet number was introduced in order to locate areas causing numerical difficulties within AMG's coarsening process.

For a plating example and using a somewhat simplified model, Fig. 10 compares SAMG's convergence history and that of an ILU-based standard one-level solver, clearly demonstrating the benefit of using SAMG.

6.3 Oil Reservoir Simulation

The numerical simulation of multiphase sub-surface flow processes plays a key role in the design of resource recovery in the oil and gas industry. The complex physics

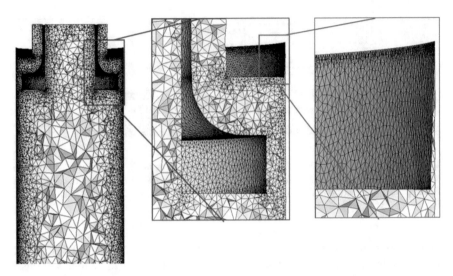

Fig. 9 Electrochemical machining example: diesel injection system, ~ 4.2 mio tetrahedra, aspect ratio up to 1:10,000. Published with kind permission of ©Fraunhofer SCAI 2016. All Rights Reserved

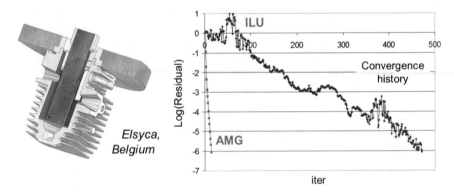

Fig. 10 Chromium plating example. *Left*: Geometry; *Right*: convergence history of SAMG in comparison to a standard solver (simplified model). Published with kind permission of ©Fraunhofer SCAI 2016. All Rights Reserved

of corresponding models—ranging from the Black Oil model for multiphase flows to different models for multicomponent flows, including thermal and mechanical influences—results in linear systems that are challenging for linear solvers. Both the increasing complexity and size of the systems makes their solution by far the most time consuming part of today's reservoir simulators. Nowadays, systems with more than a billion variables are not uncommon in real-world applications and scalable AMG-based solvers are of great practical importance.

Unfortunately, straightforward AMG approaches, although often efficient, in general lack robustness. In contrast to the applications described in the previous two subsections, the linear systems from reservoir simulation systems require certain preconditioning to allow a robust AMG employment. This section summarizes corresponding results of [29]. Regarding details of reservoir simulation itself we refer to [3].

6.3.1 The Reservoir Simulation Models

Basic reservoir models take only fluid flow into account, assuming oil to be one 'black' fluid and assuming temperature and mechanical forces to be constant. Such *Black Oil models* are still widely used in the oil industry. More sophisticated *compositional models* decompose the oil into its chemical components. In their full generality, all of these models additionally take *temperature* and *geomechanics* into account.

In order to describe the current situation regarding the application of AMG, let us start with the Black Oil model which simulates multiphase flows in porous media based on the conservation of mass for the three phases oil, gas and water, formally denoted by α, β and γ, respectively. For each phase ℓ the *mass conservation* (*continuity equation*) is given by

$$0 = \frac{\partial}{\partial t} \varphi S_\ell \rho_\ell + \nabla \rho_\ell \mathbf{u}_\ell + \rho_\ell q_\ell \quad (\ell = \alpha, \beta, \gamma). \tag{8}$$

Here, $\mathbf{u}_\ell$ denotes the velocity, S_ℓ the saturation, φ the porosity, ρ_ℓ the density distribution and q_ℓ a pressure-dependent external source. Source terms not only result from *injection* and *production wells* (with $q_\ell < 0$ and $q_\ell > 0$, respectively), also *fractures* and *chemical reactions* can be modeled via source terms.

The velocity-pressure dependence for flows in porous media is described by *Darcy's law*,

$$\mathbf{u}_\ell = -\lambda_\ell \mathbf{K}(\nabla p_\ell - g \nabla d) \quad (\ell = \alpha, \beta, \gamma), \tag{9}$$

where d is the depth and g the gravitational constant. Furthermore, p_ℓ denotes the pressure, λ_ℓ the mobility and $\mathbf{K}$ the permeability tensor. The absolute permeability varies in space by, typically, several orders of magnitude in a strongly discontinuous manner. In Fig. 11, the color scale indicates the variation of the permeability as a function of space for a typical case.

By combining the previous two equations, we obtain the *mass conservation equation* for phase ℓ,

$$0 = \frac{\partial}{\partial t} \varphi S_\ell \rho_\ell - \nabla \rho_\ell \lambda_\ell \mathbf{K}(\nabla p_\ell - g \nabla d) + \rho_\ell q_\ell \quad (\ell = \alpha, \beta, \gamma). \tag{10}$$

From this set of equations, pressures and saturations for all phases can be computed if one takes into account that $\sum_\ell S_\ell \equiv 1$ and that the individual phase pressures are interrelated by means of simple non-PDE relations involving known (but saturation-dependent) capillary pressures,

$$p_{cap}^{\alpha\beta} = p_\alpha - p_\beta, \quad p_{cap}^{\alpha\gamma} = p_\alpha - p_\gamma \quad \text{and} \quad p_{cap}^{\beta\gamma} = p_\beta - p_\gamma. \tag{11}$$

Exploiting these non-PDE relations, the three equations (10) can be used to compute one pressure and two saturations. Usually, these are the oil-pressure, p_α, and the gas and water saturations, S_β and S_γ, respectively. Throughout this section we assume these settings.

6.3.2 Fully Implicit Methods

In contrast to *streamline-based* IMPES approaches (see Sect. 4.2.2), *fully implicit methods* (FIM) treat all unknowns implicitly. The linearization is typically realized by Newton's method. Let

$$M(x) = (M_\alpha(x), M_\beta(x), M_\gamma(x))^T = 0 \tag{12}$$

denote the discretized mass balance equations (10) and $x = (p_\alpha, S_\beta, S_\gamma)^T$ the discrete pressure and saturations. Starting from an initial guess x_1, the $(k + 1)$-st Newton iteration reads as

$$x_{k+1} = x_k - \delta \quad \text{with} \quad \mathbf{J}(x_k)\delta = M(x_k), \tag{13}$$

where $\mathbf{J}(x_k)$ is the *Jacobian matrix* evaluated with the current solution vector, for simplicity also referred to as $\mathbf{J}$, i.e.

$$\mathbf{J} := \begin{pmatrix} \dfrac{\partial M_\alpha}{\partial p_\alpha} & \dfrac{\partial M_\alpha}{\partial S_\beta} & \dfrac{\partial M_\alpha}{\partial S_\gamma} \\[2ex] \dfrac{\partial M_\beta}{\partial p_\alpha} & \dfrac{\partial M_\beta}{\partial S_\beta} & \dfrac{\partial M_\beta}{\partial S_\gamma} \\[2ex] \dfrac{\partial M_\gamma}{\partial p_\alpha} & \dfrac{\partial M_\gamma}{\partial S_\beta} & \dfrac{\partial M_\gamma}{\partial S_\gamma} \end{pmatrix}. \tag{14}$$

Hence, in each Newton step we have to solve linear systems

$$\mathbf{J}x = f. \tag{15}$$

Scalar AMG for Solving Pressure Equations (CPR-AMG)

A classical, widely used approach for solving (14)–(15) is the *Constrained Pressure Residual* (CPR) method [58, 59]. Since each CPR iteration consists of two successive preconditioning stages, it is also referred to as *two-stage preconditioning method*.

Rewriting $\mathbf{J}$ in the following block form

$$\mathbf{J} = \begin{pmatrix} \mathbf{A}_{pp} & \mathbf{A}_{ps} \\ \mathbf{A}_{sp} & \mathbf{A}_{ss} \end{pmatrix} \quad \text{with} \quad \mathbf{A}_{pp} := \frac{\partial M_\alpha}{\partial p_\alpha}, \tag{16}$$

one CPR iteration, computing $x^{k+1} = (p^{k+1}, S^{k+1})$ from $x^k = (p^k, S^k)$, reads as:

1. Based on the scalar pressure sub-problem given by $\mathbf{A}_{pp}$, compute a temporary pressure $p^{k+\frac{1}{2}}$ by approximately solving

$$\mathbf{A}_{pp}p^{k+\frac{1}{2}} = -\mathbf{A}_{ps}S^k + f_p. \tag{17}$$

2. Compute a new approximation, $x^{k+1} = (p^{k+1}, S^{k+1})$, by performing one (or more) ILU(0) iterations to improve the temporary approximation, $(p^{k+\frac{1}{2}}, S^k)$, based on the full system.

Usually, CPR is not applied directly to (15) but rather to a transformed system,

$$\tilde{\mathbf{J}}x = g \quad \text{where} \quad \tilde{\mathbf{J}} = \mathbf{C}\mathbf{J}, \quad g = \mathbf{C}f, \tag{18}$$

with some non-singular matrix $\mathbf{C}$. The purpose of (18) is to approximately decouple pressure and saturation in order to help the 'outer' CPR convergence. Various choices of $\mathbf{C}$ are described in the literature, the most common ones being *alternate block factorization*, *quasi-IMPES* and *true-IMPES*, all of which approximate the

Schur complement by eliminating the dominating couplings in one way or another. The methods are only approximative in order to be applicable efficiently and to (widely) maintain the sparsity structure of the matrix. For more details, we refer to [29] and the references given therein.

As early as the late 1990s, AMG—more precisely amg1r5 (see Sect. 3.2)—had been used for solving (17). Since then, this CPR-AMG approach became increasingly popular. When the development of SAMG started in 2000, the academic code amg1r5 was replaced by SAMG both in research on reservoir simulation and in its industrial use.

Unfortunately, while working very efficiently in many simulations, a failure of CPR-AMG was observed in a significant number of practical simulations. This failure was nearly always caused by AMG-related issues in solving (17). A closer analysis shows that essentially two aspects are responsible for such failures:

- **A1:** Under real conditions, well settings are becoming increasingly complex with today's well-bore capabilities. The well source terms, q_ℓ in (8), can have a drastic impact on the properties of the pressure related matrix $\mathbf{A}_{pp}$. This may even result in $\mathbf{A}_{pp}$ becoming strongly indefinite so that the reliable applicability of AMG is no longer ensured. As the well-settings may change in time, the strength of such impacts can even drastically change during a complete simulation run.
- **A2:** Because of the transformation (18), AMG has to be applied with some $\tilde{\mathbf{A}}_{pp}$ rather than $\mathbf{A}_{pp}$. While the underlying approximate decoupling may, in principle, improve the outer CPR iteration, it is often counter-productive for the efficient applicability of AMG.

There have been attempts to improve the robustness of AMG. Krylov-deflation methods and Schwarz approaches, together with stronger AMG-smoothers, have been considered in [17, 34], respectively. These approaches indeed improved the robustness, but did not cure the deeper reason for the above-mentioned difficulties. In the following we will get back to the solution of this problem and, at the same time, consider the use of AMG for the *full problem* rather than just the pressure part.

System AMG for Solving Fully-Coupled Systems

It is very important to have a pressure description which corresponds to a nicely elliptic, ideally positive definite M-matrix type discrete problem. We recall that there is no well-defined single pressure equation as, for instance, in CFD pressure-correction type approaches (cf. Sect. 4.2.1). In fact, choosing the pressure derivative of the oil phase α as $\mathbf{A}_{pp}$ in (16) was somewhat arbitrary; we could as well have chosen the pressure derivative of any of the other phases. Since the source term of each phase has a different and locally changing influence on the respective pressure part, it appears to be reasonable, to use a dynamic and locally adapted combination instead. This allows to cope with both above mentioned difficulties (**A1, A2**).

We outline this approach for the most general reservoir model where, besides *oil pressure* and *phase saturations*, further unknowns are considered such as *concentrations of chemical components, temperature* and/or *displacements*. In either case, the Jacobian $\mathbf{J}$, now ordered point-wise, takes the general form

$$\mathbf{J} = \begin{pmatrix} \mathbf{J}_{11} & \cdots & \mathbf{J}_{1n_p} \\ \vdots & \ddots & \vdots \\ \mathbf{J}_{n_p1} & \cdots & \mathbf{J}_{n_pn_p} \end{pmatrix}. \tag{19}$$

Here, n_p denotes the number of grid points and the $\mathbf{J}_{ij}$ are small matrix blocks of dimension $\leq n_u$, describing the block couplings between points i and j where n_u denotes the (maximum) number of unknowns per point. Due to the potential disappearance of phases and components during a simulation, as well as due to different well models, the real number of unknowns per cell may vary.[6]

Rather than solving $\mathbf{J}x = f$, we solve (18) with a block-diagonal $\mathbf{C}$,

$$\mathbf{C} = \begin{pmatrix} \mathbf{C}_1 & & \\ & \ddots & \\ & & \mathbf{C}_{n_p} \end{pmatrix}. \tag{20}$$

However, in contrast to the preconditionings used in the CPR context, our goal here is mainly to ensure the most robust applicability of AMG. Motivated by the pressure-related arbitrariness mentioned above, we dynamically select

$$\mathbf{C}_i = \begin{pmatrix} \delta^i_1 & \delta^i_2 & \cdots & \delta^i_{n_u} \\ & 1 & & \\ & & \ddots & \\ & & & 1 \end{pmatrix} \quad (i = 1, 2, \ldots, n_p). \tag{21}$$

Such preconditioning with dynamic weights $\delta^i_j \geq 0$ is called *Dynamic Row Summing* (DRS), see [29]. The concrete selection of δ^i_j will not only avoid that this matrix transformation introduces 'artificial' difficulties for AMG (cf. **(A2)**), but it will also ensure matrix properties which are well-suited for AMG. Roughly speaking, we will shield AMG from the negative impacts that may result from the complex physics in industrial simulations, which adresses **(A1)**.

The complete system, preconditioned this way, can now be solved by SAMG. Essential for the desired robustness and efficiency is the selection of suitable SAMG components and their optimal adaptation to the requirements of the different simulation models. In particular, the standard use of AMG for the fully-coupled problem $\tilde{\mathbf{J}}x = g$ is neither robust nor efficient. One reason is the very different—from a

---

[6]We assume the sorting of unknowns at points to be so that the first column of $\mathbf{J}_{ij}$ corresponds to pressure derivatives.

mathematical point of view, both elliptic and hyperbolic—physical processes in a model. The respective sub-systems of the full system of equations are characterized by very different properties and scales.

A few remarks regarding the SAMG configuration (for details, also on the concrete selection of δ_j^i, we refer to [29, 30]):

- **Elliptic equations:** Elliptic equations are coarsened and interpolated in an *unknown-based* fashion. Only for the geomechanics sub-component a *point-based* approach might be considered if it gives a better performance. Depending on the respective boundary conditions, one might even need an aggregation approach for the geomechanics component. That is, a hybrid coarsening approach may be the safest approach in the long run. However, for all of the investigated test cases, this was not necessary.
- **Hyperbolic equations:** Hyperbolic equations are not coarsened at all, their main matrix blocks are typically strongly diagonally dominant. However, in order to resolve the corresponding cross couplings sufficiently well, ILU(0) is used as smoother on the finest level.
- **Further aspects:** An adaptive coarsening is employed so that coarsening is done only where beneficial. The different scaling between pressure and displacements usually require more sophisticated Uzawa-type smoothing processes. Apart from this, Gauss-Seidel can be used as smoother on all coarser levels.

Remark 6.1 Further performance improvements can be achieved without influencing the suitability for AMG. This includes adaptations of the AMG components, as well as the exploitation of further left and right preconditionings in certain situations. We refer to [29, 30] for details.

The developed System-AMG approach can be used like a black-box, provided certain basic information (such as the type of model and the involved physical functions) are known to the solver. It adapts itself to concrete problems, such as the reservoir properties, the number of considered components, etc. The efficiency and the robustness of the overall process have been verified for a suite of applications, many of which are from industrial users. Various applications can now be solved efficiently, which was not possible before. In our experience, without preconditioning by dynamic rowsuming (21) it seems difficult to ensure robustness for AMG.

Remark 6.2 SAMG applied directly to fully-coupled simulation models has several advantages over CPR-AMG-like approaches. On the one hand, it is very easy for a user to integrate the solver in his existing simulator. Second, one can extend this approach relatively easily to further physical quantities. Particularly important is the fact that the entire range of available AMG components is available and can be used to optimize the solution process. All physical information is available in the solution process "in its original form". In particular, no artificial algebraic scalings have been applied, which contain the risk of destroying the efficient applicability of AMG. The latter was the main reason for the lack of robustness of CPR-AMG.

7 Summary, Conclusions and Lessons Learned

In this paper we reviewed the path of AMG from an interesting academic idea in the early 1980s to a flexible methodology to create scalable solvers urgently needed by industrial simulation software since the mid 1990s. Before this, virtually all iterative linear solvers used in industrial/commercial simulation packages were based on Krylov methods, combined with some standard one-level preconditioner such as Jacobi or ILU. To a large extent, this is true still today. In many technical engineering disciplines, however, physical models in the 1990s had become so large and sophisticated that classical linear solvers were too slow to make 'real-life' simulations practicable. To remove their performance bottlenecks, throughout Europe commercial codes were migrated to parallel computing platforms, extensively funded by the European Commission (Europort project). HPC indeed gave a boost to the practicability of the participating simulation software, but with further increasing model complexities the linear solver bottleneck was back soon, simply because of the lack of numerical scalability.

Interestingly, at about the same time (around 1995), extensive research on algebraic multigrid restarted. Since AMG-based solvers promised both numerical scalability and an easy integration into existing simulation software, a real performance breakthrough was to be expected by replacing classical one-level solvers with AMG-based solvers. However, AMG was not a fixed method but rather a methodology, allowing the composition of specific solvers by combining application-dependent multilevel components. No general and robust off-the-shelf AMG-solvers existed. Moreover, the methodology was rapidly developed further by investigating new components, alternative processes and algorithms. Except for simple scalar (close to M-matrix) problems, it was not at all obvious, let alone settled, how to compose robust and efficient solvers for the wide range of industrial PDE problems.

There was a huge gap between the research progress on the academic side and the urgent requirements on the industrial side. However, there was no attempt to actively bridge this gap by consolidating the academic progress and to make relevant results usable for industry. Hence, except for simple applications, it was virtually impossible for an industrial developer of simulation software to set up an efficient AMG environment on his own.

When SCAI became a Fraunhofer institute in 2001, generally speaking, its new mission was to make R&D for and in the interest of industry. Obviously, this was a perfect environment to focus on bridging the above gap. In the course of this development, most industrial requests related to AMG's applicability to different types of coupled PDE systems. Since there was no settled AMG approach for coupled PDE systems at that time, the goal was to develop and step-by-step extend an AMG framework, SAMG, general and flexible enough to be usable in as many applications as possible. To maximize SAMG's practical usefulness, industrial partners have been involved in key technical decisions, either directly or by providing typical applications. The main focus was on industrial applications from computational fluid dynamics (Euler and Navier-Stokes), linear elasticity, electro-chemistry, casting and molding, reservoir simulation (oil and groundwater), circuit simulation, and semi-conductor process and device simulation. Today, SAMG is included as linear solver package in many commercial/industrial simulators of these application areas.

The importance of better linear solvers for industrial software developers can be seen from the fact that SCAI's return-of-investment (through licensing and/or focused industrial projects) on the average has allowed a continuous support of 6–7 researchers on SAMG for nearly 15 years now. We point out, however, that the decisive factor was not the software as such, it was the accompanying service, support and maintenance which was continuously provided, and which took by far most (more than 90%, say) of the total work. Many clients have effectively 'outsourced' the numerical work on linear solvers. Clearly, to allow for this in a commercial context, contracts have to be concluded with detailed action plans, rights and duties, responsibilities and—most importantly—liabilities. This, in turn, requires not only extensive administrative and legal support, but also a technical and scientific staff which can permanently take these responsibilities. Most probably here lies the main reason, why none of the freely available AMG packages has become a serious competitor (yet).

Looking back in time, there was a long and stony way from the first 'business plan' to the current role of SAMG in industrial simulation. This success definitely had not been possible without the support (and also the 'pressure') of Fraunhofer through its consequent focus on industrial needs. The biggest market in the near future will be industrial software developers which up to now have been happy with one-level solvers (because their models were still of limited size). It is only a matter of time before most of them will also need scalable solvers. New types of applications and/or discretizations will require further basic research to develop robust and efficient AMG components.

Already for quite some time there has been a serious request for 'intelligent solver interfaces', that is, for some kind of automatic control mechanisms which attempt to find and adjust solver strategies 'on the fly' during run time of a simulation. This is a very important and demanding topic of R&D for the near future. For instance, in time-dependent reservoir simulation processes (oil or water), the properties of discretization matrices may abruptly change in time, requiring a change of solution parameters—or even the solver itself—to maintain overall efficiency. Unfortunately, an optimal (that is, robust *and* efficient) solver strategy generally cannot be defined a priori but rather requires an automatic process at run time. It is an open question, how general such an automatism can be. It will certainly be possible to define efficient automatic processes as long as the range of reasonable choices is known in a given application area (in fact, SAMG already contains such an automatism for groundwater applications). However, although there is a big market for something like a black-box 'super solver'—which always finds the optimal solver and automatically applies it independently of the application environment—there is some doubt, whether such a super solver can ever be also efficient.

References

1. R.E. Alcouffe, A. Brandt, J.E. Dendy Jr., J.W. Painter, The multigrid method for the diffusion equation with strongly discontinuous coefficients. SIAM J. Sci. Stat. Comput. **2**, 430–454 (1981)
2. AMG Summit. http://grandmaster.colorado.edu/summit/. Accessed 31 Mar 2017
3. K. Aziz, A. Settari, *Petroleum Reservoir Simulation* (Applied Science Publishers, London, 1979)
4. N.S. Bahvalov, Convergence of a relaxation method under natural constraints on an elliptic operator. Ž. Vyčisl. Mat. i Mat. Fiz. **6**, 861–883 (1966)
5. P. Bastian, K. Birken, K. Johannsen et al., UG – a flexible software toolbox for solving partial differential equations. Comput. Vis. Sci. **1**, 27–40 (1997)
6. R.P. Batycky, M.J. Blunt, M.R. Thiele, A 3d field-scale streamline-based reservoir simulator, SPE Reservoir Engineering (1997)
7. W.N. Bell, L.N. Olson, J.B. Schroder, PyAMG: Algebraic multigrid solvers in Python v3.0 (2015). Release 3.2
8. M. Blatt, P. Bastian, The iterative solver template library, in *Applied Parallel Computing. State of the Art in Scientific Computing: 8th International Workshop, PARA 2006*, Umeå, June 18–21, 2006. Revised Selected Papers, ed. by B. Kågström, E. Elmroth, J. Dongarra, J. Waśniewski (Springer, Berlin, Heidelberg, 2007), pp. 666–675
9. A. Brandt, Multi-level adaptive technique (MLAT) for fast numerical solution to boundary value problems, in *Proceedings of the Third International Conference on Numerical Methods in Fluid Mechanics: Vol. I General Lectures. Fundamental Numerical Techniques*, July 3–7, 1972, ed. by Universities of Paris VI and XI, H. Cabannes, R. Temam (Springer, Berlin, Heidelberg, 1973), pp. 82–89
10. A. Brandt, Multi-level adaptive solutions to boundary-value problems. Math. Comp. **31**, 333–390 (1977)
11. A. Brandt, Algebraic multigrid theory: the symmetric case. Appl. Math. Comput. **19**, 23–56 (1986). Second Copper Mountain Conference on Multigrid Methods (Copper Mountain, CO, 1985)
12. A. Brandt, J. Brannick, K. Kahl, I. Livshits, Bootstrap AMG. SIAM J. Sci. Comput. **33**, 612–632 (2011)
13. A. Brandt, S. McCormick, J. Ruge, Algebraic multigrid (amg) for automatic multigrid solution with application to geodetic computations. Technical report, Institute for Computational Studies (1982)
14. A. Brandt, S. McCormick, J. Ruge, Algebraic multigrid (AMG) for sparse matrix equations, in *Sparsity and Its Applications (Loughborough, 1983)* (Cambridge University Press, Cambridge, 1985), pp. 257–284
15. M. Brezina, A. J. Cleary, R.D. Falgout et al., Algebraic multigrid based on element interpolation (AMGe). SIAM J. Sci. Comput. **22**, 1570–1592 (2000)
16. T. Clees, *AMG Strategies for PDE Systems with Applications in Industrial Semiconductor Simulation*. Fraunhofer Series in Information and Communication Technology (Shaker Verlag, Aachen, 2005)
17. T. Clees, L. Ganzer, An efficient algebraic multigrid solver strategy for adaptive implicit methods in oil reservoir simulation. SPE J. **15** (2010)
18. T. Clees, K. Stüben, Algebraic multigrid for industrial semiconductor device simulation, in *Challenges in Scientific Computing—CISC 2002*. Lecture Notes in Computational Science and Engineering, vol. 35 (Springer, Berlin, 2003), pp. 110–130
19. T. Clees, T. Samrowski, M.L. Zitzmann, R. Weigel, An automatic multi-level solver switching strategy for PEEC-based EMC simulation, in *2007 18th International Zurich Symposium on Electromagnetic Compatibility* (2007), pp. 25–28
20. K. Coats, A note on IMPES and some IMPES-based simulation models. SPE J. **5**, 245–251 (2000)

21. Copper Mountain Conferences. http://grandmaster.colorado.edu/~copper/. Accessed 31 Mar 2017
22. P.M. de Zeeuw, Matrix-dependent prolongations and restrictions in a blackbox multigrid solver. J. Comput. Appl. Math. **33**, 1–27 (1990)
23. J.E. Dendy Jr., Black box multigrid. J. Comput. Phys. **48**, 366–386 (1982)
24. J.E. Dendy Jr., S.F. McCormick, J.W. Ruge, T.F. Russell, S. Schaffer, Multigrid methods for three-dimensional petroleum reservoir simulation, in *Proceedings of the 10th SPE Symposium on Reservoir Simulation* (Society of Petroleum Engineers, Richardson, 1989)
25. R.D. Falgout, U.M. Yang, hypre: a library of high performance preconditioners, in *Computational Science — ICCS 2002: International Conference, 2002 Proceedings, Part III*, Amsterdam, April 21–24, ed. by P.M.A. Sloot, A.G. Hoekstra, C.J.K. Tan, J.J. Dongarra (Springer, Berlin, Heidelberg, 2002), pp. 632–641
26. R.P. Fedorenko, On the speed of convergence of an iteration process. Ž. Vyčisl. Mat. i Mat. Fiz. **4**, 559–564 (1964)
27. T. Füllenbach, K. Stüben, Algebraic multigrid for selected PDE systems, in *Elliptic and Parabolic Problems (Rolduc/Gaeta, 2001)* (World Scientific Publishing, River Edge, NJ, 2002), pp. 399–410
28. T. Füllenbach, K. Stüben, S. Mijalkovic, Application of an algebraic multigrid solver to process simulation problems, in *2000 International Conference on Simulation Semiconductor Processes and Devices (Cat. No.00TH8502)* (2000), pp. 225–228
29. S. Gries, System-AMG Approaches for Industrial Fully and Adaptive Implicit Oil Reservoir Simulations, PhD thesis, Universität zu Köln, 2016
30. S. Gries, K. Stüben, G.L. Brown, D. Chen, D.A. Collins, Preconditioning for efficiently applying algebraic multigrid in fully implicit reservoir simulations. SPE J. **19** (2014)
31. W. Hackbusch, *Multi-Grid Methods and Applications*. Springer Series in Computational Mathematics, vol. 4 (Springer, Berlin [u.a.], 1985)
32. P.W. Hemker, A note on defect correction processes with an approximate inverse of deficient rank. J. Comput. Appl. Math. **8**, 137–139 (1982)
33. M.A. Heroux, R.A. Bartlett, V.E. Howle et al., An overview of the Trilinos Project. ACM Trans. Math. Softw. **31**, 397–423 (2005)
34. H. Klie, M. Wheeler, T. Clees, K. Stüben, Deflation AMG solvers for highly ill-conditioned reservoir simulation problems, in *SPE Reservoir Simulation Symposium* (Society of Petroleum Engineers, Richardson, 2007)
35. O.E. Livne, A. Brandt, Lean algebraic multigrid (LAMG): fast graph Laplacian linear solver. SIAM J. Sci. Comput. **34**, B499–B522 (2012)
36. B. Metsch, Algebraic Multigrid (AMG) for Saddle Point Systems. Dissertation, Institut für Numerische Simulation, Universität Bonn, July 2013
37. W.A. Mulder, A new multigrid approach to convection problems. J. Comput. Phys. **83**, 303–323 (1989)
38. N.H. Naik, J. Van Rosendale, The improved robustness of multigrid elliptic solvers based on multiple semicoarsened grids. SIAM J. Numer. Anal. **30**, 215–229 (1993)
39. A. Napov, Y. Notay, An algebraic multigrid method with guaranteed convergence rate. SIAM J. Sci. Comput. **34**, A1079–A1109 (2012)
40. M. Naumov, M. Arsaev, P. Castonguay et al., AmgX: a library for GPU accelerated algebraic multigrid and preconditioned iterative methods. SIAM J. Sci. Comput. **37**, S602–S626 (2015)
41. Y. Notay, P.S. Vassilevski, Recursive Krylov-based multigrid cycles. Numer. Linear Algebra Appl. **15**, 473–487 (2008)
42. H.-J. Plum, A. Krechel, S. Gries et al., Parallel algebraic multigrid, in *Scientific Computing and Algorithms in Industrial Simulation — Projects and Products of Fraunhofer SCAI*, ed. by M. Griebel, A. Schüller, M.A. Schweitzer (Springer, New York, 2017)
43. S. Reitzinger, Algebraic Multigrid Methods for Large Scale Finite Element Equations, PhD thesis, Johannes Keppler Universität Linz, 2001

44. J. Ruge, K. Stüben, Efficient solution of finite difference and finite element equations, in *Multigrid Methods for Integral and Differential Equations (Bristol, 1983)*. Institute of Mathematics and Its Applications Conference Series, New Series, vol. 3 (Oxford University Press, New York, 1985), pp. 169–212

45. J.W. Ruge, K. Stüben, Algebraic multigrid, in *Multigrid Methods*. Frontiers Applied Mathematics, vol. 3 (SIAM, Philadelphia, PA, 1987), pp. 73–130

46. Y. Saad, H.A. van der Vorst, Iterative solution of linear systems in the 20th century. J. Comput. Appl. Math. **123**, 1–33 (2000). Numerical analysis 2000, vol. III. Linear algebra

47. K. Stüben, Europort-D: commercial benefits of using parallel technology, in *Parallel Computing: Fundamentals, Applications and New Directions*, ed. by E. D'Hollander, F. Peters, G. Joubert, U. Trottenberg, R. Völpel. Advances in Parallel Computing, vol. 12 (North-Holland, Amsterdam, 1998), pp. 61–78

48. K. Stüben, A review of algebraic multigrid. J. Comput. Appl. Math. **128**, 281–309 (2001). Numerical analysis 2000, vol. VII, Partial differential equations

49. Super Matrix Solver SMS-AMG. http://www.vinas.com/en/seihin/sms/SMS_AMG_index. html. Accessed 31 Mar 2017

50. M.R. Thiele, R.P. Batycky, M.J. Blunt, A Streamline-Based 3D Field-Scale Compositional Reservoir Simulator. SPE-38889-MS (Society of Petroleum Engineers, 1997)

51. P. Thum, *Algebraic Multigrid for the Multi-Ion Transport and Reaction Model - A Physics-Aware Approach* (Logos-Verlag, Berlin, 2012)

52. P. Thum, T. Clees, Towards physics-oriented smoothing in algebraic multigrid for systems of partial differential equations arising in multi-ion transport and reaction models. Numer. Linear Algebra Appl. **17**, 253–271 (2010)

53. P. Thum, T. Clees, G. Weyns, G. Nelissen, J. Deconinck, Efficient algebraic multigrid for migration–diffusion–convection–reaction systems arising in electrochemical simulations. J. Comput. Phys. **229**, 7260–7276 (2010)

54. U. Trottenberg, C.W. Oosterlee, A. Schüller, *Multigrid* (Academic, San Diego, CA, 2001). With contributions by A. Brandt, P. Oswald and K. Stüben

55. P. Vaněk, J. Mandel, M. Brezina, Algebraic multigrid by smoothed aggregation for second and fourth order elliptic problems. Computing **56**, 179–196 (1996)

56. P. Vaněk, M. Brezina, J. Mandel, Convergence of algebraic multigrid based on smoothed aggregation. Numer. Math. **88**, 559–579 (2001)

57. M. Wabro, Algebraic Multigrid Methods for the Numerical Solution of the Incompressible Navier-Stokes Equations, PhD thesis, Johannes Keppler Universität Linz, 2003

58. J.R. Wallis, Incomplete Gaussian elimination as a preconditioning for generalized conjugate gradient acceleration, in *SPE Reservoir Simulation Symposium* (Society of Petroleum Engineers, Richardson, 1983)

59. J.R. Wallis, R. Kendall, T. Little, J. Nolen, Constrained residual acceleration of conjugate gradient acceleration, in *SPE Reservoir Simulation Symposium* (Society of Petroleum Engineers, Richardson, 1985)

60. T. Washio, C.W. Oosterlee, Flexible multiple semicoarsening for three-dimensional singularly perturbed problems. SIAM J. Sci. Comput. **19**, 1646–1666 (1998)

Parallel Algebraic Multigrid

Hans-Joachim Plum, Arnold Krechel, Sebastian Gries, Bram Metsch, Fabian Nick, Marc Alexander Schweitzer, and Klaus Stüben

1 Introduction

Typical problem sizes in industrial applications today range from several million to billions of degrees of freedom which cannot be handled by a single CPU core or single compute node in acceptable time. Very large problems can in fact not be solved at all on a single node due to memory restrictions. Thus, there is a principle demand for parallel software tools that utilize modern multi- and many-core systems as well as cluster computers. Typically, performance measurements with respect to parallelization focus on so-called strong and weak scaling properties which, however, are somewhat more of academic interest. Since the development focus of SAMG is on industrial use, we rather need to be concerned with the question: How to minimize the time-to-solution on fixed and limited (but arbitrary) hardware resources for a small range of problem sizes? Hence, we need to provide more flexibility within SAMG to allow for a trade-off between e.g. memory requirements and parallel scalability versus overall time-to-solution for a fixed problem size using a particular hardware configuration. This flexibility together with the coverage of virtually all algorithmic directions that are state of the art in AMG [3] and their

H.-J. Plum (✉) • A. Krechel • S. Gries • B. Metsch • F. Nick • K. Stüben
Fraunhofer Institute for Algorithms and Scientific Computing SCAI, Schloss Birlinghoven, 53757 Sankt Augustin, Germany
e-mail: hans-joachim.plum@scai.fraunhofer.de

M.A. Schweitzer
Fraunhofer Institute for Algorithms and Scientific Computing SCAI, Schloss Birlinghoven, 53757 Sankt Augustin, Germany

Institute for Numerical Simulation, Rheinische Friedrich-Wilhelms-Universität Bonn, Wegelerstr. 6, 53115 Bonn, Germany
e-mail: marc.alexander.schweitzer@scai.fraunhofer.de

© Springer International Publishing AG 2017
M. Griebel et al. (eds.), *Scientific Computing and Algorithms in Industrial Simulations*, DOI 10.1007/978-3-319-62458-7_6

parallelization (multi-core, distributed parallelism or a hybrid combination) are truly unique selling points of SAMG.

In [3] the importance of combining HPC with *scalable* (optimal) numerical solvers was already stressed. However, when HPC became a major issue for industrial software developers in Europe (in the mid 1990s, when virtually all major European simulation software companies joined the Europort parallelization project [2]), numerically scalable solver software was not yet available. The major numerically scalable solver technology of those days was geometric multigrid which was not appropriate for industrial software development due to its lack of robustness and its non-plugin character [3]. Moreover, existing AMG-based solvers were pure research codes and far from being general and robust enough to be suitable for industrial use. In this situation, parallelization of industrial simulation codes necessarily had to rely on standard one-level solvers only. Fortunately, with the further development of the AMG technology, industrially relevant scalable solvers became available whose combination with HPC then promised a real breakthrough in industrial solver technology.

How important it is to combine HPC and scalable numerical solvers is illustrated by the following simplifying, somewhat amusing scenario in Table 1. There we compare the typical application parameters of the very first AMG in the early 1980s with those of SAMG in the 2010s in a pressure correction CFD simulation context. Ignoring other hardware and software details (like operating system, compilers, ...), the increase in compute power is 10^7-fold which is roughly 20 times larger than the increase in the user demand at $0.5 \cdot 10^6$. Hence, we anticipate to find an improvement by a factor of 20 in the overall AMG run time, which we can roughly observe. Although this comparison is extremely simplified, it highlights the crucial benefit of AMG over standard solution techniques. Only through its numerical scalability can AMG (in principle) keep pace with the rapid development of the hardware; while any non-hierarchical solver will show an explosion of the run times with increasing problem size.

The growth factors in Table 1 pretty much match the famous *Moore's Law* which predicted, already in 1965, an exponential growth in complexity of integrated circuits with a doubling period of only 18–24 months [1]. These growth factors, however, are the joint result of many technological developments. For the ease of notation, we refer to the entirety of all hardware components and their organization as (compute) architecture.

Nowadays high-performance computers, operating systems and compilers offer a large range of features and components for optimizing the run time of applications.

Table 1 Comparison of typical AMG use-case in the 1980s and 2010s

	AMG (1980s)	AMG (2010s)	Factor
CPU clock	≈ 1 MHz	≈ 10 THz[a]	10^7
Main memory	≈ 48 KBytes	≈ 100 TBytes[b]	$> 2 \cdot 10^9$
'Customer demand'	$N = 256$	$N = 130,000,000$	$0.5 \cdot 10^6$
Solution time	$O(1)$ min	$O(1)$ s	$O(10)$

[a]Sum of all cores' frequency
[b]Sum of all cores' main memory

However, the more intricate the architecture is, the more sophistication in software development is needed to attain a satisfactory utilization of the computing system. This is particularly true for AMG and its implementation for different flavors of parallelism (multi-core, many-core, distributed memory parallelism). Besides parallelization, also an effective use of the floating point architecture (loop optimization, vectorization, cache usage) is a challenging task for sparse matrix operations which dominate the performance of AMG.

Note, however, that it is virtually impossible to achieve the theoretical limit in parallel performance with AMG on modern compute architectures due to their complexity and AMG's involved algorithmics. Yet, in SAMG we have gone a long way in order to get as close as possible to optimal performance while maintaining the necessary robustness required by industrial users. The substantial amount of programming effort spent on pure parallel performance optimization today pays of for our customers and provides a strong basis for future hardware developments.

In the remainder of this paper we will introduce various challenges and issues encountered in the parallelization of any hierarchical iterative linear solver and AMG in particular. Then, we will shortly summarize the steps we took in the parallel implementation of SAMG to cope with these challenges. In particular we will focus on the key components of SAMG (smoothing, coarsening, coarse grid solving) and sketch our approaches to the successful utilization of HPC features within these components. Finally, we present some performance indicators of SAMG in real-world application settings and summarize the current status and philosophy of parallel SAMG, including a reference to a special "autoparallel" version of SAMG: XSAMG.

2   Challenges Imposed by Parallel Computer Architectures

In the following we are concerned with the impact of three fundamental hardware developments on linear solver technology and AMG in particular. Here, we focus on single-core floating point performance, multi-core/shared memory parallelism and distributed memory parallelism.

2.1   Single Core Performance: CPU Clock Speed and Memory Frequency

Until the mid 2000s, CPU clock speeds grew rapidly, pretty much at the rate of Moore's Law, ending up at speeds of approximately 3–4 GHz. This period was called the "GHz era". For operation intensive applications, it provided a "free ride": Without changing anything, the performance doubled in speed every 2 years. However, this is not the complete story: While the CPU clock speed and the size of

Table 2 Impact of memory efficiency on AMG performance

	Base CPU frequency	α by "stream"	SAMG run time
Machine A	3.3 GHz	≈7	17.7
Machine B	2.6 GHz	≈2	13.7

the main memory was growing rapidly, the memory access speed did not increase at the same rate. When looking at prominent chip series, one finds a 50-fold increase in CPU frequency between the mid 1990s and mid 2000s (matching Moore's Law), but only about a 10-fold increase in the memory frequency. That is, the relation of CPU and memory performance, expressed by α, which is defined as the number of CPU cycles elapsed during a memory read of one word, increased by a factor of 5. Thus, intensive memory access has become a potential bottleneck.

This has a direct impact on serial SAMG performance. The basic operation type in AMG is a sparse *matrix × vector* multiplication or structurally similar operations. Such operations are largely memory bound. Roughly speaking, the whole matrix has to be loaded from main memory, but only two floating point operations are executed per entry. Since, memory loads are (much) more expensive than floating point operations on current architectures, AMG performance is dominated by memory performance. On top of this, the typical memory access patterns in sparse matrix operations are neither cache- nor vectorization friendly.

Thus, the parameter α defines a performance barrier for sparse matrix operations and thus also for AMG. We manifest this claim by a concrete benchmark on two current machines, where one of the machines has a higher CPU frequency but also a much slower memory access, i.e. a significantly higher α, see Table 2. Here, we give the run time of SAMG and the results of the well-established "stream" [6] benchmark which directly measures α. From these numbers, we can clearly observe that AMG (i.e. sparse matrix operations in general) cannot benefit from an increase in CPU clock speed if the memory performance does not improve accordingly. In fact, the execution of SAMG on the slower CPU with much better memory performance is 25% faster. Thus, AMG did benefit from the GHz era but not to the fullest extent.

2.2 Multi-Core CPUs and Shared Memory Parallelism

With the advent of "multi-core", the GHz era came to an abrupt end. Important arguments for stopping the CPU frequency boost were heat production and energy efficiency of computers. Physics dictates that e.g. a dual core CPU running both cores at 80% of the clock speed consumes the same amount of energy as a single core CPU at full speed. Similarly, an eight core CPU running its cores at 50% of the clock speed requires the same energy. Therefore, multi-core has the potential to deliver a largely improved "performance per Watt" balance. This way, multi-core became a success model which could no longer be circumvented. As a consequence, the free ride of the GHz era was over for everyone and almost all codes had to be parallelized with multi-core compliant new programming paradigms to improve run time performance.

This was thoroughly done for SAMG via the directive based programming paradigm OpenMP [7], which (usually) spans so-called parallel threads across the multi-core architecture. There are almost 1000 algorithms and code blocks in SAMG that have all been equipped with OpenMP parallelism, most of them in an intricate manner far beyond simple implicit OpenMP programming. Besides this effort, multi-core has also introduced a couple of new difficulties into SAMG. The memory barrier discussed in Sect. 2.1 has stayed or even worsened so that the multi-core speedups are (indispensably) below optimum (see Sect. 2.2.1). Many components of AMG are intrinsically serial by nature and have to be replaced by parallel variants which, however, usually change the numerical behavior and entail non bit-wise identical results with the serial version (see Sect. 2.2.2). Moreover, OpenMP parallelization (by definition in its standard) allows for so called race conditions which may render results non reproducible even between two runs with the same number of threads (Sect. 2.2.3). Finally, most HPC platforms offer multiple multi-core-CPU s (multiple "sockets") which still, from the logical point of view, can be handled by OpenMP like a homogeneous shared memory system. This, however, may introduce additional performance losses which are difficult to detect (see Sect. 2.2.4).

For non commercial software packages and research codes, such issues are usually accepted and considered negligible. But, SAMG has a different positioning with its industrial user focus. The lack of perfect multi-core speedups is difficult to sell, race conditions are usually not accepted, while results depending on the degree of parallelism, given for virtually all non trivial algorithms, is meanwhile accepted also by many industrial users.

2.2.1 Multi-Core and Memory Access

Regarding the memory bottleneck and the factor α described in the previous section, two effects come together in multi-core. On the one hand, hardware vendors have substantially improved memory links so that the absolute memory throughput has improved. However, this throughput does not scale with the number of cores. Thus, the sustainable overall throughput saturates at a certain point, as clearly shown by "stream" on a typical eight core CPU of today, see Table 3. This means that the sustainable α increases significantly when going from 1 to 8 cores, i.e. relative memory performance deteriorates. It is clear that this effect is not beneficial to SAMG's performance. Yet, the numbers given in Table 4 clearly show that SAMG shows an almost identical scaling behavior as a simple Jacobi relaxation, i.e. sparse matrix operations. Thus, SAMG actually performs at the practical limit it can reach due to the underlying memory access patterns.

Table 3 Memory throughput on typical eight-core CPU

Number cores	Throughput (stream), GB/s	Speedup
1	14.1	1
2	24.4	1.73
4	36.2	2.57
8	37.3	2.65

Table 4 Comparison of parallel speedup of SAMG and simple Jacobi-smoothing

Number cores	Speedup SAMG	Speedup Jacobi
1	1	1
2	1.84	1.85
4	3.3	3.4
8	5.1	5.5

2.2.2 Intrinsically Serial Components

Many key code blocks consist of loops that are recursive, i.e. iterations depend on the outcome of preceeding ones. As said, a Gauß–Seidel smoother or the standard Ruge Stüben coarsening process are key examples of this issue. Chopping up such loops by simple implicit OpenMP is just wrong. One has to create parallelizable variants of such components by blocking strategies where the respective algorithm is applied in parallel to each block, combined with some sort of correction handling the dependencies between the blocks (e.g. by data or domain decomposition techniques). Unfortunately, there is no simple general rule how to select these blocks and the respective corrections to account for inter-block dependencies. In fact, most of these blocking strategies implemented in SAMG are rather complicated and each one is designed for a single specific task.

Let us point out one crucial consequence of introducing such parallel variants or generalizations of intrinsically serial algorithms: The overall computation becomes dependent on the number of parallel instances (usually threads in the OpenMP case). Thus, computational results are subject to deviations between different runs of the same simulation case on different numbers of threads. This observation holds true for block-variants of any intrinsically serial algorithm and not only for AMG.

2.2.3 Race Conditions

A race condition occurs when multiple threads that share a common memory segment run into a read/write conflict at particular memory locations. Note that such events may render results to be not reproducible: Two different runs with exactly the same code, input, number of threads and compute architecture may yield different results, depending on the order how the conflict is resolved. Moreover, OpenMP (per standard definition) may introduce such race conditions by itself, e.g. when thread local floating point data get summed up across the threads (via so-called OpenMP reductions). Another issue for an OpenMP enabled library is of course that the calling user program may itself contain race conditions which may effect the input to the library and thereby introduce these effects into an otherwise race condition free library.

2.2.4 Multiple Sockets

Implicit data sharing between sockets may affect the performance, as data movements between sockets are usually more expensive than within a physically shared memory on a single socket.

Such effects are (unfortunately) related to the memory management of the user program (and the operating system). To give an example, a data array allocated and initialized on a certain socket will be persistently placed in this socket's memory (at least on Linux machines). Any access to this array by another socket will therefore result in such implicit and unwanted data movements. To avoid this issue or at least reduce its impact on the run time, it is advisable to initialize a shared memory location by the very thread which is mainly using it (a so-called "first touch" policy).

2.3 Distributed Memory Parallelism

Since the late 1980s, long before multi-core, distributed memory parallelism has been an unavoidable means to improve on excessive run times. However, this kind of parallelism requires even more programming effort since data is no longer shared among processors but must be exchanged explicitly, usually by MPI [5]. Nowadays, industrial clients use SAMG on large compute clusters, where many "nodes" of multi-socket multi-core CPUs are linked together by efficient communication interconnects.[1] This imposes another significant level of complexity in AMG programming, especially when MPI based communication is employed only for inter-node data exchange while intra-node parallelism is attained via OpenMP. Note also that MPI-parallel AMG is in fact more an algorithmic variant of AMG rather than an exact parallelization, compare Sect. 2.2.2.[2] New parallelizable variants of existing algorithms, also based on certain blocking strategies, have to be implemented for MPI. Note that for this, the programming effort is another add on, since OpenMP and MPI implementations are separate and cannot profit from each other. Compared to OpenMP, where we have already claimed that about 1000 code blocks in SAMG are equipped with OpenMP statements, MPI-parallelism is much more involved. The MPI version of SAMG requires about 50% of its code lines just to organize the MPI-parallelism.

---

[1] SAMG is in fact used very successfully by our customers with more than 3000 cores.

[2] Note, however, that due to customer demand parallel SAMG can be run in a (low efficiency) mode which comes very close to a serial algorithm.

The performance difficulties imposed by distributed parallelism are similar to those of the multi-core context, see Sect. 2.2. The limited intra-node speedup discussed in Sect. 2.2.1 is, of course, not removable by MPI, so that the per-node performance figures of SAMG are pretty much the same for OpenMP and MPI. Benefits of MPI over OpenMP are that no race conditions exist and that multiple sockets pose no additional challenge. Unfortunately, the "coarse grid challenge" becomes much harder for MPI-parallelism than with pure OpenMP parallelism. Often 100s or 1000s of processes are involved in MPI computations, so that on very coarse grids, there is hardly enough work remaining for all the processes. In such a situation, the decision of where to stop coarsening and declare a given level as the coarsest one, becomes crucial. Recall that (usually) the coarsest level has to be solved (almost) exactly since alternative coarse grid solves, like executing only a few iterations of an iterative solver, can seriously affect the numerical robustness of the overall AMG process and thus need to be employed with care. Therefore, one is faced with the problem of solving a sparse linear system of moderate and non negligible size on many processes when stopping the coarsening too soon. While continuing coarsening for too long yields more and more idling processes and a deteriorating operation to communication ratio. In any event, somewhere AMG incurs an "Amdahl Law" on coarse grids which cannot be completely avoided. Note that this phenomenon, however, is completely due to distributed parallelism and not present in serial or pure OpenMP parallel AMG.

3   How SAMG Counters the HPC Challenges

Now that we introduced the fundamental challenges arising from parallelization in general, let us shortly sketch the strategies we employed in the parallelization of SAMG's core components: Smoothing, coarsening and the coarse grid solution. As indicated earlier, this covers only a small section of all the OpenMP and MPI organization that has been implemented in SAMG, but provides good role models for the strategies used throughout SAMG. Before we go into the details, let us summarize the overall effect of our parallelization efforts:

- SAMG has a close to 100% parallelization for OpenMP and MPI; i.e., there are no crucial serial code blocks in SAMG which would induce a strong "Amdahl effect". Moreover, SAMG provides hybrid parallelization combining OpenMP and MPI-parallelism which gives more flexibility to the user to reduce the time-to-solution on a particular hardware configuration. To our knowledge, this is a unique feature among all available AMG packages.
- We have eliminated all potential OpenMP race conditions in SAMG (at the cost of a tiny additional overhead). Round off affected OpenMP reductions were avoided and manually re-programmed. This and a diversity of other measures have entailed a fair amount of extra programming work in SAMG which may

serve as an example that industrial demands require additional quality and programming compared to academic software.

- We employ a first touch policy (see Sect. 2.2.4) wherever sensible in SAMG to minimize the likelihood for multi socket performance losses.
- Our focus was always on robustness and generality for a broad range of industrial use cases. Thus, extremely tuned variants for very special cases are not included in SAMG. In such cases, we sacrificed a few percent of performance in favor of generality and in the best interest of our customers which obtain a satisfactory parallel performance for all of their employed parameter settings. Moreover, we provide substantial support to our customers in finding the optimal settings to minimize the time-to-solution for their respective application and hardware resources.

3.1 Tuning and Parallelization of Smoothing

The classical smoothers, "Jacobi" and "Gauß–Seidel", consist of seemingly very simple compact six line double loops. However, in these loops (if selected as smoother), easily 50% (if not more) of SAMG's solution time are spent. Thus, optimizing and parallelizing these code blocks, i.e. tackling all of the HPC challenges described in Sect. 2, is a crucial task for SAMG.

The said loops are pretty much accepted as the standard way of formulating these smoothers. Thus, very simply, everything is left to the compiler and a thorough check of the selected flags.

However, one can go beyond this. The standard formulation uses the standard data structures for sparse matrices. However, there is a whole bunch of interesting alternative formats for storing sparse matrices. In the SAMG development, we have quite thoroughly checked such re-formulations and finally came to the conclusion that, for certain cases, it is worthwhile to sacrifice some memory and transform the user matrix into a new beneficial format. In these cases, the run time overhead for the transformation amortizes and the overall run time is reduced at the cost of extra storage. In some use cases from our industrial customers we find a performance gain of roughly 20% in the overall run time, see Table 5, while the parallel scaling

Table 5 Comparison of parallel run times of relaxation schemes

Number cores	Jacobi	Gauß–Seidel (with race conditions)	Gauß–Seidel (optimized)
1	1.70	1.74	1.36
2	0.79	0.82	0.72
4	0.46	0.47	0.40
5	0.32	0.32	0.25
16	0.18	0.17	0.14

behavior is maintained. Even better is the situation when the user matrix has a sub-structuring into dense 2×2 or 3×3 blocks, as it is for instance the case for matrices arising in elasticity applications. By using a respective block data structure to store the system matrix, the execution time for a single Jacobi smoothing step improves by a factor of 2–3, see Table 6 and Fig. 1.

In many cases, intrinsically serial Gauß–Seidel smoothing provides a substantially shorter time-to-solution than naturally parallel Jacobi smoothing due to its better smoothing. For instance, it took SAMG 35 iterations to converge for the problem considered in Table 5 when using the Jacobi smoother whereas it took only 24 iterations when using Gauß–Seidel smoothing (running on a single core). Thus, our optimized Gauß–Seidel smoother in SAMG yields an overall reduction in the total time-to-solution of over 50% compared to SAMG with Jacobi smoothing in this example. Unfortunately, extra programming effort is again necessary to maintain this improved convergence behavior also in parallel, at least to a large

Table 6 Comparison of parallel run times of relaxation schemes for matrices with internal block structure		Jacobi	Jacobi
	Number cores	(scalar)	(block)
	64	3.59	1.49
	128	1.84	0.55
	256	0.92	0.30
	512	0.46	0.17
	1024	0.23	0.09
	2048	0.14	0.07

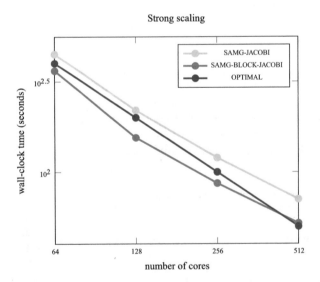

Fig. 1 Strong scaling behavior of parallel SAMG solver in complex elasticity application with $N = 46,875,000$ degrees of freedom (Architecture: Intel (R) E5-2650 v2 @ 2.60 GHz, 16 core nodes, hybrid OpenMP/MPI-parallelism)

extent. Gauß–Seidel smoothing updates the solution vector one component after the other, with any update depending on the preceding ones. This process can in general not be parallelized by any blocking strategy. For sparse matrices, each of the elementary component updates (i.e. the inner loop), involves only a thin amount of data which doesn't deliver enough mass for parallelization either. Thus, a literal parallelization of this algorithm is virtually impossible.

A standard approach to this issue is a domain decomposition technique: Apply block local Gauß–Seidel and Jacobi for the inter-block couplings. However, this approach is prone to loose some of the improved smoothing quality along the block boundaries and asymptotically will yield Jacobi smoothing behavior. In SAMG, we developed several refinements of this approach which aim at reducing the "Jacobi-share" of the smoothing process while maintaining the race condition free implementation in OpenMP. Due to this effort, our Gauß–Seidel smoothing is rather robust with respect to the number of cores. For instance, while SAMG took 24 iterations to converge with Gauß–Seidel smoothing on a single core in the previous example, it required 29 iterations with 16 cores which is very much acceptable compared with the 35 iterations required with pure Jacobi smoothing.

3.2 Ruge-Stüben Coarsening

Let us now briefly indicate some of the parallelization ideas for one of the coarsening strategies in SAMG: the classic Ruge-Stüben coarsening process. The original process is recursive and employs a global view of the system matrix. Again, restricting the coarsening into sub-blocks is an obvious idea. However, when doing this in a simple brute force fashion by completely localizing AMG and ignoring all inter-block dependencies, one ends up with a naive domain decomposition approach with local AMG which does not converge in multigrid quality. Early improvements of this approach, also in SAMG, were "sub-domain blocking" techniques (see also [4]) which handle the inter-block couplings in a sligthly more refined way. However, a version of this initial sub-domain blocking approach which was truly robust, memory friendly and efficient was not found at the time.

Interestingly, it was one of the early commercial users of SAMG who requested a substantial improvement here. At this point, a significant development phase began in SAMG and led to the realization of a suite of strategies that essentially fix this inter-block issue. These generalizations of the sub-domain blocking approach allow for trading memory or communication requirements against numerical properties. With a reasonable increase of communication requirements alone one of these strategies allows for (almost) re-capturing the numerical behavior of the serial code in many cases.

Recall, however, that our SAMG library provides virtually all relevant flavors of AMG and coarsening procedures which all have their specific issues when it comes to parallelization. Even though we have invested a substantial amount of research and programming efforts into the parallelization of all these techniques over the years, the work on parallel coarsening schemes continues to further improve on the scalability and time-to-solution.

3.3 Coarse Grid Solution

Finally, let us take a closer look at the infamous "coarse grid challenge" including a potential "Amdahl effect" which we already mentioned in Sect. 2.3. The fundamental issue here is that after a number of coarsening steps the resulting linear systems become too small in size to be distributed efficiently among the participating cores; this holds true for any multigrid scheme.

To cope with this issue, SAMG provides a number of strategies since there is no one-size-fits-all solution to this problem. First of all the impact of the quality of the coarse grid solver on the convergence behavior is problem-dependent and thus this issue may be safely ignored in some applications while in others it will dominate the overall performance. Note also that the available hardware resources, e.g. communication network and available memory, need to be considered in the design of an appropriate coarse grid approach to minimize the overall time-to-solution. Therefore, we restrict ourselves in this treatise to demonstrating the effectiveness of our counter measures in an industrial use case, see also Fig. 2. Recall that in serial SAMG, the grids are coarsened down to a negligible size of roughly 100 degrees of freedom. Using a similar local coarse grid load in parallel,

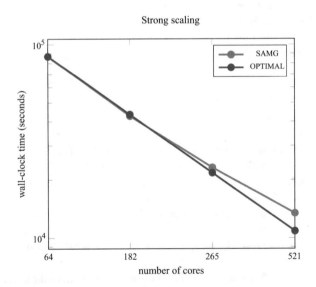

Fig. 2 Strong scaling behavior of parallel SAMG solver in OpenFoam with $N = 130,000,000$ degrees of freedom (Architecture: Intel (R) E5-2650 v2 @ 2.60 GHz, 16 core nodes, pure MPI-parallelism)

however, yields a much larger total coarse grid system size. For instance, when we use 512 processes, as given in this example, the total coarse grid problem is (roughly) a system of dimension $50{,}000 \times 50{,}000$ and the computational effort associated with its solution is no longer negligible. If we were to perform the direct solution of this coarse grid system serially on a single core, its solution would easily amount to more than 20% of the total run time in this concrete example. Thus, a serial coarse grid solve would entail a harsh barrier for parallel scaling which is not acceptable to our customers. With one of the coarse grid strategies available in SAMG which takes the available hardware configuration into account, however, we realize a perfectly acceptable parallel scaling behavior, see Fig. 2. From the plots depicted in Figs. 2 and 1 we can observe the near optimal strong scaling behavior of SAMG on 64–2048 cores. The respective measurements come from the solution of a customer fluid flow application from automotive industry with $N = 130{,}000{,}000$ degrees of freedom and an elasticity application with $N = 46{,}875{,}000$ degrees of freedom.

4 SCAI's Parallel SAMG Solver Library

In summary, the parallelization of multigrid methods and AMG in particular is by no means a trivial task and many of the issues presented in this paper will have to be revisited time and again with every new hardware development. Moreover, the usability of AMG in the industrial context imposes a number of additional challenges that we have also touched upon. The key point being that the only two relevant measures for industrial users are robustness and time-to-solution—with respect to their problem size and hardware resources. Therefore, SAMG provides multiple flavors of parallelism: OpenMP, MPI and hybrid, which allows the user to choose the optimal settings with respect tp the available hardware.

Finally, we want to point to a special version of MPI-parallel SAMG, XSAMG, which provides a unique selling point by equipping completely serial user applications by full-blown hybrid MPI-OpenMP parallelism—at least for the linear solver. This product line of SAMG is utilized by customers whose simulation code is non MPI-parallel, but uses a high share of the run time in the linear solver.

In such a case, when the XSAMG branch of SAMG is used, MPI-parallelism is created "on the fly". Internally, the user data are re-distributed among the available MPI processes, and MPI-parallel SAMG is executed. In the end, the distributed solution is gathered back. Note that the calling sequence to XSAMG is identical to the serial interface—the user code does not have to be modified (except for two marginal initialization and finalization calls). Of course, the data distribution and gathering is not (at all) for free and the run time speedups are limited to the linear solver part only. Thus, perfect speedups cannot be expected, and this approach is worthwhile only when the overall share of time in the linear solver is significant.

XSAMG is of particular interest for users who have a serial application but multiple cluster nodes available. With XSAMG, they are enabled to span the solver

Table 7 Scaling of XSAMG across multiple MPI nodes

Number nodes	XSAMG runtime
1	24.3
2	12.2
4	7.3

part over multiple nodes of the cluster, a feature which is not possible with any other approach. For instance, one of our customers of XSAMG obtains a speedup of 3.3 within the linear solver part when going from 1 to 4 nodes using XSAMG which clearly outperforms pure single node parallelism, see Table 7.

We conclude by briefly commenting on SAMG's philosophy regarding special architectures. Apart from partial successes for selected modules, the attempt to port SAMG in full generality to special chips like GPUs or many-core co-processors has failed so far. Enormous efforts have to be invested in order to make even simple code structures run effectively on such chips. Special AMG versions are around, but these cannot at all stand up to the generality and robustness required for industrial use. However, SAMG will be prepared for a new generation of chips of that kind, once their user friendliness and effective programmability have been enhanced.

References

1. G.E. Moore, Cramming more components onto integrated circuits. Electronics **8**, 114–117 (1965)
2. K. Stüben, Europort-D: commercial benefits of using parallel technology, in *Parallel Computing: Fundamentals, Applications and New Directions*, ed. by E. D'Hollander, F. Peters, G. Joubert, U. Trottenberg, R. Völpel. Advances in Parallel Computing, vol. 12 (North-Holland, Amsterdam, 1998), pp. 61–78
3. K. Stüben, J.W. Ruge, T. Clees, S. Gries, Algebraic multigrid — from academia to industry, in *Scientific Computing and Angorithms in Industrial Simulation — Projects and Products of Fraunhofer SCAI*, ed. by M. Griebel, A. Schüller, M.A. Schweitzer (Springer, New York, 2017)
4. U.M. Yang, Parallel algebraic multigrid methods — high performance preconditioners, in *Numerical Solution of Partial Differential Equations on Parallel Computers*, ed. by A.M. Bruaset, A. Tveito (Springer, Berlin, Heidelberg, 2006), pp. 209–236
5. MPI Forum. http://mpi-forum.org. Accessed 31 Mar 2017
6. STREAM: Sustainable Memory Bandwidth in High Performance Computers. https://www.cs.virginia.edu/stream/. Accessed 31 Mar 2017
7. The OpenMP API Specification for Parallel Programming. http://www.openmp.org. Accessed 31 Mar 2017

MpCCI: Neutral Interfaces for Multiphysics Simulations

Klaus Wolf, Pascal Bayrasy, Carsten Brodbeck, Ilja Kalmykov, André Oeckerath, and Nadja Wirth

1 Introduction

In various research and engineering fields, there is a growing demand for more realistic simulations covering all relevant aspects from different simulation disciplines—the multiphysical or multidisciplinary simulations. Fluid-structure interaction (FSI), magneto-hydro dynamics, thermal radiation or manufacturing process chains define only a subset of multiphysics applications.

To this purpose, Fraunhofer SCAI has developed flexible vendor neutral interfaces since 1996 in order to transfer simulation data from one tool to another—either at run-time or file-based. The tools provide methods and algorithms to translate (i.e. map) the data to the software-specific syntax and to the problem-specific discretization. The mapping algorithms use the finite element or finite volume formulation (i.e. the shape functions) of the source mesh in order to interpolate the data onto the target mesh, cf. the MpCCI documentation [4].

The main software product MpCCI CouplingEnvironment provides a framework for co-simulations: the data is repeatedly exchanged at run-time in a bi-directional way. This allows to combine specialized simulation tools in order to create a commonly converged multiphysical result. The tool is applied for multiphysical phenomena which require a high degree of interactions.

For application examples, where the influence of the fluid dynamics or the electromagnetics to the thermal or mechanical behavior of a structure is higher than vice versa, it is sufficient to map the load conditions once. To this purpose the file-based software MpCCI FSIMapper has been developed.

K. Wolf (✉) • P. Bayrasy • C. Brodbeck • I. Kalmykov • A. Oeckerath • N. Wirth
Fraunhofer Institute for Algorithms and Scientific Computing SCAI, Schloss Birlinghoven, 53757 Sankt Augustin, Germany
e-mail: klaus.wolf@scai.fraunhofer.de

© Springer International Publishing AG 2017
M. Griebel et al. (eds.), *Scientific Computing and Algorithms in Industrial Simulations*, DOI 10.1007/978-3-319-62458-7_7

Simulation of complex manufacturing processes to predict structural product characteristics does often require a serialized chain of distinct simulation disciplines. To achieve realistic simulation results, local material properties have to be mapped as initial conditions to each step of the virtual process chain [6]. The MpCCI Mapper provides a vendor neutral interface for such simulation workflows. Typical application areas are passive safety for automotive vehicles, optimization of forming tools, and design of composites manufacturing process and products.

The following sections describe the MpCCI concepts and selected multidisciplinary applications. Parts of these product and application descriptions are excerpts from other sources [2, 5–8, 10, 11, 14].

2   MpCCI CouplingEnvironment

MpCCI CouplingEnvironment has been developed for the simulation of multiphysical phenomena. For simulation disciplines like finite element analysis (FEA), computational fluid dynamics (CFD), multibody systems (MBS), electromagnetics (EM), etc., a lot of specialized commercial or open-source simulation tools are available on the market. However, in many cases the combination of different simulation disciplines cannot be realized within a closed software environment from a single vendor [10].

On this account, an application independent interface for the direct coupling of different simulation codes has been designed: MpCCI CouplingEnvironment. It has been accepted as a 'de facto' neutral standard for simulation code coupling. Its multiphysics framework provides a complete and ready-to-use co-simulation environment with a dedicated user front end and visualizer system.

To ensure best interoperability between the codes, MpCCI CouplingEnvironment has established a standardization of coupling procedures independent from the utilized codes and coupling quantities definition. The supported simulation codes are shown in Fig. 1.

MpCCI CouplingEnvironment provides a coupling manager which automatically organizes the communication of the coupled codes [10]. The software employs a staggered approach for all co-simulation problems which can be defined as

- a globally *explicit* coupling method: the coupled fields are exchanged only once per coupling step. This approach is applicable to problems with weak physics coupling.
- an *implicit* iterative coupling method: the coupled fields are exchanged several times per coupling step until an overall stabilized solution is achieved before advancing to the next coupling step. This approach is applicable to problems with strong physics coupling.

Complementary to the coupling method, MpCCI CouplingEnvironment offers two coupling algorithms—Gauss-Seidel and Jacobi (Fig. 2):

- The *Gauss-Seidel* coupling scheme is also known as serial or "Ping-Pong" algorithm where one code waits while the partner code proceeds.

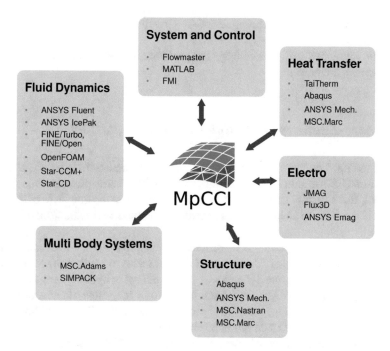

Fig. 1 MpCCI CouplingEnvironment's list of supported simulation codes (December 2016). Published with kind permission of ©Fraunhofer SCAI 2016. All Rights Reserved

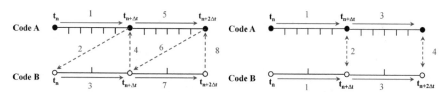

Fig. 2 Flow diagram for the Gauss-Seidel coupling scheme (*left*) and the Jacobi coupling scheme (*right*). $t_{n+\Delta t}$ represents the physical time t at time step $n + \Delta t$. Δt is the coupling time step size [10]. Published with kind permission of ©Fraunhofer SCAI 2016. All Rights Reserved

- The *Jacobi* coupling scheme is also known as a parallel algorithm where both analysis codes run concurrently.

MpCCI CouplingEnvironment will automatically exchange the data between the meshes of two or more simulation codes by using interpolation methods and considering the nature of the quantities exchanged. The co-simulation application can exchange nearly any kind of data between the coupled codes; e.g. energy and momentum sources, material properties, boundary condition values, mesh definitions, or global quantities.

To provide a stabilization of the coupling iteration, relaxation algorithms (fixed factor, ramping or automatic) are implemented. For problems with different time scales or adaptive time stepping, MpCCI CouplingEnvironment offers a time-interpolation.

In the following, various multiphysical problems solved with MpCCI Coupling-Environment are presented.

2.1 Aero-Elasticity and Fluid-Structure-Interaction

One of the most common examples of multiphysics simulations is a fluid-structure interaction (FSI) simulation: when a surrounding fluid exerts pressure on a flexible structure, this structure deforms which leads to changes in the flow field of the fluid. The fluid mesh has to adapt to the new position computed by the structure simulation code [8].

This effect can be observed in different application areas, for example when studying the aerodynamic performance of aircraft wings or spoilers of racing cars or while investigating machine dynamics of valves, pumps or hydraulic engine mounts.

Although FSI simulations have been in use for a long time already, depending on the fluid and solid properties, the choice of the right coupling algorithm and the mesh motion on the CFD side might be challenging.

The two following selected applications highlight different aspects of FSI with deformable structures.

2.1.1 Wing and Spoiler Design

Aircraft wings and racing car spoilers are surrounded by fast moving air, as shown in Fig. 3. The pressure building up on the surface of the wings and spoilers leads to deformation of the solid parts. To predict the behaviour of the wing during flight a coupled FSI simulation is necessary [15].

The deformation of the wing or spoiler elements is calculated with an FEA simulation software, e.g. Abaqus or MSC.Nastran. This FEA code is then coupled—via MpCCI CouplingEnvironment—with a CFD code (Fluent, Star-CCM+ or OpenFOAM) calculating the fluid flow. Depending on the investigated driving or

Fig. 3 CFD model of racing car. Rear wings are investigated using FSI simulations [8]. Published with kind permission of ©Fraunhofer SCAI 2016. All Rights Reserved

flight conditions, different coupling algorithms can be used: either steady state or transient simulations can be coupled. Only with a coupled simulation, the behaviour of wings (e.g. flutter) or the performance of a racing car spoiler using drag and lift coefficients can be predicted.

2.1.2 Hydraulic Pump Layout

With the help of numerical simulation a new high-pressure hydraulic axial pump has been developed at Gdansk University [19], see Fig. 4. To find an optimal layout of the different pump chambers, coupled FSI simulations using Abaqus, Fluent and MpCCI CouplingEnvironment were used.

The pump is equipped with a compensation chamber with an elastic membrane. During phases with high pressure, this membrane deforms, giving the fluid more room and thereby reducing harmful pressure peaks.

The CFD model consists of all pump chambers—including the compensation chamber—and implements the movements of the pistons. The hydraulic oil is assumed to be slightly compressible to achieve a stable coupling. Abaqus/Explicit is used to compute the deformation of the elastic wall of the compensation chamber. Very small time steps are necessary for the coupled simulation.

While a CFD stand-alone simulation is not capable of reproducing the experimental measurements for the pump, the coupled simulation results show a very good agreement with the experiments and thus have been used to optimize the geometric layout of the compensation chamber and the materials for the elastic wall.

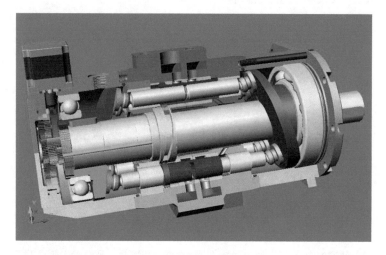

Fig. 4 Constant displacement PWK pump with four chambers   [16]. Published with kind permission of ©Fraunhofer SCAI 2016. All Rights Reserved

2.2 Thermal and Vibration Loads in Turbomachinery

The development of highly efficient turbo-machines makes the detailed knowledge of the mechanical, thermal and fluid dynamic processes indispensable. They serve for optimization of the flow geometry, thermal stresses, lifetime, etc. Realistic simulation of a turbomachinery system often requires the knowledge of boundary conditions, which usually are the results of other simulation disciplines or are even strongly interrelated with each other [11].

The following selected applications highlight different solution strategies for turbomachinery design.

2.2.1 Thermal Loads on Ceramic Impeller

In order to increase the efficiency of micro gas turbines by increasing the gas temperatures, it is necessary to consider new material concepts for the high temperature loaded parts. The solution strategy targets to compute a realistic temperature distribution in turbo-machines. For the analysis of the heat transfer, well chosen thermal boundary conditions are necessary. This includes, among other things, the heat input to the structure resulting from the flow field (cf. Fig. 5). The heat flux to solid parts is influenced by the difference between the fluid and the structural wall temperature. The wall temperature is in turn dependent on the effective heat flux. Only if both entities are at equilibrium, the actual component temperatures are reached. Fraunhofer SCAI used a thermally coupled FINE/Turbo—Abaqus solution to create a new ceramic impeller design [24].

Fig. 5 Turbomachinery flow paths colored by the temperature [11]. Published with kind permission of ©Fraunhofer SCAI 2016. All Rights Reserved

2.2.2   Life-Time Estimation of Turbine Blades

Flow-induced vibrations can lead to a high noise emission and to blade fatigue which can endanger the integrity of the whole system. Excitations are caused by pressure fluctuations in the flow field generated mainly by interactions between rotating and stationary blade rows. To estimate the long term behavior and high-cycle fatigue in operation, it is necessary to know the periodic pressure oscillations of the flow and thus the excited oscillations of the turbine blades. A transient coupling of the fluid pressure and the blade deformation delivers stress oscillations and thus gives the basis of fatigue analyses [11].

2.3   Vehicle Dynamics and Nonlinear Component Behavior

Analysis of multibody systems (MBS), finite element analysis (FEA), and computational fluid dynamics (CFD) are well established practices in computational engineering. The simulation of multibody systems is mainly used for the analysis of mechanisms consisting of rigid components connected with joints to represent the whole system dynamics. The FEA and CFD methods allow much more detailed investigations of the system behavior. They need, however, more computational resources and are time consuming compared to the MBS simulation. The following selected examples highlight some applications where co-simulation with MBS codes provides an accurate result and is a good strategy compared to standard computational engineering methods [2, 14].

2.3.1   Driving Over Obstacles

For some critical situations, e.g. driving over an obstacle (Fig. 6), the misuse of components over the vehicle dynamics needs to be investigated. Various automotive original equipment manufacturers (OEMs) use a combination of Abaqus and MSC. Adams to model the nonlinear behavior of single critical components, e.g. transverse links, and their interaction with the complete vehicle system model. All critical components are modeled with Abaqus in order to calculate the nonlinear response from the multibody system in MSC.Adams which provides the kinematic constraints of the whole system. This solution strategy has provided a considerable reduction of the total amount of simulation time compared to a full FEA analysis. Especially analysis types with a strong dependency on accuracy, like e.g. fatigue life calculations, can benefit from co-simulation [2].

2.3.2   Wading Simulation for Off-Road Vehicles

Vehicle wading refers to a situation where a vehicle traverses through water at different speeds, cf. Fig. 7. One of the major challenges is computing the inertial

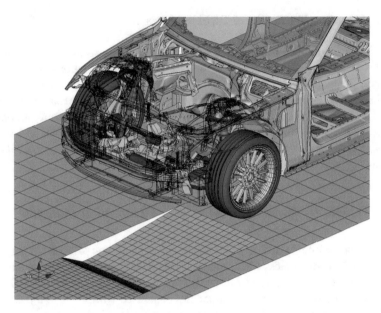

Fig. 6 Coupled FEA-MBS model of a car driving over an obstacle [2]. Published with kind permission of ©Fraunhofer SCAI 2016. All Rights Reserved

Fig. 7 Wading of a car through deep water modelled by a simplified block in STAR-CCM+ [14]. Published with kind permission of ©Fraunhofer SCAI 2016. All Rights Reserved

field of a vehicle while wading. In cooperation with an automotive OEM, Fraunhofer SCAI has developed a new method of co-simulation between CFD (STAR-CCM+) and MBS (SIMPACK) [14]. This solution strategy provides a new level of design and analysis capabilities to the industrial users.

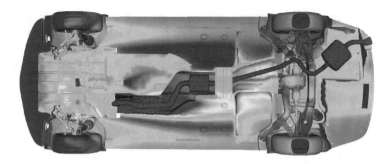

Fig. 8 Temperature distribution on the underbody of a full vehicle [1]. Published with kind permission of ©Fraunhofer SCAI 2016. All Rights Reserved

2.4 Automotive Thermal Management

The calculation of underhood component temperatures of entire passenger cars requires the consideration of different heat transport phenomena: convection, conduction and radiation. In addition to conjugate heat transfer models, the application of separate software tools dedicated to certain transport mechanisms is still demanded by users due to applicability, computing time or other company specific reasons. For these coupled approaches, a co-simulation environment like MpCCI is yet needed. The following applications highlight different aspects of thermal management for automotive engineering (Fig. 8).

2.4.1 Automotive Thermal Management for Full Vehicles

Calculation of the thermal behavior of automotive vehicles requires simulations for the full complexity of a vehicle's geometry and transport phenomena of heat including convection, radiation and conduction in fluids and solid bodies [5]. With regard to an accurate prediction of the temperature distribution of the entire car, radiation plays an important role in the overall heat management calculation. In areas with a relevant temperature influence (e.g. engine compartment, gear box or exhaust system), convective heat transfer and radiation are calculated in a coupled environment. As transient problems like dynamic drive cycles gained lately more importance in the development stage, the implementation of fast but accurate quasi-transient simulation approaches is further advanced [1]. For these specific tasks, OEMs frequently use a combination of TAITherm with STAR-CCM+ or in-house CFD solvers.

2.4.2 Automotive Thermal Management for Vehicle Manifolds

An important issue in automotive industry is to provide temperature distributions for vehicle components as input for following stress analyses. The transfer of temperature fields, heat transfer coefficients and film temperatures in an engine exhaust manifold may illustrate the importance of the thermal coupling in the transient heating due to the flow of the internal hot exhaust gas stream. The temperature distribution may be used to calculate the temperature expansion and resulting stresses as accomplished in [17].

2.5 Component Design in Electrical Engineering

The prediction of heating and cooling processes is of eminent importance in the development of electrical devices. The alternating current induces heating due to losses by Ohm's law, and the usual mechanism for cooling is free convection. The increasing tendency of miniaturization requires to fully exploit the thermal potential of the materials involved. The following two selected applications highlight different aspects of electrical component design [7].

2.5.1 Cooling of a 3-Phase Transformer

The thermal performance of an oil-immersed power transformer is governed by the oil flow for the transfer of heat generated in the windings and core towards the tank and the surrounding air. A coupled JMAG-Fluent model has been used to detect the local hot spots [22] within the device. The Joule losses calculated by JMAG are used as source terms for Fluent which conducts a thermal analysis taking into account the transport phenomena of heat including convection and conduction in fluids (coolant and air) and solid bodies (core and coils). The resulting temperature distribution for the core and coils will affect the electrical property of the copper material, which is temperature dependent, and will induce a new distribution of Joule losses. The temperature distribution in coils is very important because heat resistant designs are required for safety.

2.5.2 Electric Arc in Switching Devices

Switching arcs, as shown in Fig. 9, can be modeled using ANSYS EMAG to solve the magnetic field problem and Fluent to solve the fluid dynamics problem—coupled in volume through MpCCI CouplingEnvironment [21]. Based on magneto-hydrodynamic equations, a 3D model for a switching arc considering Lorentz forces, ohmic heating and radiation transport can be developed using a co-simulation approach. Beside the co-simulation solution, which allows to subdivide a complex problem in smaller problems, the challenge still remains in modeling the plasma. The modeling of the electric arc behavior is governed by different stages, e.g. the arc motion, the arc elongation, the arc commutation and arc cutting process. The inclusion of all these phenomena in the analysis requires to consider

Fig. 9 Plasma temperature of
an electric arc in a switching
device [7]. Published with
kind permission of
©Fraunhofer SCAI 2016. All
Rights Reserved

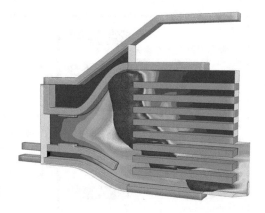

additional models like the arc root, the material erosion, etc. The resulting physical
and material model of the electric arc provides methods and references for the
optimization work of switching devices.

3   MpCCI FSIMapper

For many problems the influence of the CFD on the FEA solution is more significant
than vice versa, e.g. when structural deformation caused by thermal expansion
or pressure loading does not affect the flow field. Thus, a one-way transfer of
the stationary fluid solution to the solid solver as boundary condition is a time-
saving and good approximation to a co-simulation. Therefore, Fraunhofer SCAI
has designed a file-based tool to address application cases where a single one-way
transfer is sufficient.

MpCCI FSIMapper allows to read data of various CFD result formats as well
as an EM result format (see Fig. 10). The universal EnSight Gold format can be
exported by diverse CFD tools which enlarges the practicability.

Thermal and mechanical loads can be transferred to an FEA model to be used
in a subsequent structural analysis. The tool exports a file including the mapped
boundary conditions using the syntax of Abaqus, ANSYS Mechanical, or Nastran.

The quantities that can be transferred are volume temperature, film temperature,
wall heat transfer coefficient, wall heat flux, pressure and forces. The two meshes,
between which the interpolation of physical entities shall take place, have to be
either surface meshes or volume meshes. Robust and efficient interpolation schemes
allow the data transfer for different discretization accuracy or even in non-matching
model regions using extrapolation.

The mapping of static, transient (only MagNet .vtk, EnSight Gold .case) and
harmonic (only FINE/Turbo .cgns, EnSight Gold .case) results is offered. A Fourier
transformation of transient force or pressure data is provided in order to create the
loading for NVH (noise vibration harshness) analyses.

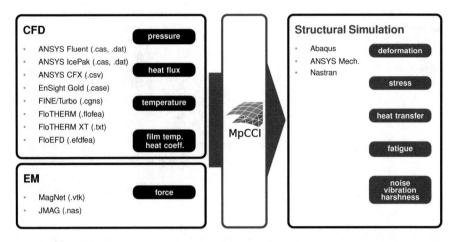

Fig. 10 Schematic view of the capabilities of MpCCI FSIMapper. In the *left*, the importable file formats are listed, in the *right* the available solver formats for exporting the mapping result (December 2016). Published with kind permission of ©Fraunhofer SCAI 2016. All Rights Reserved

If the models, between which the mapping shall take place, are defined in different unit systems or if their position and orientation differ from each other, MpCCI FSIMapper provides on the one hand an automatic and on the other hand a user defined transformation (translation and rotation) in order to generate geometrically coinciding models. Also, the mapping between periodic models which differ only with respect to their section shape is possible.

In the context of frozen-rotor analyses, the simple mapping of the thermal or mechanical loading would lead to an unbalanced structural behavior. In order to produce a blade-wise average, MpCCI FSIMapper offers the possibility to build an "average over rotation", where the mean is built over sections defined by a certain pitch angle.

Furthermore, with MpCCI FSIMapper it is possible to compare the geometric shape of used models to locate and evaluate differences in modeling the particular domain.

The tool has a graphical user interface but can also be used in batch.

4 MpCCI Mapper Solution for Integrated Simulation Workflows

Simulation of complex manufacturing processes to predict structural product characteristics does often require a chain of distinct simulation disciplines. Each simulation step typically requires a specific problem discretization to handle the physical effects. To achieve realistic simulation results, local material properties have to be specified as initial condition at each step of the virtual process chain.

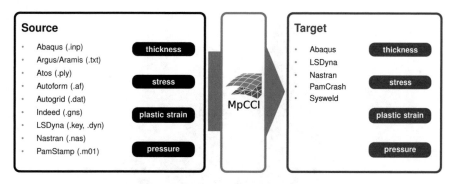

Fig. 11 MpCCI Mapper list of file interfaces supported (December 2016). Published with kind permission of ©Fraunhofer SCAI 2016. All Rights Reserved

In addition to the transfer of local material properties along a manufacturing chain, simulation models have to be validated by comparison with experimental test results [9].

In order to transfer local material properties between consecutive simulation steps, Fraunhofer SCAI has developed MpCCI Mapper to supply a link between the different computer-aided engineering (CAE) tools involved in the process chain.

For instance, the mechanical properties, like material thinning or plastic strain from a sheet metal forming process, can be integrated in the structural design process of crash relevant automotive body components.

MpCCI Mapper provides advanced and robust methods to map, compare and transfer simulation results and experimental data in integrated simulation work-flows. It supports a growing number of native file formats (Fig. 11) and can be used in a variety of engineering applications.

The MpCCI Mapper software is a standard tool in the engineering departments of most German automotive OEMs and has been validated in the VDA/FAT working group *Formed Chassis Parts* for Forming to Crash workflows.

In the subsequent sections, some application areas are presented, where the manufacturing history is significant for the component behavior.

4.1 Passive Safety

For the accurate prediction of the structural behavior of metal sheet car body components, the local manufacturing history must be taken into account. Local reduction of material thickness, stresses, plastic strain and other material properties, e.g. the local crystalline structure of high-strength steel resulting from single manufacturing steps such as deep drawing, immersion lacquering, and welding, may have significant influence on the resulting car body component. As local material properties may vary during a manufacturing process, result information has to be transferred downwards the process chain. In [12, 18, 23] it was shown that only the

Fig. 12 Car seat crash simulation [18]. Published with kind permission of ©Fraunhofer SCAI 2016. All Rights Reserved

consideration of the manufacturing history of high strength steels leads to a good correlation of simulation and experimental testing. Figure 12 shows the application in the crash simulation of a seat system.

4.2 Forming Tools and Material Properties

4.2.1 Lightweight Stamping Tools: Use Forming Loads in Structural Optimization

The combination of increased diversity of automotive parts and the pressure for decreased tool development times results in the need for optimization of the structural layout of stamping tools. A number of German OEMs have used MpCCI Mapper to transfer the maximal pressure loads from the stamping process into a structural optimization environment. The optimization process thus can consider local stamping loads to determine improved designs with less total mass but the same stability [20].

4.2.2 Validation of Material Model Parameters: Compare Forming Results and Experimental Data

Due to the stringent requirements with respect to feasibility, stability and crash performance, exact models for the specific material behavior are required. This validation process is supported by the comparison of different simulation or experimental results with each other. MpCCI Mapper has been used to obtain information about the deviation of results either in a section or over the whole geometry of a component [3].

4.3 Composite Structures and Plastic Components

4.3.1 CFRP Workflows: From Draping via Mulling and Curing to Structural Analysis

The excellent mass-specific properties of carbon-fiber reinforced plastics (CFRP) can be tailored to the actual requirements and make CFRP well qualified for use in lightweight constructions. However, the economical exploitation of these theoretical potentials is currently limited by insufficiencies of manufacturing processes, by lack of knowledge of the material behavior and by insufficient prediction of the structural performance. These weaknesses can only be solved by establishing a close collaboration between the three disciplines of methods, materials and processes. Another important precondition for improving CFRP applications is an integrated simulation of the entire CFRP process chain, where all significant process parameters and process results are transferred between the single simulation steps.

In a research project, KIT Karlsruhe has used the MpCCI Mapper technology to link the process steps from draping via molding and curing to the final structural analysis of a prototype trunk lid geometry, see Fig. 13 [13].

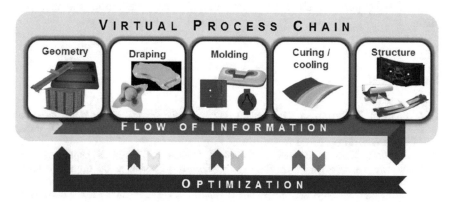

Fig. 13 Virtual process chain combining design, process and structural simulation for a CFRP workflow [13]. Published with kind permission of ©Fraunhofer SCAI 2016. All Rights Reserved

4.3.2 Structural Integrity of Blow Moulded Plastic Components

Within a research project, MpCCI Mapper was used to transfer local material properties and orientations from the BSim simulation as initial conditions for a subsequent structural analysis. This simulation workflow is essential for a range of standard products: from plastic bottles to complex automotive components like fuel tanks.

5 Conclusion

Coupled multiphysics simulation and simulation workflows consisting of different simulation steps can help to implement more realistic models. For each of the relevant physical effects, most suited simulations tools can be applied and combined in a larger application scenario. The benefits of the interface solution MpCCI are its neutrality and openess to all commercial software vendors. MpCCI Coupling-Environment provides a flexible way to run co-simulations of two or more codes at once; the MpCCI Mapper solutions give a simpler path for a 1-way data transfer.

While the MpCCI interface solutions are already used in many different application domains and industrial sectors, there are still open issues which need to be solved during the next years: The combination of full 3D models (fluid dynamics, structural analysis) with complex system models (e.g. a complete 1D vehicle model) or the seamless transfer of local material properties in a multi-disciplinary simulation workflow still require more sophisticated synchronisation algorithms and standardised export and import facilities for the different simulation tools in a CAE chain.

References

1. C. Brodbeck, P. Bayrasy, in *Co-Simulationsmethoden für das thermische Management von Gesamtfahrzeugen bei Betrachtung dynamischer Fahrzyklen*, VDI-Berichte 2279 (VDI-Verlag, Düsseldorf, 2016), pp. 31–43
2. J. Christl, S. Kunz, P. Bayrasy, I. Kalmykov, J. Kleinert, FEA-MBS-coupling-approach for vehicle dynamics, in *Proceedings of 2nd European Conference on Coupled MBS-FE Applications* (NAFEMS, Turin, 2015), pp. 29–32
3. G. Eichmüller, M. Meywerk, Stochastische simulation - versuchsabgleich der deformation eines vierkantrohres. NAFEMS Mag. **27**, 75–87 (2013)
4. Fraunhofer Institute for Algorithms and Scientific Computing SCAI, *MpCCI 4.4.1 Documentation, Part V User Manual* (Fraunhofer Institute for Algorithms and Scientific Computing SCAI, Sankt Augustin, 2015)
5. Fraunhofer Institute for Algorithms and Scientific Computing SCAI, Automotive Thermal Management (2016), https://www.scai.fraunhofer.de/en/business-research-areas/multiphysics/application-areas/automotive-thermal-management.html

6. Fraunhofer Institute for Algorithms and Scientific Computing SCAI, Manufacturing processes and passive safety (2016), https://www.scai.fraunhofer.de/en/business-research-areas/multiphysics/application-areas/manufacturing-process-chains.html

7. Fraunhofer Institute for Algorithms and Scientific Computing SCAI, MpCCI - Electrical Components (2016), http://www.mpcci.de/en/application-areas/electrical-components.html

8. Fraunhofer Institute for Algorithms and Scientific Computing SCAI, MpCCI - Flexible Structures in Aerodynamics and Machinery Design (2016), http://www.mpcci.de/en/application-areas/flexible-structures-in-aerodynamics-and-machinery-design.html

9. Fraunhofer Institute for Algorithms and Scientific Computing SCAI, MpCCI - Manufacturing Processes and Passive Safety (2016), http://www.mpcci.de/en/application-areas/manufacturing-process-chains.html

10. Fraunhofer Institute for Algorithms and Scientific Computing SCAI, Mpcci couplingenvironment (2016), http://www.mpcci.de/en/mpcci-software/mpcci-couplingenvironment.html

11. Fraunhofer Institute for Algorithms and Scientific Computing SCAI, Turbomachinery applications (2016), https://www.scai.fraunhofer.de/en/business-research-areas/multiphysics/application-areas/turbomachinery-applications.html

12. M. Hunkel, UmCra - Werkstoffmodelle und Kennwertermittlung für die industrielle Anwendung der Umform- und Crash-Simulation unter Berücksichtigung der mechanischen und thermischen Vorgeschichte bei hochfesten Stählen (AiF379ZN). FAT-Schriftenreihe **273**, 1–123 (2015)

13. L. Kärger, A. Bernath, F. Fritz, et al., Development and validation of a CAE chain for unidirectional fibre reinforced composite components, Compos. Struct. **132**, 350–358 (2015)

14. P. Khapane, U. Ganeshwade, J. Senapathy, et al., Deep water wading simulation of automotive vehicles, in *Proceedings of NAFEMS World Congress* (NAFEMS, San Diego, 2015)

15. B. Landvogt, Coupled fluid-structure-interaction simulations for aero-elastic benchmark cases, in *Proceedings NAFEMS European Conference on Multiphysics Simulation, Manchester, UK* (2014)

16. B. Landvogt, L. Osiecki, T. Zawitowski, B. Zylinski, Numerical simulation of fluid-structure interaction in the design process for a new hydraulic axial pump, in *ANSYS Automotive Simulation World Congress* (2012)

17. F. Mendonca, Approaches to industrial FSI - an overview with case studies, in *Proceedings of the 7th MpCCI User Forum* (2006)

18. M. Meyer, A. Oeckerath, Simulation of local material characteristics at Faurecia. ATZ worldw. eMag. **113**, 42–45 (2011)

19. L. Osiecki, P. Patrosz, B. Landvogt, et al., Simulation of fluid structure interaction in a novel design of high pressure axial piston hydraulic pump. Arch. Mech. Eng. **60**, 509–529 (2013)

20. C. Pfeiffer, D. Assmann, R. Canti, et al., Gewichtseinsparungen im Automobil-Werkzeugbau durch Topologieoptimierung, in *Tagungsband Deutsche Simulia-Konferenz, Bamberg* (2011)

21. C. Rümpler, Low-voltage circuit breaker arc interruption – a multi-physics modeling challenge, in *Proceedings NAFEMS European Conference on Multiphysics Simulation, Manchester, UK* (2014)

22. M. Salari, P. Bayrasy, K. Wolf, Thermal analyis of a three-phase transformer with coupled simulation, in *Proceedings of NAFEMS UK Conference, Oxford* (2014)

23. C. Steinbeck-Behrens, T. Menke, J. Steinbeck, et al., *Gemeinsamer FuE-Abschlussbericht des Verbundprojektes "Durchgängige Virtualisierung der Entwicklung und Produktion von Fahrzeugen (VIPROF)"* (2011)

24. N. Wirth, Development of a micro-gas-turbine with ceramic impeller: SCAI tools for stress and vibration analyses, in *Proceedings NAFEMS European Conference on Multiphysics Simulation, Manchester, UK* (2014)

Cooling Circuit Simulation II: A Numerical Example

Tanja Clees, Nils Hornung, Detlef Labrenz, Michael Schnell,
Horst Schwichtenberg, Hayk Shoukourian, Inna Torgovitskaia,
and Torsten Wilde

1 Introduction

This is the second of two associated articles on the simulation of cooling circuits. The first article conveys information on the underlying simulation model, whereas this second article focuses on the application to a real test case. In practice, several questions arise, such as what network information is available with respect to the circuit and its elements, how this information can be automatically or manually transferred into a description of the simulation, and which accuracy can be achieved by the simulation.

The main goal of this article is to give a detailed description of numerical results for a given network, part of a real cooling system, and to compare these results to measurement data. Our aim is also to specify the background of our example circuit, determine different sources of network data—for example, the network layout itself, device characteristics, measurements from operating practice, as well as data inferred from combining all other sources—and discuss their immanent issues. We want to shed light on all steps necessary to finally predict the behavior of the cooling circuit, where we particularly try to show the limitations of a computer simulation as to quality, not so much due to numerical issues, but because of the uncertainty and heterogeneity of the data and system description involved. Where possible, our objective is to propose ways to deal with this situation.

T. Clees (✉) • N. Hornung • M. Schnell • H. Schwichtenberg • I. Torgovitskaia
Fraunhofer Institute for Algorithms and Scientific Computing SCAI, Schloss Birlinghoven, 53757
Sankt Augustin, Germany
e-mail: tanja.clees@scai.fraunhofer.de

D. Labrenz • H. Shoukourian • T. Wilde
Leibniz Supercomputing Centre of the Bavarian Academy of Sciences and Humanities,
Boltzmannstr. 1, 85748 Garching, Germany

© Springer International Publishing AG 2017
M. Griebel et al. (eds.), *Scientific Computing and Algorithms in Industrial
Simulations*, DOI 10.1007/978-3-319-62458-7_8

153

In order to achieve our aim, we describe a real cooling system and one selected subcircuit which will represent our simulation example in Sect. 2. Section 3 introduces the overall concept of the simulation framework. It specifically addresses the issues of data collection and discusses, with a focus on automation, in which way the different data sources can be combined into meaningful scenario descriptions. Finally, numerical results are presented for a simplified subcircuit in Sect. 4, where the uncertainty in the initial data is spotlighted by a comparison of simulation results to measurements.

2 Application

In the following sections, we describe the cooling circuit that will represent our test case. The case is part of a real cooling system deployed at the Leibniz Supercomputing Centre of the Bavarian Academy of Sciences and Humanities, Garching, Germany.

2.1 Cooling System

The Leibniz Supercomputing Centre has $9554\,m^2$ ($102,838\,ft^2$) of floor space with a redundant power feed of $10\,MW$. The IT equipment floor space is $3160.5\,m^2$ ($34,019\,ft^2$) distributed over six rooms on three floors. The floor space for the data center infrastructure amounts to $6393.5\,m^2$ ($68,819\,ft^2$). Figure 1 shows an overview of the cooling infrastructure at the Leibniz Supercomputing Centre.

As stated in [9], the Leibniz Supercomputing Centre uses a mix of free cold water cooling (well water), chiller-supported cold water cooling, chiller-less cooling, and air cooling. For instance, the cooling capacity for the latest extension of the data center is stated in Table 1. This mix represents a real challenge for the infrastructure control system and for an energy efficient operation since all cooling technologies need to operate in accord but each requires different optimization techniques.

Figure 2 shows the overview map of the chiller-less cooling circuit. This cooling technology is available for two floors of the data center (HRR and NSR). Both connect to one water distribution which connects to four roof cooling towers (KLT11–KLT14).

2.2 Circuit Basics and Example

The cooling circuit discussed in this article is called KLT72 and provides hot water cooling to NSR. An overview picture from the Johnson Controls (JCI) building automation system is shown in Fig. 3. For another overview of the chiller-less cooling circuit and KLT72, as used in our simulation software, see Fig. 14 later on. As can be seen, KLT72 consists of one pump pair, one heat exchanger, which

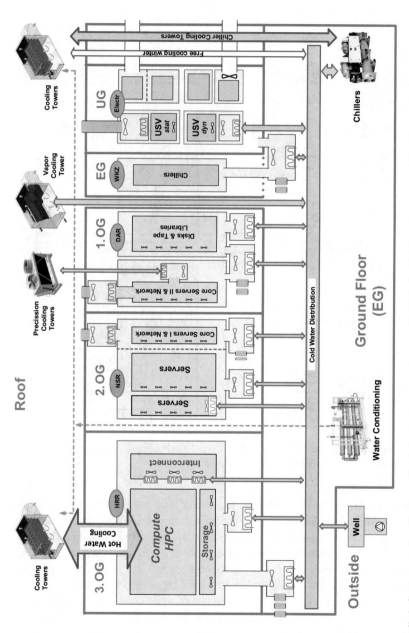

Fig. 1 Leibniz Supercomputing Centre cooling infrastructure, overview map.

Table 1 Cooling capacities at Leibniz Supercomputing Centre, Garching, Germany

Cooling	Capacity (MW)
Vapor cooling	2
Well water	0.6
Chiller-supported	3.2
Evaporative cooling towers	8

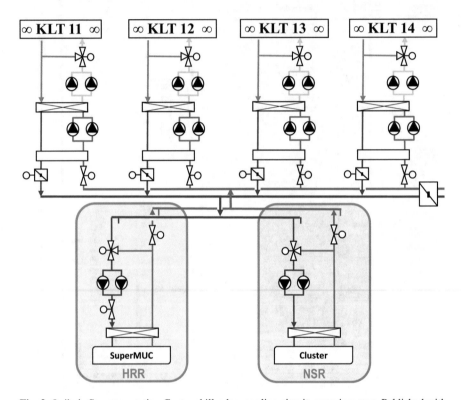

Fig. 2 Leibniz Supercomputing Centre chiller-less cooling circuit, overview map. Published with kind permission of ©Fraunhofer SCAI 2016. All Rights Reserved

transfers heat from the IT systems to the circuit, and a three-way valve used for circuit control. It is important to note that in standard mode of operation only one pump is active. Each pump, identical in construction, is active in weekly turns. Since we consider a partial network, sources and sinks represent the boundary to the rest of the network. Sources also introduce thermal energy to the circuit. Cold water is heated by energy that enters via the heat exchanger, which is, in turn, used to cool a neighboring circuit. Furthermore, the considered circuit KLT72 encompasses a closed loop. Within this loop, the flow is mainly induced by the pump. Via controllable pumps or valves, circuit operators try to achieve given temperature, pressure, or flow goals. Actually, the combination of the pump and mixing valve produces a circular flow in order to keep pressure difference and inflow temperature

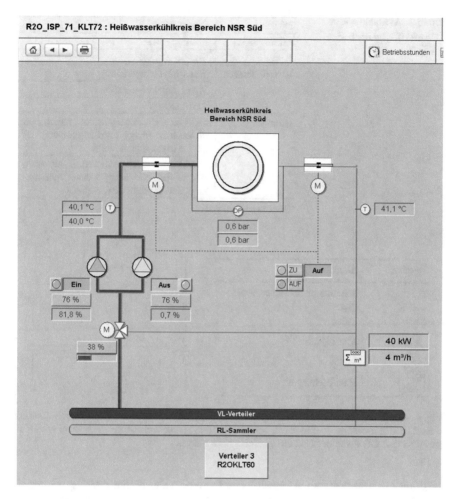

Fig. 3 Leibniz Supercomputing Centre hot water cooling circuit KLT72, overview map. Published with kind permission of ©Fraunhofer SCAI 2016. All Rights Reserved

at user-defined values by feedback control. A certain amount flows back to the circuit through the mixing valve, while a small amount leaves the circuit.

The available sensor data is given in Table 2, starting from left bottom of Fig. 3 and proceeding in clockwise rotation. Circuit control parameters are marked in Table 2 and are depicted with a cyan background color in Fig. 3.

Figure 4 (left) lists common symbols used to depict elements in circuit visualizations. Pipe directions, if shown, represent topological information, but do not necessarily coincide with the directions of flow. However, the topology is chosen in such a way that it conveys intended flow directions. A node that does not coincide with any incoming pipes is typically a source, a node without outgoing pipes is a sink.

Table 2 Available sensor data and control goals at KLT72

Device	Sensor data
Mixing valve	Position (lift)
Pump	Rotational speed
Heat exchanger	Water inlet temperature[a]
	Water outlet temperature
	Pressure loss[a]
Circuit	Flow as measured by a flow meter, part of the heat energy transfer measurement device (right side, bottom)
	Total heat power transferred to the water, part of the heat energy transfer measurement device (right side, bottom)

[a]Used as control goal

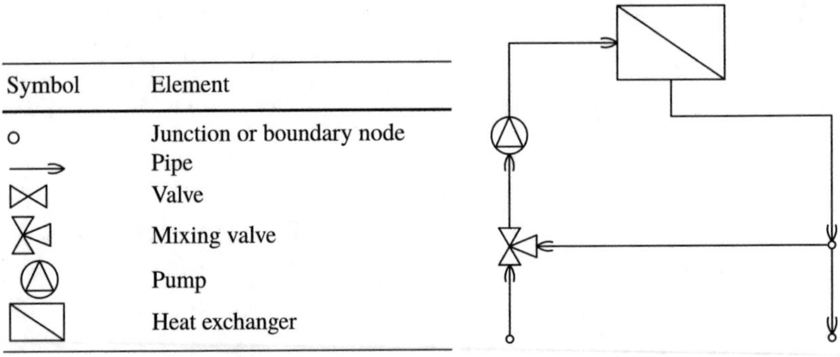

Symbol	Element
o	Junction or boundary node
—⇒	Pipe
⋈	Valve
⋈	Mixing valve
⊕	Pump
⊠	Heat exchanger

Fig. 4 *On the left*: Element symbols used in circuit layout visualizations. *On the right*: Cooling circuit KLT72 including mixing valve, pumping station, and heat exchanger. Published with kind permission of ©Fraunhofer SCAI 2016. All Rights Reserved

Figure 4 (right) shows a simplified schematic view of the cooling subcircuit KLT72 from Fig. 3, which is used for simulation. This view renders the topology of the circuit more simple. Physically, though, the new topology is almost equivalent to the original one, also cf. Fig. 14.

Boundary conditions at the source are pressure (taken to be constant 6 bar) and temperature (dotted curve in Fig. 5). At the sink, the boundary condition is volumetric flow (Fig. 6). During operation, sensors measure the inflow and outflow temperatures (dashed curves in Fig. 5) and pressure differences (Fig. 7) at the heat exchanger. In correspondence to the measured data, rotational frequencies of the pump and lift values of the mixing valve are known (Fig. 8). Via the exchanger, heat is externally supplied (Fig. 9). Notice that *appointed* control values and their *measured* equivalents—of rotational pump speed and valve lift—differ significantly. We only apply measurements here and ignore the values that were originally appointed to the components. This choice in favor of measured control data is arbitrary.

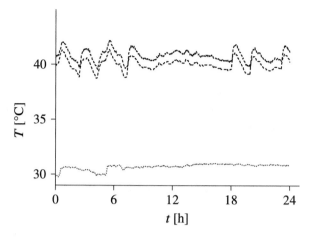

Fig. 5 Cooling circuit KLT72, measured temperature curves on April 29th, 2014—low circuit inflow temperature (*dotted*) and, at the heat exchanger, medium entrance and higher exit temperature (*both dashed*). Published with kind permission of ©Fraunhofer SCAI 2016. All Rights Reserved

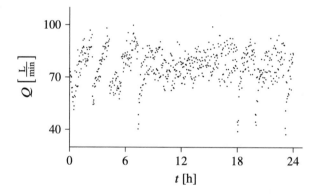

Fig. 6 Cooling circuit KLT72, volumetric flow at the sink on April 29th, 2014. Published with kind permission of ©Fraunhofer SCAI 2016. All Rights Reserved

If we assume perfect insulation, i.e., no temperature losses along pipes, the supplied amount of heat corresponds exactly to the difference of inflow and outflow of the full circuit. Since certain temperature losses occur in practice, we approximate the temperature at the outlet of the heat exchanger by that at the outflow of the entire circuit. In agreement with this approximation we equate the external heat supply of the circuit to that of the heat exchanger. Most measurements are only captured if their change in quantity exceeds a given threshold. Notably, the threshold of 0.1 bar for pressure differences accounts for the two seemingly different curves in Fig. 7.

While the temperatures, the pressure differences, the total outflow from the network, the relative pump speed and valve aperture, as well as the goals for automatic control are captured as explained above, the pipe length and the geodesic

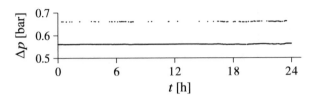

Fig. 7 Cooling circuit KLT72, pressure drop at the heat exchanger on April 29th, 2014—only changes of more than 0.1 bar are captured. Published with kind permission of ©Fraunhofer SCAI 2016. All Rights Reserved

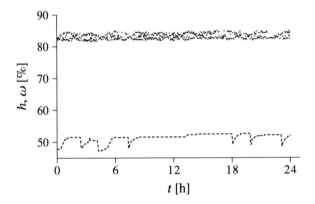

Fig. 8 Cooling circuit KLT72, relative mixing valve aperture h (*dashed*) and pump speed ω (*dotted*) on April 29th, 2014. Published with kind permission of ©Fraunhofer SCAI 2016. All Rights Reserved

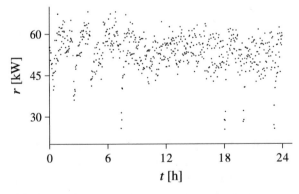

Fig. 9 Cooling circuit KLT72, supplied thermal power on April 29th, 2014. Published with kind permission of ©Fraunhofer SCAI 2016. All Rights Reserved

height can only be approximated. Inner diameters are known from the specification of the system. Typically, vendors deliver certain characteristic data for their devices via product sheets. All information gathered from such data sheets is displayed in Table 3. There are several reasons why measurements and characteristic data might

Table 3 Characteristic data of devices from cooling circuit KLT72

Element	Data type	Data
Mixing valve	Rangeability	100
(Hora BR316GF)	Form of characteristic curve	Equal percentage (A→AB), linear (B→AB)
	Resistance k_{vs}	125 m³/h
	Inner diameter	105.3 mm (DN100)
Pump	Characteristic height curve	See Fig. 10 (left)
(Grundfos NKE 150-250/243)	Characteristic power curve	See Fig. 10 (right)

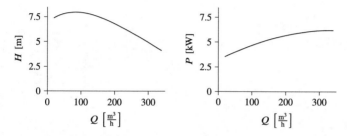

Fig. 10 Pump—height and power characteristics Q vs. $H_{100\%}$ and $P_{100\%}$, respectively, for Grundfos NKE 150-250/243. Published with kind permission of ©Fraunhofer SCAI 2016. All Rights Reserved

not coincide in practice, such as fatigue, corrosion, calcification within devices, or measurement, transmission, storage, and quantification problems within the system of sensor data collection.

3 Concept and Software

In the following, we are going to address the general framework of our simulation software MYNTS. Section 3.1 gives a general overview of how all components interact, while the subsequent sections describe how data is collected and processed.

3.1 Framework and Components

In order to set up a model which can be used for analysis, simulation, and optimization of the infrastructure network considered, the following components are necessary (see also [9]):

C1   A layout of the network (topology).
C2   Physical and numerical models of the network components (elements).

Table 4 Sources of information for different components

Source	For components	Description
S1	C1	Schematics (for example, AutoCAD data)
S2	C2	Technical documentation of devices including parameters and characteristic maps
S3	C2 and C3	Measurement data stemming from sensors installed in the considered network
S4	C3	For each sensor, a mapping of its tag in the measurement system to the corresponding name of the element (in the network description) which refers to this sensor
S5	C3 and C4	Discussions with the relevant stakeholders

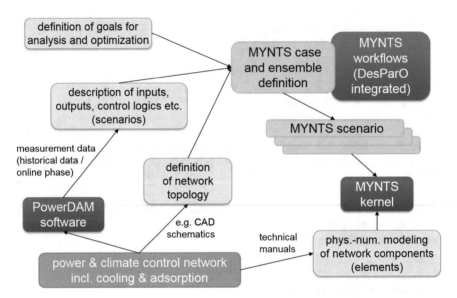

Fig. 11 The overall architecture of the system developed. Published with kind permission of ©Fraunhofer SCAI 2016. All Rights Reserved

C3   A description of inputs, outputs, control logics etc. (scenarios).

C4   A definition of goals for analysis and, if desired, optimization.

Sources for these components are summarized in Table 4.

The overall concept for targeting cooling circuits is depicted in Fig. 11 by means of a high-level description of its components and their connections. In the following sections, we explain the main components and sources in more detail.

3.2 Semi-Automatic Model Creation with Schemparser

Ideally, the topology of a cooling circuit can be obtained or reconstructed by data used in an infrastructure or building management system (for example, data given in HTML, XML, or OPC UA format with appropriate contents). Quite often, though, only computer aided design (CAD) data is available for the specific circuits considered. A schematic representation of the cooling and support infrastructure of the Leibniz Supercomputing Centre, useful for our purposes, is only available as a CAD drawing in the Drawing Exchange Format (DXF). Pipes and components like pumps or valves are constructed by simple primitives, for example, by lines, polygons, and circles. More that 40,000 lines and polygons are contained in the Leibniz Supercomputing Centre schematics. In order to create a realistic model for the simulation, a parsing software is required that can identify these components automatically and generate a model in the intrinsic format of our simulation software package [1]. Therefore, the tool "Schemparser" was developed to meet these demands and to simplify the model creation process.

A schematic design usually follows certain rules, which can be used to simplify the identification of the different components. We benefit from these rules by merging dashed lines that describe the same pipe. Figure 12 (left) shows an example of such a case, where a single pipe is composed of many different lines. At the start and at the end point of every line the surroundings are searched for compatible lines nearby. Certain requirements must be met before the lines are merged. Among other criteria,

- merged lines must be located in the same layer,
- the slopes of the lines must be compatible with that of the source line,
- and the lines must have the same line properties (for example, color and width).

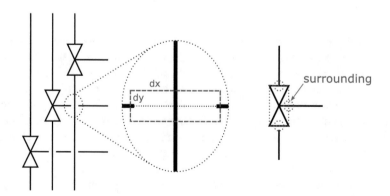

Fig. 12 *On the left*: Illustration of the pipe merging algorithm. *On the right*: Three-way valve with identification of its connecting pipes. Published with kind permission of ©Fraunhofer SCAI 2016. All Rights Reserved

Since this comparison task must be executed for every primitive, the algorithm needs to be refined to speed up the process. In a first step, all lines are filtered not fulfilling typical engineering rules for drawing pipes. The remaining lines are sorted by their orientation (horizontal and vertical). Only lines with the same orientation are processed in the line merging process. A run time reduction by a factor of ten is achieved using this approach for the Leibniz Supercomputing Centre schematics.

After the pipe identification and merging process, a component-finding algorithm is needed: A database containing all common representations for each technical component is used in combination with a template matching algorithm. When a component is found, its connections to the pipes must be identified. An algorithm, similar to the pipe merging described above, is used to determine the connections to the pipes, and the database is queued to identify the type of the connection (for example, input or output of an element). This is important, for instance, if the flow direction of a pump has to be determined. In Fig. 12 (right) the identification is illustrated for a three-way valve.

Once all pipes and components are identified, the result is translated to the internal format of our simulation software. Despite the good recognition capabilities of Schemparser certain information has to be added manually like the enumeration and model types of some components. These are not specified in the Leibniz Supercomputing Centre schematics in a standard way.

Due to the tolerance in every parsing step of the program, Schemparser is, in principle, suited to convert scans of schematics drawn on paper that were created with other drawing programs. Although an additional parsing is required to convert the input document to DXF format, this approach can significantly simplify the model creation process.

3.3 Device Modeling and Sensor Mapping

To model the physical behavior of the network components, data contained in technical documentation is required. For the overall cooling system of Leibniz Supercomputing Centre, the important components include regulated or unregulated pumps, heat exchangers, cooling towers, hydraulic separators, additional resistors, regulators, one-way or three-way valves, chillers, and pipes. Technical specifications of decisive parameters, settings, and characteristic maps can be obtained by digitizing the respective information from the device manuals and from web pages.

In particular, our simulation software creates a characteristic map by reading a table of data points of that map and by approximating the dependencies. Hence, for each characteristic map which is needed, appropriate points on the respective plots in technical documentation must be identified, manually adding the obtained values to the table.

Although vendors use functional dependencies (data-fitting) to create such plots, only printed versions of the plots are usually available, and neither the original data points nor the functional dependencies.

Since the device names could not be read automatically from the infrastructure plan or management system, a mapping from model generated device names to original device names was generated semi-automatically, based on the respective naming systems of the SCADA system vendors.

3.4 Collection of Measurement Data with PowerDAM

Measurement data and descriptions of control logics serve the purpose to

- define inputs or starting points for the operating scenarios to be analyzed,
- define boundary conditions, describing important events from the past, in order to set up valid simulation models,
- validate simulation results (historical data for offline validation, online data for dynamic validation),
- and define input for calibration processes.

The latter is necessary if a decisive device acts as a "black box" and cannot be described based on technical documentation. This can happen if the model, the type, or the specifications of the device are not known, or if the device represents a more complex system. Instead of characteristic maps from vendors, measurement data from nearby elements might be used to train approximation models. A software tool box for the interactive exploration and analysis of parametrized data sets (DesParO [3]) can be used for this training task, for instance. In our exemplary subcircuit, the heat exchanger and the remaining network are treated as black boxes.

The software PowerDAM (Fig. 13, more details in [7]) was developed at Leibniz Supercomputing Centre to collect important information from different data center systems and transform it into a standard format that supports cross-system analysis and reporting, independent of system specific data layouts. PowerDAM generates, from the collected data, the XML data file required by the modeling and solver framework (see Sect. 3.5).

A plug-in infrastructure, together with the support of publish-subscribe communication (under development), allows for easy integration of data from new systems. Currently, PowerDAM collects data from two data center infrastructure monitoring and control systems,

- JCI, a cooling infrastructure building automation system,
- and WinCC, an electrical infrastructure monitoring system.

PowerDAM also collects data from two high performance computing systems and their scheduling systems (CooLMUC, SuperMUC, Slurm, LoadLeveler), and from one special technology system (SorTech adsorption chiller).

The quality of the data is not guaranteed, which is one challenge of this approach. The main reason for this is that the current infrastructure management systems do not consider further data processing by third party tools as part of their supported use cases. For example, some sensor changes are not recognized in JCI because, for

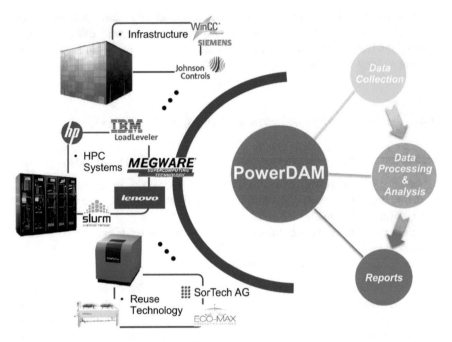

Fig. 13 Overview of the PowerDAM software architecture at Leibniz Supercomputing Center.
Published with kind permission of ©Fraunhofer SCAI 2016. All Rights Reserved

the circuit control, only differences bigger than a given threshold are considered
important. Additionally, even though one can set a data collection frequency
(currently 1 min for JCI and 15 min for WinCC), some sensors are connected via
technologies that have a longer read-out interval, meaning, one might see the same
value for multiple time stamps for some sensors.

3.5   Nonlinear Problem Setup and Solution with MYNTS

In detail, the relevant equations of continuum mechanics are explained in [4],
with application to water dynamics in pipe networks. Possible simplifications such
as incompressibility of the fluid, quasi-stationarity, and assumptions regarding
thermodynamics are made explicit and concepts of regularization as well as
discretization on a non-collocated grid are carefully discussed. For a network of
pipelines, a graph-like topology is introduced, for which the non-collocated grid is
adapted in a reasonable way. Furthermore, simplified equations are introduced for a
subset of devices that are typically found in cooling circuits.

The considerations of [4] always refer to an implementation within an optimization approach based on an interior point method where nonlinear equality

constraints are satisfied via a Newton-type algorithm. In the application case described here, the role of the optimization framework is played by IpOpt [8]. Optimization and nonlinear root finding are linked via a file interface in NL format [5]. This interface expects mathematical formulae to be conveyed in an algebraic modeling language, which allows for automatic differentiation. As a result, the Jacobian matrices needed by any Newton-type approach are not calculated numerically, which would introduce considerable error. Instead of that, they are dealt with algebraically and calculated directly.

Our Multiphysical Network Simulation Framework (MYNTS, cf. [1, 2, 6]) has been developed as software for modeling and simulating as well as for analyzing and optimizing energy networks such as gas and power grids. It is used, for instance, by Open Grid Europe, a company running Germany's largest long-distance gas transport network. MYNTS already offers an implementation for gas and power transport. For the case study discussed in this article, it was extended to handle cooling circuits as well.

Actually, MYNTS is built upon the fact that such networks can be modeled in a very similar manner as systems of differential-algebraic equations. The respective nonlinear system to be solved—simulated or optimized—is set up based on analytic formulations. MYNTS' converter to the NL format creates all data necessary for runs with an underlying solver for nonlinear problems (NLPs). MYNTS supports user-defined elements (custom physical formulations and constraints). In order to facilitate coding, MYNTS features the following components:

- a human-readable description of the system of equations derived automatically from a model of the circuit,
- an automatic translation of that description into the NL format,
- a workflow involving a series of NLP setups and solver calls to IpOpt (for stationary cases, a series is helpful to ramp up physics, meaning, increasingly more complex problems are solved; for time-dependent cases, such a workflow is used at startup, as well as at subsequent steps progressing in time),
- and a final remapping of the results to the network components.

On top of this process, an automatic handling of a full set of scenarios is realized, including the comparison to given values. MYNTS supports batch-style processing of scenarios and ensemble runs (for the analysis of parameter variations), for which the MYNTS graphical user interface offers a convenient configuration and run environment.

In combination with net'O'graph (SCAI's graph analysis library built into MYNTS) and DesParO [3], graph analysis, statistical analysis, calibration, and robust optimization tasks are supported. The available features in the field of graph analysis include graph reduction, graph matching, analysis of supply and demand scenarios, and network (de-)composition as well as automatic layout methods. DesParO provides metamodeling (efficient interpolation by means of response surfaces with adaptively built models), statistical analysis of parameter-criteria dependencies (including our own nonlinear correlation, tolerance, and quantile estimators, and Pareto (i.e., multi-objective) optimization.

Figure 14 sketches the network with its elements and sensor tag references, as used for MYNTS. The enlarged area, see Fig. 15, shows the KLT72 circuit and its original topology before the physically equivalent simplifications, see Sect. 2.2.

Usually, optimizing a complex infrastructure in combination with possibly conflicting operating requirements leads to a multi-objective optimization. The specific selection of an optimum is not unique as in single-objective optimization. It is strongly influenced by goals of the data center and of other stakeholders. A common approach is to weight goals against each other to form a single optimization criterion. An alternative approach is multi-objective optimization in order to compute Pareto-optimal solutions, i.e., best compromises. Due to its coupling to IpOpt and DesParO, MYNTS offers both methods.

4   Numerical Tests

We are now going to describe several test results, obtained using MYNTS, for the simulation of the cooling circuit introduced in Sect. 2.2, see also Fig. 15. Data sheets are available for the pumping station and mixing valve. The model type and, particularly, the resistance k_{vs} of the heat exchanger are not known, though.

4.1   Simplified Heat Exchanger

In a first step, a simulation problem is set up to determine possible values for the resistance k_{vs} at the heat exchanger. Strictly speaking, pressures, temperatures, and flows are independent variables wherever defined. As introduced in [4] (see pages 61–79 in this book), a system of equations derived from the laws of continuum mechanics is accounted for to determine these variables. Resistance k_{vs} is an additional free parameter at the heat exchanger influencing the relation between flow and pressure there. During our calibration step, it is determined by a given pressure difference between inlet and outlet at the heat exchanger. The aperture at the mixing valve, the rotational speed of the pump, the power transferred to the circuit at the heat exchanger, as well as the inflow and outflow of the circuit are treated as given parameters. They are assigned values measured during the operation of the real system. The measurements represent a time series taken over an entire day, April 29th, 2014, where the captured values seem to be reliable enough. Pump and valve characteristics are determined from characteristic data supplied by the corresponding manufacturers or vendors.

Figure 16 shows the resulting k_{vs} values where convergence is achieved. Solutions are accepted as converged when the right hand side of the system of equations exhibits a relative error of less than 0.1. As will be seen from the subsequent tests, such a vague criterion is necessary and we must take the results with caution. For simple heat exchangers a constant resistance would be expected, which can in fact

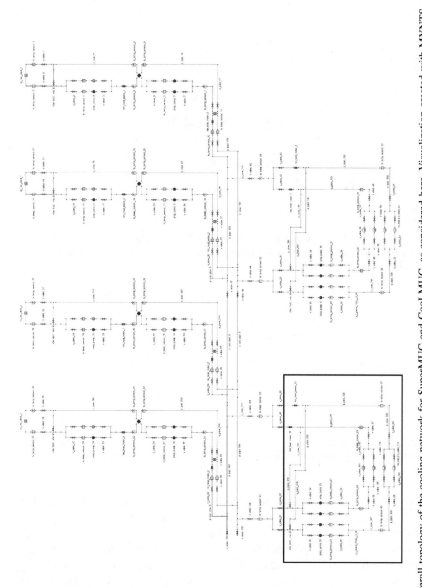

Fig. 14 The overall topology of the cooling network for SuperMUC and CoolMUC, as considered here. Visualization created with MYNTS. An enlarged version of the *boxed part in the lower left corner* is shown in Fig. 15. Published with kind permission of ©Fraunhofer SCAI 2016. All Rights Reserved

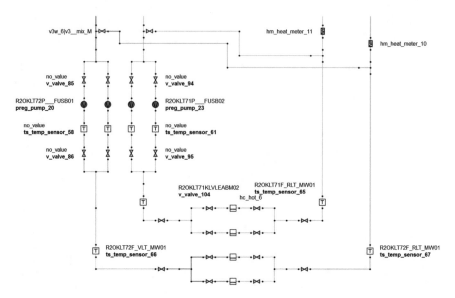

Fig. 15 NSR part of the overall network with the KLT72 circuit (outer circuit, pipes drawn in *color*). Visualization created with MYNTS. Published with kind permission of ©Fraunhofer SCAI 2016. All Rights Reserved

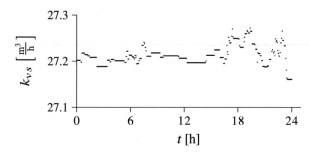

Fig. 16 Cooling circuit KLT72, k_{vs} values determined by optimization for April 29th, 2014. Published with kind permission of ©Fraunhofer SCAI 2016. All Rights Reserved

be observed in our case. A clear concentration of realizations at $k_{vs} \approx 27.2\,\mathrm{m^3/h}$ motivates our adopting values of similar magnitude in a second step.

This second step consists of a series of simulations with $k_{vs} = 30\,\mathrm{m^3/h}$ as well as with $k_{vs} = 50\,\mathrm{m^3/h}$ and boundary data from April 30th, 2014, intentionally different from the data used for calibration. The pressure difference goal, previously used to determine resistance, is omitted now and we apply a stricter relative threshold of 10^{-5} for accepting a solution as converged. Not all scenarios of the time series do converge under these conditions, though. As a matter of fact, if lower resistance value are chosen, less scenarios yield acceptable solutions, which is why we do not present results for $k_{vs} < 30\,\mathrm{m^3/h}$. The computed temperatures and

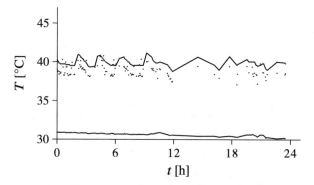

Fig. 17 Cooling circuit KLT72, temperatures at the entrance to the heat exchanger—measured (*upper solid line*) and computed with $k_{vs} = 30\,\mathrm{m^3/h}$ (*dots*)—vs. at the entrance to the circuit—measured (*lower solid line*)—for April 30th, 2014. Published with kind permission of ©Fraunhofer SCAI 2016. All Rights Reserved

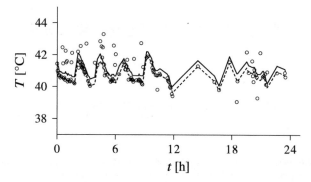

Fig. 18 Cooling circuit KLT72, temperatures at the exit of the heat exchanger—measured (*solid line*) and computed with $k_{vs} = 30\,\mathrm{m^3/h}$ (*circles*)—vs. at the exit of the circuit—measured (*dashed line*)—for April 30th, 2014. Published with kind permission of ©Fraunhofer SCAI 2016. All Rights Reserved

pressure differences at the heat exchanger are then compared to measured data, see Figs. 17, 18, and 19 for resistance values of $30\,\mathrm{m^3/h}$. Results for $k_{vs} = 50\,\mathrm{m^3/h}$ are shown in Figs. 20, 21, and 22. Notice that scenarios not converged are not depicted in the figures.

The temperature results exhibit strong variance. However, a concentration of values near measurements can be observed. For $k_{vs} = 30\,\mathrm{m^3/h}$, these values deviate from measurements by a fixed distance of not more than $2\,^{\circ}\mathrm{C}$ at the inlet of the heat exchanger. At the outlets, the deviation amounts to about $1\,^{\circ}\mathrm{C}$. The latter observation can be perfectly explained, since the simulation deliberately ignores temperature losses along pipes. Furthermore, the power transferred to the circuit at the heat exchanger is fixed in such a way as to account for the measured power gain between the inflow and outflow point, which is correct for lossless systems. As can be seen

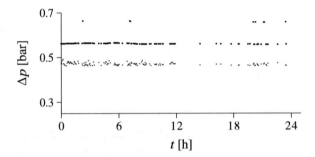

Fig. 19 Cooling circuit KLT72, heat exchanger—pressure drop measured (*bold dots*) and computed with $k_{vs} = 30\,\mathrm{m}^3/\mathrm{h}$ (*regular dots*) for April 30th, 2014. Published with kind permission of ©Fraunhofer SCAI 2016. All Rights Reserved

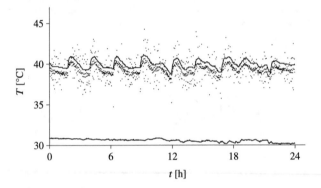

Fig. 20 Cooling circuit KLT72, temperatures at the entrance to the heat exchanger—measured (*upper solid line*) and computed with $k_{vs} = 50\,\mathrm{m}^3/\mathrm{h}$ (*dots*)—vs. at the entrance to the circuit—measured (*lower solid line*)—for April 30th, 2014. Published with kind permission of ©Fraunhofer SCAI 2016. All Rights Reserved

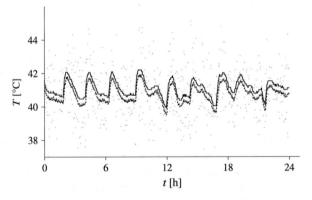

Fig. 21 Cooling circuit KLT72, temperatures at the exit of the heat exchanger—measured (*upper solid line*) and computed with $k_{vs} = 50\,\mathrm{m}^3/\mathrm{h}$ (*gray dots*)—vs. at the exit of the circuit—measured (*lower dashed line*)—for April 30th, 2014. Published with kind permission of ©Fraunhofer SCAI 2016. All Rights Reserved

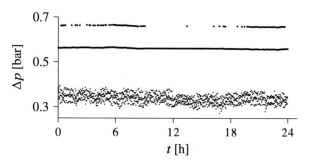

Fig. 22 Cooling circuit KLT72, heat exchanger—pressure drop measured (*bold dots*) and computed with $k_{vs} = 50\,\text{m}^3/\text{h}$ (*regular dots*) for April 30th, 2014. Published with kind permission of ©Fraunhofer SCAI 2016. All Rights Reserved

from the solid and dashed lines in Figs. 18 and 21, the real system does, however, exhibit energy losses between the outlet of the heat exchanger and the outflow node. The simulation results are, therefore, in line with our basic assumptions and boundary conditions: The observed concentration of temperature values coincides with those measured at the outflow point. Interestingly, the higher k_{vs} value yields more accurate temperature results at the inlet to the heat exchanger.

This last observation is remarkable because the higher k_{vs} value gives less accurate pressure results, as displayed in Figs. 19 and 22. A comparison of both plots clearly shows that the computed pressure drop behaves similar to the measured data, except for a deviation of slightly more than 0.1 bar. The deviation is larger for the higher resistance value. Notice again that the measured pressure data has an accuracy of 0.1 bar and can be understood to have arbitrary values between and beyond the maximum and the minimum in Fig. 7. However, a lack of accuracy in the measurements cannot be held fully accountable for the simulated pressure drop, which is too small throughout the entire time series.

In fact, Fig. 10 shows a maximum pump height of less than 8 m at a flow rate of approximately $90\,\text{m}^3/\text{h}$. With a typical rotational speed of 85 % a maximum pressure increase of $(85\,\%)^2 \cdot 8\,\text{m} \cdot \rho g \approx 0.67\,\text{bar}$ can occur at a flow rate of approximately $(85\,\%)^{-1} \cdot 90\,\text{m}^3/\text{h} \approx 106\,\text{m}^3/\text{h}$.

Figure 5 shows a resulting temperature at the three-way valve which is strongly influenced by the mixing inlet. Let us assume a situation where temperatures of 30 and 41 °C result in a mixed temperature of 40 °C. The energy balance law at the three-way valve, $(m_1 + m_2) \cdot 40\,°\text{C} = m_1 \cdot 30\,°\text{C} + m_2 \cdot 41\,°\text{C}$, yields $m_1 = m_2 \cdot 41\,°\text{C} - 40\,°\text{C}/40\,°\text{C} - 30\,°\text{C} = 10^{-1} m_2$. We, therefore, have to expect flows m_2 through the mixing entry about ten times higher than flows m_1 measured at the inlet to the circuit. The typical boundary flows of Fig. 6 range between $40\,\text{L/min} \approx 2.4\,\text{m}^3/\text{h}$ and $100\,\text{L/min} = 6\,\text{m}^3/\text{h}$. Hence, the typical flow rates at the pump must lie within a range between 24 and $60\,\text{m}^3/\text{h}$.

Flow rates of less than $(85\,\%)^{-1} \cdot 60\,\text{m}^3/\text{h} \approx 70.6\,\text{m}^3/\text{h}$ produce a pressure increase of not more than $(85\,\%)^2 \cdot 7.95\,\text{m} \cdot \rho g \approx 0.56\,\text{bar}$ at the pump and a pressure decrease of about

$$\frac{1}{k_v^2}\left((60 - 6)\,\text{m}^3/\text{h}\right)^2 \text{bar} \approx 0.75\,\text{bar}$$

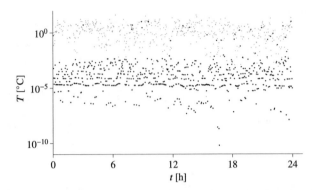

Fig. 23 Cooling circuit KLT72, scatter analysis for simulation temperature error $T = |T_s - T_m|$ (April 30th, 2014, $k_{vs} = 50\,\mathrm{m}^3/\mathrm{h}$)—index s denotes the simulation results, m the measurements, *gray dots* mark scenarios with zero flow change, *bold black dots* mark all other scenarios. Published with kind permission of ©Fraunhofer SCAI 2016. All Rights Reserved

at the mixing entry of the three-way valve with $k_v = \frac{k_{vs}}{2} = \frac{125}{2}\,\mathrm{m}^3/\mathrm{h}$ for typical apertures of $h = \frac{1}{2}$. In order for the pressure to decrease by 0.5 bar at the heat exchanger, the mixing entry needs to be wide open given the characteristics of the three-way valve from Table 3.

If there does not exist another source for a pressure increase within the closed loop of KLT72, typical measurements of a pressure drop of about 0.5 bar at the heat exchanger are not in accord with what we know about the pump and valve characteristics. In conclusion, this observed contradiction can be held accountable for the above mentioned observation that the simulated pressure drop at the heat exchanger is too small throughout the entire time series, compared to measurements.

Before we proceed to more involved simulations, let us briefly return to the scatter observed within temperature results from our simulation, see again Fig. 21, for instance. A simple analysis shows that constant inflow measurements during consecutive time steps correlate with large temperature errors. Since inflow measurements usually exhibit strong variance, one may argue that constant values for consecutive time steps should not occur at all within the measured data. This observation suggests the idea that strong variance may be a result of flow measurement errors, i.e., of flow values erroneously not updated in the data base upon variation.

Figure 23 provides numerical evidence of this interpretation for the simulation cases of Figs. 20 and 21. It depicts the absolute distances, on a logarithmic scale, between temperature simulations and measurements at the outlet of the circuit. Simulations corresponding to scenarios without flow change (with respect to the previous time step) are responsible for temperature errors around 1 °C (gray dots), while only few other simulation results (bold black dots) exhibit errors of a similar magnitude. This observation supports the notion that temperature simulation scatter is induced by measurement errors.

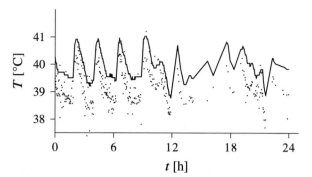

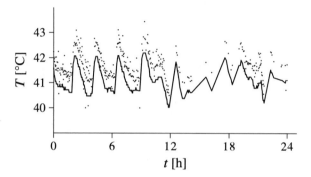

4.2 Logarithmic Mean Temperature Difference

We are now going to apply a more accurate model of the heat exchanger. In the
previous computations, heat was supplied as a given amount r determined from past
measurements. This is, of course, not possible for predictions of future behavior.
Section 4.3 in [4] (see page 73 in this book) provides a model of a heat exchanger,
where r is rendered more precise via the logarithmic mean temperature difference
(see Eq. (4) in [4] on page 73 in this book). The use of this more detailed model
with coefficient $\frac{c_{ht}L}{D} = 40\,\frac{\mathrm{kW}}{\mathrm{K}}$, roughly estimated from data measurements, leads to
slightly different results, see Figs. 24, 25, and 26, as well as Figs. 27, 28, and 29.
As before, all figures only portrait successful solver runs. The implementation of
Eq. (4) in [4] fails substantially more often, but yields less variance and comparable
systematic error upon completion.

In particular, Figs. 27 and 28 give a much more accurate account of the de facto
temperature conditions around the heat exchanger. These results refer to k_{vs} values
of $50\,\mathrm{m^3/h}$. Figure 24, displaying results for k_{vs} values of $28\,\mathrm{m^3/h}$, shows simulated
temperatures clearly lower than the measured ones at the inlet to the heat exchanger,
whereas the simulated outlet temperatures in Fig. 25 are too high. Large temperature
differences at a heat exchanger may indicate low flow rates as explained in Sect. 4.1.
Arguing via the measured temperatures, we again come to the conclusion that $k_{vs} = 28\,\mathrm{m^3/h}$ induces, together with the measured pump speed, a too low flow rate.

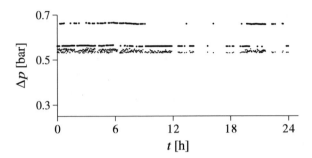

Fig. 26 Cooling circuit KLT72, heat exchanger—pressure drop measured (*bold dots*) and computed with $k_{vs} = 28\,\text{m}^3/\text{h}$ (*regular dots*, more accurate model of the heat exchanger) for April 30th, 2014. Published with kind permission of ©Fraunhofer SCAI 2016. All Rights Reserved

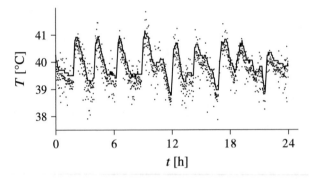

Fig. 27 Cooling circuit KLT72, entrance to the heat exchanger—temperatures measured (*solid line*) and computed (*dots*, more accurate model of the heat exchanger with $k_{vs} = 50\,\text{m}^3/\text{h}$) for April 30th, 2014. Published with kind permission of ©Fraunhofer SCAI 2016. All Rights Reserved

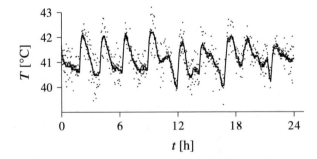

Fig. 28 Cooling circuit KLT72, exit of the heat exchanger—temperatures measured (*solid line*) and computed (*dots*, more accurate model of the heat exchanger with $k_{vs} = 50\,\text{m}^3/\text{h}$) for April 30th, 2014. Published with kind permission of ©Fraunhofer SCAI 2016. All Rights Reserved

On the other hand, $k_{vs} = 28\,\text{m}^3/\text{h}$ gives pressure conditions almost in agreement with the measured pressure drop at the heat exchanger, see Fig. 26. Figure 29 exhibits simulated pressure differences too low in comparison to the situation for lower k_{vs} values.

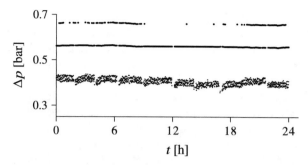

Fig. 29 Cooling circuit KLT72, heat exchanger—pressure drop measured (*bold dots*) and computed (*regular dots*, more accurate model of the heat exchanger with $k_{vs} = 50\,\mathrm{m}^3/\mathrm{h}$) for April 30th, 2014. Published with kind permission of ©Fraunhofer SCAI 2016. All Rights Reserved

We can, therefore, conclude that, under the given physical assumptions and with the pump and valve characteristics provided by the respective manufacturing companies, we cannot determine a pressure resistance at the heat exchanger that accounts for measured temperatures and flows at the same time. Either the simulated temperatures *or* the pressures contradict real system data. A situation in which characteristic data and measurements are not in accordance with each other seems to be typical when one works with real systems.

5   Conclusion

For a real-world application, cooling circuit simulation has been applied in an exemplary way. The numerical solution of equations from continuum mechanics, already introduced in [4] (see pages 61–79 in this book), represents the background of this application case. Models mostly based on common engineering knowledge are presented for a range of devices in order to enable the reader to apply the methods to the most typical examples of cooling circuits from practice.

After a brief survey of the numerical libraries involved to solve the system and a description of all developed software components and their interfaces, we particularly discussed problems and possible solutions related to the definition of the circuits themselves. There are three main data sources: The network layout itself, documentation from the manufacturers of the different components, and measurements conducted with the real system under load. Not only do we, therefore, dispose of a multitude of possibly conflicting data sources, we also have seen that errors and uncertainties within the data need to be anticipated. In fact, all three main sources of documentation exhibit certain issues specific to their nature: The network layout usually represents the design of the system and may vary significantly from its actual implementation, mostly due to changes during construction or due to later adaption. Manufacturer documentation may, on the one hand, be scarcely available and, on the other hand, refer to an ideal situation. Experience shows that characteristic values of components integrated into and aged within a real system can show considerable differences to this ideal situation. Eventually, measurements may suffer from defective or badly calibrated sensors, transmission delays or failures, as well as from all possible issues of digital data storage.

In lieu of a general and exhaustive discussion of data related uncertainties, this article has demonstrated these problems in an exemplary manner. We have proposed approaches how to collect, process, and apply data from the mentioned sources. Ideas for an automation of the data collection and processing are suggested in many subcases. Automatic data processing is far from trivial and two main areas remain where we apply manual work: Most documentation of single components is only available as printed data sheets, sometimes even including hand-drawn plots. These are collected and incorporated manually by now. The second prominent area of manual intervention is network simplification. The analysis of a subcircuit, as presented here, is often easier if only the most important components and effects are displayed. Knowing that only one of two parallel pumps is used in standard operational mode, for example, we have discarded the second pump. Such a simplification may partly be automatized. However, implicit knowledge of operational practice must again be collected and incorporated by the user.

Finally, this article has presented numerical simulation results of the most frequent type of subcircuits which can be found within our example application. The chosen subcircuit possesses a source, a sink, as well as a closed loop. It is regulated using the rotational speed of a pump and the opening of a three-way mixing valve. Energy is transferred from a neighboring circuit via a heat exchanger. Our focus was on simulation, not on automatic regulation or optimization. However, this is not at all a restriction of the described method, which is specifically designed to be able to couple simulation to optimization. Here, some kind of regulation process was only used in order to complete component characteristics where documentation was not available. Then a comparison of numerical results to previous measurements from the real system was conducted and discussed. Apart from scatter, attributed to flow measurement errors, the simulation is found to reproduce the main trends in temperature values and in pressure decrease. A systematic displacement with respect to the measured values can be observed, though. Depending on chosen component characteristics, this displacement either occurs for temperatures *or* pressures, which we see as an indication that the documented component characteristics do not exactly match the measurements from the real system. The simulated temperature curves can be improved if a more dedicated description of the heat exchanger is applied—at the expense of numerical stability. Whereas a very simple treatment of heat exchange can be stably calculated, our implementation of the mean logarithmic temperature difference formula suffers from convergence issues in some of the tested scenarios.

This article can only give an idea of the different areas and innumerable issues a reader interested in the simulation and optimization of real cooling circuits may face. Although we hope that we can point towards solutions to several of these issues, very many areas for systematic research do naturally remain. Some questions are obvious from what has been written above, such as how to better deal with the inhomogeneous and sometimes conflicting data sources, missing documentation, etc. Other self-evident goals are to handle larger networks, more components of types not yet discussed, and cooling liquids different from water. In particular, there is a vast amount of different types of energy exchangers and cooling devices, ranging from simple cooling doors to absorption or adsorption cooling, evaporators, compressor circuits, and many more.

A closer view on the numerics will provoke questions about stability, uniqueness and, of course, existence of solutions. From a more practical standpoint, one might be interested in how to best implement nonlinear formulae such as the mentioned mean logarithmic temperature difference in order to improve convergence. Particularly, a combination of several heat exchangers does not at all seem trivial, although there is practically no real system that does not consist of at least two of them.

Eventually, we expect any coupling of optimization and simulation to raise many more practical and theoretical questions. Additionally, automatic regularization algorithms are often applied within real systems, while their definition is seldom available to the user. Besides the question how to reproduce such a system behavior, any attempt to predict such algorithms within a simulation *may*, if a patent exists, even bring up legal issues. Nevertheless, it *is* certain that the problem of cooling circuit simulation and optimization does not only span the different fields of computer science and engineering, mathematics, classical physics, water and thermal engineering. Also within many of these fields, a multitude of areas is touched, such as data management, data basis engineering, data processing, visualization, image processing, nonlinear root finding, and the huge area of optimization, among others, which definitely makes the application a rich, complex, and very interesting object of study.

Acknowledgements Work reported in this publication has received funding from the German Federal Ministry of Education and Research (BMBF) under grant agreement no. 01IH13007A/B (SIMOPEK project).

References

1. K. Cassirer, T. Clees, B. Klaassen, I. Nikitin, L. Nikitina, MYNTS User's Manual, Release 3.7. Fraunhofer SCAI, Sankt Augustin (2015). http://www.scai.fraunhofer.de/mynts
2. T. Clees, MYNTS – Ein neuer multiphysikalischer Simulator für Gas, Wasser und elektrische Netze. Energ. Wasser Prax. **09**, 174–175 (2012)
3. T. Clees, N. Hornung, I. Nikitin, L. Nikitina, D. Steffes-Lai, DesParO User's Manual, Release 2.4. Fraunhofer SCAI, Sankt Augustin (2014). http://www.scai.fraunhofer.de/desparo
4. T. Clees, N. Hornung, È.L. Alvarez et al., *Cooling Circuit Simulation I: Modeling* In: (Springer, Berlin, 2017)
5. D.M. Gay, Writing .nl files, in *Optimization and Uncertainty Estimation* (Sandia National Laboratories, Albuquerque, 2005)

6. S. Grundel, N. Hornung, B. Klaassen, P. Benner, T. Clees, Computing surrogates for gas network simulation using model order reduction, in *Surrogate-Based Modeling and Optimization*, ed. by S. Koziel, L. Leifsson (Springer, New York, 2013), pp. 189–212
7. H. Shoukourian, T. Wilde, A. Auweter, A. Bode, Monitoring power data: a first step towards a unified energy efficiency evaluation toolset for HPC data centers. Environ. Model. Softw. **56**, 13–26 (2014). Thematic issue on modelling and evaluating the sustainability of smart solutions
8. A. Wächter, L.T. Biegler, On the implementation of an interior-point filter line-search algorithm for large-scale nonlinear programming. Math. Program. **106**, 25–57 (2006)
9. T. Wilde, T. Clees, N. Hornung et al., Increasing data center energy efficiency via simulation and optimization of cooling circuits – a practical approach, in *Energy Informatics, 4th D-A-CH Conference, Proceedings*, ed. by S. Gottwalt, L. König, H. Schmeck. Information Systems and Applications, vol. 9424. Internet/Web, and HCI, Karlsruhe (Springer, Berlin, 2015), pp. 208 ff

The LAMA Approach for Writing Portable Applications on Heterogenous Architectures

Thomas Brandes, Eric Schricker, and Thomas Soddemann

1 Introduction

At present, most heterogeneous computing systems are composed of CPUs with accelerator boards. Those boards incorporate powerful computing chips like GPU[1] (predominantly), various many-core processors (e.g. Intel Xeon Phi—most used in this field), or even FPGA[2] and DSP[3] for more dedicated tasks in bio-informatics or geophysical applications.

In November 2008, the first GPU accelerated super-computer (Tsubame [6]) hit the Top500 list. Today[4] 104 systems make use of acceleration by GPU and Intel® Xeon Phi™. Four of them can even be found in the Top 10. By having a look at the road-maps of the prevailing hardware vendors, it is by far more than just conceivable that heterogeneity in the hardware landscape is going to grow in the future.

Programming those systems and achieving reasonable code performance is a challenge. Programming as well as memory models exhibit tremendous differences between architectures. Optimizing code on non-traditional and novel CPU architectures is a challenge in itself due to the initally small user community and

---

[1]Graphics Processing Unit, can also be used as accelerator for scientific applications.

[2]Field Programmable Gate Array, integrated circuit that enables designers to program customized logic.

[3]Digital Signal Processor, takes real-world signals and processes them by performing mathematical functions.

[4]Status is the Top500 list from November 2015 [7].

T. Brandes (✉) • E. Schricker • T. Soddemann
Fraunhofer Institute for Algorithms and Scientific Computing SCAI, Schloss Birlinghoven, 53757 Sankt Augustin, Germany
e-mail: thomas.brandes@scai.fraunhofer.de

© Springer International Publishing AG 2017
M. Griebel et al. (eds.), *Scientific Computing and Algorithms in Industrial Simulations*, DOI 10.1007/978-3-319-62458-7_9

the small experience using those. Addressing the performance portability aspect across hardware architectures is a challenge in itself. In addition, work balancing approaches and task (co-)scheduling on those hardware architectures is in its infancy. However, these novel architectures are offering high processing element densities and memory bandwidths. Application domains with a huge demand for computing power are going to make use of those architectures.

Today, we face the task of porting legacy codes to benefit heterogeneous computing infrastructures. That process reveals design weaknesses which should be avoided in new developments at all costs. That was the motivator and the incentive for creating LAMA

LAMA started out as a linear algebra library for sparse matrices. It supports an abstraction scheme which allows a separation of concerns regarding implementing numerical algorithms vs. implementing clever and highly tuned code segments for utilizing modern and future hardware architectures. The main design goals can be formulated as

- being easily extensible as far as supporting novel hardware architectures are concerned,
- being flexible regarding its application domains.

This paper describes the general design of the LAMA library. First, we want to outline the main concepts of our implementation (Sect. 2). Later on, we perform a comparison especially with PETSc [1] (Sect. 3). Finally we sum up the presented approaches and draw some conclusions for future work.

2   LAMA

Extensibility and flexibility are main design goals for LAMA. We see extensibility as a main goal in order to answer user demands for creating new algorithms and augmenting LAMA with new features. This also holds for the underlying subsystem with respect to supporting next generations of hardware architectures. Flexibility was a design criterion in order to be able to support different hardware architectures and meet user requirements at the same time.

Therefore, a modular software stack was created. In its layering scheme, levels of abstractions are stacked on top of each other from a strong hardware dependence layer to a complete hardware agnostic implementation layer for numerical algorithms. Each of these layers is organized as a separate library. This allows any part of the library to be employed or reused independently from the ones on the higher levels of abstraction.

The dependencies to internal and external libraries are reduced to a minimum, and interfaces are kept as simple as possible. Figure 1 illustrates the design of the software stack with its dependencies.

On the bottom, all external libraries are listed that are (optionally) used by LAMA for the support of the different architectures and inter-node communication.

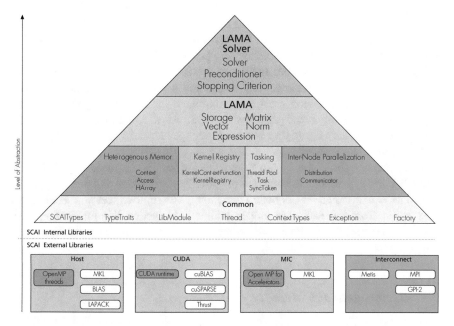

Fig. 1 Hierarchical design of LAMA showing the software stack of LAMA with major concepts in the separate libraries as well as internal and external dependencies. External dependencies are mapped by *color*. Published with kind permission of ©Fraunhofer SCAI 2016. All Rights Reserved

Primarily, these are the runtime libraries for supported devices (OpenMP® thread, CUDA®[5] driver and runtime, OpenMP[6] for Accelerator)[7] as well as vendor specific BLAS,[8] sparse BLAS and LAPACK[9] implementations. Currently we provide interfaces to Intel MKL[10]/BLAS, cuBLAS and cuSPARSE. For inter-node

---

[5]Compute Unified Device Architecture, an application programming interface developed by NVIDIA for GPUs.

[6]Open Multi-Processing, application programmer interface that supports multi-platform shared memory multiprocessing.

[7]In LAMA the host CPU is considered as a device that consists of multiple cores and therefore OpenMP is used for parallelization. OpenMP is also exploited for the implementation of kernels on the Intel Xeon Phi architecture using the offload support.

[8]Basic Linear Algebra Subprograms, specification that prescribes a set of low-level routines for performing common linear algebra operations.

[9]Linear Algebra Package, standard software library for numerical linear algebra on dense matrices.

[10]Math Kernel Library (Intel®) is a library of optimized math routines for science, engineering and financial applications, contains also BLAS and LAPACK implementations.

parallelization LAMA makes use of the communication standards MPI[11] (message passing programming mode) or GPI[12] (partitioned global address space programming model). Efficient implementations are provided on the respective sides.

The next level indicates our main concept for the design of portable parallel applications. A strict separation between memory management, kernel routines, tasking and communication creates a common layer for all supported devices, asynchronous execution and distributing data. As a consequence, all routines implemented on the next higher levels are independent of the underlying device and communication model.

More details about the memory and the kernel management will follow in Sects. 2.1 and 2.2. Section 2.3 explains the principles of tasking and Sect. 2.4 details how data is distributed across node boundaries. The top two levels with the most abstraction from the hardware model exhibit the functionality for dealing with matrices and vectors and finally linear equation solvers. The main functionalities are described in Sects. 2.5 and 2.6.

2.1 Heterogeneous Memory

The central concept of the LAMA design is the possibility to implement operations independent from the underlying hardware architecture. Hence, we introduce an abstraction for memory and kernels. This abstraction has independent backends (represented by a `Context`), supporting one processor kind or a particular group of hardware. In the following we are going to introduce LAMAs memory abstraction.

The goal for the memory abstraction layer is to provide means for accessing data on all supported hardware devices transparently in a unique way. Specialized device kernels on the layer below will be called for executing the data accesses on a particular hardware device. Listing 1 shows an example for the implementation of an add method operating on two arrays. The optional argument `prefCtx` specifies which `Context` this operation should ideally be executed in.

The heterogeneous memory management provides a container class `HArray` which allows memory management on any supported back end. During its lifetime it can have incarnations on different backends. From its first use on at least one incarnation is always valid. Accessing data is only possible via an explicit `Read`- or `WriteAccess` for a given `Context`. The constructor of the corresponding access object takes care that valid data is available, i.e. potentially necessary memory transfers are performed implicitly. Figure 2 demonstrates and explains the impacts of accesses for a snapshot of possible states of the involved arrays from Listing 1.

---

[11]Message Passing Interface, is a standardized and portable message-passing system, designed by a group of researchers from academia and industry, available on most parallel computing architectures.

[12]Global Address Space Programming Interface is an application programmer interface for one-sided asynchronous communication.

```
 1  template<typename T>
    void add( HArray<T>& res, const HArray<T>& a, const HArray<T>&
        b, ContextPtr prefCtx = ContextPtr() )
 3  {
        IndexType n = a.size(); /* assuming b.size() is the same
            */
 5      static LAMAKernel<UtilKernelTrait::add<T> > add;
        ContextPtr ctx = add.getValidContext( prefCtx );
 7      ReadAccess<T> readA( a, ctx );
        ReadAccess<T> readB( b, ctx );
 9      WriteOnlyAccess<T> write( res, ctx, n );
        add[ctx]( write.get(), readA.get(), readB.get(), n );
11  }
```

Listing 1 Generic implementation for an `add` method operating with two arrays: In line 5 the required kernel method is looked up from the registry and a valid `Context` for the kernel is retrieved (line 6). Afterwards the mandatory `Access` instances on the arrays are tied to the active `Context` (lines 7–9), then the device specific kernel method is called (line 10)

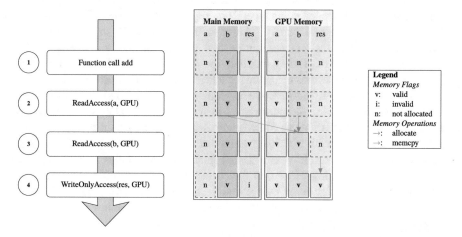

Fig. 2 Incarnations of heterogeneous arrays before and after an operation on GPU: Array *a* has only a valid incarnation on the GPU, *b* and *res* only on the CPU. The operations should be executed on the GPU. While the `ReadAccess` for *a* is directly possible, the `ReadAccess` of *b* implies the allocation of data on the GPU and the corresponding memory transfer. The `WriteAccess` for *res* invalidates the incarnation on the CPU without freeing the memory and allocates memory on the GPU. As this `WriteAccess` is declared as write-only, the previously valid data on the CPU is no more needed and no memory transfer is involved. Published with kind permission of ©Fraunhofer SCAI 2016. All Rights Reserved

The decision on which `Context` the operation will be executed is independent from the operation itself. LAMA supports preferred `Context`s that can be specified for the dedicated data structure. It can be considered as a hint where the data is needed in future. A good choice might take into account on which `Context` already valid data is available in order to reduce memory traffic.

2.2 Heterogeneous Kernel

In Listing 1 the interface to our kernel management has already silently been introduced: you can access static references to a kernel functions (see line 5) and execute them on valid Contexts (line 10). So a transparent usage of a kernel routine on various Contexts is given. This design has the advantage that by adding a new back end the developer only needs to implement really required functions.

A default Context is returned by the registry (line 6) if it is not given on the preferred Context. This is helpful in the development of new algorithms as a routine does not have to be implemented for all Contexts at once. And it can be a good solution for non time-critical routines e.g. in the setup or for routines that can not be implemented efficiently on a certain Context. For obtaining optimal performance all required methods have nevertheless to be implemented for a the preferred Context, of course. This kind of kernel implementation introduces an additional layer (the shown generic function in Listing 1) for executing kernels but exhibits the desired flexibility to run on all given devices.

In case of template functions, a registration to a Context also implies the instantiation for the needed value type. Both, registration and use of kernel routines employ so-called KernelTraits (refer to (I) in Listing 2), that are structures describing the signature of the routine and providing unique names used for registration [refer to (III)]. They facilitate the use of kernel routines but they also increase the safety by avoiding any mismatches between registration and use.

In the first version of LAMA, the implementation of the KernelRegistry exploited abstract classes and pure virtual methods. Unfortunately virtual methods cannot be template methods as overloading and overriding would be mixed up. A possible workaround with a virtual routine for each supported type does not only conflict with the extensibility as a design goal but also implies an unacceptable amount of source coding. Hence, in the current implementation, we introduced a new data structure which supports overloading and overriding by a map of function pointers. The penalty of less safety as in the previous approach is compensated by employing KernelTraits.

2.3 Task Parallelism

Heterogeneous architectures exhibit a large amount of parallelism. It is essential for the optimal usage of those architectures to employ that inter-architecture parallelism. Besides the previously described parallelism within a given kernel back end, there is additional parallelism inside a node, e.g. between host and accelerator. In the following, we sketch our task parallel execution model. Inter-node parallelism is treated later in Sect. 2.4.

For intra-node task parallelism we identified two types of tasks that can be executed independently from the each other.

```
1  // (I) kernel traits for vector add
   template<typename ValueType>
3  struct add
   {
5    typedef void ( *FuncType )(
       ValueType* z,
7      const ValueType* x,
       const ValueType* y,
9      const IndexType n ) ;

11   static const char* getId()
     {
13     return "Util.add";
     }
15 } ;

17 // (II) OpenMP implementation for vector add kernel
   template<typename T>
19 void add( T res[], const T a[], const T b[], IndexType n )
   {
21   #pragma omp parallel for
     for( IndexType i = 0; i < n; ++i )
23   {
         res[i] = b[i] + a[i];
25   }
   }
27
   // (III) registration for the vector add kernel for the host
     context
29 KernelRegistry::register( UtilKernelTrait::add<double>, &add,
       context::Host ) ;
   KernelRegistry::register( UtilKernelTrait::add<float>, &add,
       context::Host ) ;
```

Listing 2 Code snippets showing the implementation of a vector add kernel for the host: (I) The corresponding KernelTrait for the vector add kernel. It is used to identify the kernel inside the KernelRegistry. The kernel signature has to match the typedef of FuncType. (II) Shows the OpenMP kernel which implements the vector addition. (III) Defines the registration for the host for different value types corresponding to the KernelTrait of (I)

1. Memory transfers between host and device or between devices
2. Computation on host or accelerator device

Tasks may be executed asynchronously to optimize the usage of available system resources.

In our tasking library we use threads for the host back end. Threads are taken from a thread pool to avoid the overhead of thread creation. Streams are used for the CUDA back end. The Xeon Phi back end just handles asynchronous memory transfer by the corresponding offload directives so far.

SyncTokens are introduced to avoid access conflicts on HArrays. They resume the ownership of the corresponding Read- and WriteAccesses that are used within the task. Tasks which cannot be executed asynchronously return dummy tokens.

In LAMA we apply task parallelism mostly in two major cases: for hiding memory transfer to and between accelerators and inter-node communication while executing calculation tasks. E.g., a pre-fetch on an HArray can be invoked while computation is carried out to get data transferred to where it is needed in time. Especially our matrix-vector-multiplication benefits from this execution model. With accelerator support this additionally includes asynchronous memory transfer. For more details we refer to [4].

2.4 Distributed Memory Support

When it comes to larger problem sizes, the hardware resources of a single node are often not sufficient any more for handling the mathematical problem efficiently. Hence, multiple nodes will be used in parallel which requires the support for coping with halo data. Optimizing the distribution of data is vital to the performance of the calculations. Therefore in LAMA a Distribution can be calculated by graph partitioning algorithms using Metis [3]. This also allows weighting distribution sizes even for non-homogeneous systems.

With assigning a Distribution to a data structure, every compute process has knowledge of that part of the data it is responsible for and a CommunicationPlan is calculated. This *plan* describes the pattern and amount of data used for exchanging data elements at the border to neighboring processes. Than *plan* can be reused, so by setting up and storing it we save overhead with every exchange.

The pure communication layer introduces an abstraction from the underlying communication library by using a Communicator instance. It has special routines, that are optimized for its needs. Currently, LAMA supports two specialized implementations: MPI and GPI-2.[13] MPI is an established standard for distributed memory support. GPI-2 is a recent development in the field of partitioned global address space (PGAS). It is an API[14] for one-sided and asynchronous communication that allows a perfect overlap between computation and communication. Together with the support of asynchronous computations some applications take advantage of it.

---

[13]GPI-2 is an open-source GPI implementation provided by Fraunhofer ITWM.

[14]Application Programming Interface, a set of routines, protocols, and tools for building software applications.

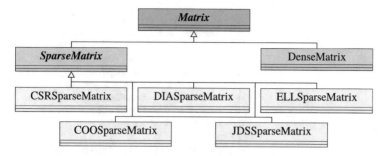

Fig. 3 Matrix inheritance hierarchy: currently LAMA contains a dense matrix format as well as COO, CSR, DIA, ELL and JDS sparse matrix formats (for more information about sparse matrix storage formats, see [5]). Every format has its own advantages and disadvantages. E.g., the CSR-format works well on CPU, while the ELL- or specialized JDS-format are usually optimal for architectures like GPUs. Published with kind permission of ©Fraunhofer SCAI 2016. All Rights Reserved

2.5 Matrices and Vectors

Generic matrices and vectors are implemented on top of `HArray`. Hence, they are independent from `Contexts`. However, a `Matrix` or `Vector` have `Distributions`. This allows a transparent use in a multi-node setup.

Several specializations of the generic `Matrix` are available in LAMA (see Fig. 3). Hence, algorithms can be formulated using the `Matrix` interface. A valid line of code for a matrix A and the vectors x and y can look like the following: `y = A * x;` This expression syntax is realized by employing expression templates (for reference on expression templates see [9]). On the basis of the underlying layers it can be executed in a single- or multi-node way on each supported `Context`.

Inside an expression the `Distributions` for matrices and vectors have to be compatible. Therefore, LAMA supports redistributing data by assigning a new distribution. Matrix data is always distributed line by line and it is split up in two parts, a local part of the matrix and its halo part. In the sparse matrix-vector multiplication for the calculation on the halo part data of the right-hand-side vector needs to be transferred before the calculation can be executed.

As mentioned before, LAMA supports overlapping communication and calculation. For this the communication of the halo data is first initiated, so the calculation on the local matrix can be executed asynchronously. Afterwards, computation on the transferred data takes place. The benefits of this approach are described for the algebraic multi-grid method in [4].

2.6 Solver Framework

At the highest level of abstraction LAMA provides a solver framework with a large set of linear `IterativeSolvers`. Each solver makes use of the textbook syntax

$$r_0 = b - Ax_0;$$
$$d_0 = r_0;$$

for $k = 0, 1, \ldots$ **to** $\|r_{k+1}\| < tol$
do

$\quad z = Ad_k;$

$\quad \alpha_k = \frac{r_k^T r_k}{d_k^T z};$

$\quad x_{k+1} = x_k + \alpha_k d_k;$

$\quad r_{k+1} = r_k - \alpha_k z;$

$\quad \beta_k = \frac{r_{k+1}^T r_{k+1}}{r_k^T r_k};$

$\quad d_{k+1} = r_{k+1} + \beta_k d_k;$

```cpp
   void cg( Vector& x, const Matrix& A,
            const Vector& b, const Scalar tol )
 2 {

     // inherit type, context of b
 4   RuntimeVectors tmpVectors(b, 3);
     Vector& r = tmpVectors[0];
 6   Vector& d = tmpVectors[1];
     Vector& z = tmpVectors[2];
 8   r = b - A * x;
     d = r;
10   Scalar rOld = r.dotProduct(r);
     L2Norm norm;
12   for (int k = 0; norm(r) < tol; k++)
     {
14     z = A * d;
       Scalar alpha = rOld /
16                      d.dotProduct(z);
       x = x + alpha * d;
18     r = r - alpha * z;
       Scalar rNew = r.dotProduct(r);
20     Scalar beta = rNew / rOld;
       d = r + beta * d;
22     rOld = rNew;
     }
24 }
```

Fig. 4 Comparison of the mathematical description (*left*) of the CG method with its implementation (*right*) in LAMA textbook syntax. The matrices and vectors involved can have any Distribution, any Storage format or any precision and might have valid data on any Context. The temporary Vectors r, d, and z are allocated with the constructor and are freed with the destructor of RuntimeVectors, they will have the same type and Context as the input Vector b. The operations will be executed (preferably) on the Context set for Matrix A and Vector b

with distributed matrices and vectors. Employing LAMA a user can implement solvers almost precisely as formulated in mathematical syntax as shown in Fig. 4.

The implementation of a solver does not make any assumption about multi-nodal distribution or execution device. As a result a Solver only has to be written once and can be executed in all heterogeneous environments including multiple nodes and different accelerators. Furthermore, the user can explore with LAMA different target hardware architectures as well as different fields of target problems with the right kind of solver and sparse matrix format.

Solvers are specializations of the abstract class Solver. IterativeSolver in turn is a specialization for solvers that are composed of single iterations, where a StoppingCriterion decides about the number of iterations in the solve method. The iterative solvers are categorized into splitting, Krylov subspace and multigrid methods. A full list of all available iterative solvers is shown in Fig. 5.

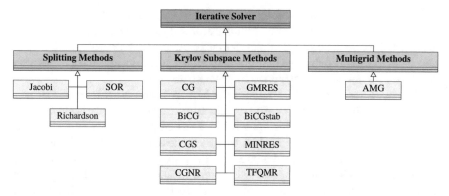

Fig. 5 Iterative solvers inheritance hierarchy: Currently LAMA supports the three categories of splitting, Krylov subspace and multigrid method with various solvers. Published with kind permission of ©Fraunhofer SCAI 2016. All Rights Reserved

```
 1  CG cgSolver( "CGSolver"/*, logger*/ );
 2  cgSolver.intialize( matrix );
    cgSolver.setStoppingCriterion( new ResidualThreshold( new
      L2Norm(), Scalar( 0.0001 ), ResidualThreshold::Absolute )
      );
 4  // optional preconditioning:
    // cgSolver.setPreconditioner( otherSolverPtr );
 6  cgSolver.solve( solutionVector, rhsVector );
```

Listing 3 Usage of an iterative solver: Set up the solver (line 1), initialize it with the system matrix (line 2), set a stopping criterion (line 3), perform optional preconditioning (line 5) and call solve with a starting solution and a right hand side vector (line 6). When the routine returns, the solutionVector holds the solution of the last performed iteration. In order to define when the iteration has to stop, different kind of stopping criteria are provided that might also be combined

Additionally an inverse solver[15] can be used as direct solver on the coarsest grid of the AMG. In principle, each solver can serve as a preconditioner for every other solver.

Which kind of solver should be used for a specific kind of problem is in the hands of the user. Also the chosen sparse matrix format should suite the matrix structure. Using a solver is really easy as shown in Listing 3 for a CG solver.

2.7 Extensibility and Maintainability

LAMA makes use of widely applied design patterns, among them templates and factories. Here, we briefly want to motivate our choices.

---

[15]The inverse solver uses an interface to the ScaLAPACK (Scalable LAPACK, a subset of LAPACK routines parallelized for distributed memory machines, e.g. by using MPI) library and can be used only for block-cyclic distributions and the MPI communicator.

Templates are employed throughout LAMA at various places, e.g. to introduce some kind of abstraction from arithmetic data types. Data structures can be instantiated for float, double, complex or double complex. The use of quad precision data types like long double and long double complex is restricted to the CPU as these data types are not directly supported by accelerators, yet.

Various factories are also offered in LAMA. All base class instances of general concepts like `Matrix`, `Vector`, `Communicator`, `Distribution` and `Solver` are created using the appropriate `Factory` for obtaining a particular implementation. This allows LAMA to be seamlessly extended with new class specializations without the need of touching the code that uses the general types.

New functionality can be added by loading a separate library module at runtime, very similar to the import feature of Python. Technically this is realized by the factories where new derived classes are registered in the application at runtime. This is also very convenient for a dual license model, as commercial extensions of LAMA (e.g. highly optimized kernels or very efficient solvers) can be added dynamically to an already installed free installation.

Furthermore, our design enhances the quality of tests within LAMA. Each extension with a new derived class can be tested directly for correct functionality and only class specific tests have to be added.

3   Performance Comparison

In this section we briefly compare LAMA's performance to PETSc which is a linear algebra library with similar focus and similar functionalities as LAMA. It is written in C and exists for quite some time. PETSc enjoys a large and vivid user community. Furthermore, in one case a basic MKL BLAS (native) implementation is used for comparison, as well.

The following benchmark results show the total run times for a CG solver after 1000 iterations.[16] The testing environment (physical system, used compilers and software) is explained in the appendix. The matrices used in the following benchmarks were taken from the University of Florida Collection [8].

Figures 6 and 7 show the single node performance for CPU and GPU. The weak scaling for both libraries is evaluated in Fig. 8 by using three Poisson matrices on one to four nodes.

For the case of the CPU-only use we have just measured for the CSR format. Here, both libraries employ Intel's highly performant MKL BLAS implementation "under the hood". This gives a good base line for our design overhead as well as for PETSc's. Looking at the results, the native BLAS implementation is (of course) the fastest. Comparing with PETSc, we see a minimal benefit for our competitors in this case. However, the results demonstrates for both libraries that their designs

---

[16]The time for one CG iteration is mainly given by one sparse matrix-vector multiplication, four dot products and three *axpy* operations.

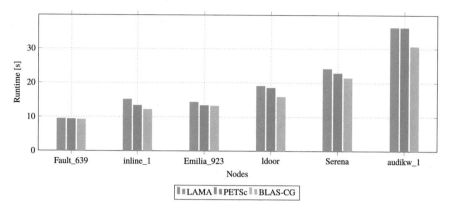

Fig. 6 CPU comparison LAMA—PETSc (CSR): Single node CPU performance utilized by six MPI-processes. Runtime is proportional to the number of non-zeros—only the irregular structure of *inline_1* and *audikw_1* show remarkably higher runtime. Published with kind permission of ©Fraunhofer SCAI 2016. All Rights Reserved

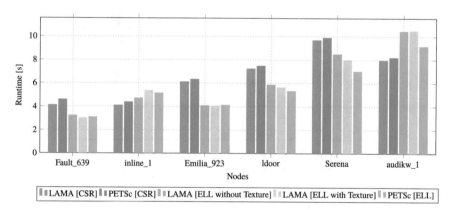

Fig. 7 GPU comparison LAMA—PETSc (CSR, ELL): Single GPU performance for CSR- and ELL sparse matrix format. PETSc relies for both formats on the cuSPARSE kernels, LAMA for CSR, too, for ELL on its own kernel (with the possibility to use the texture cache for the right-hand-side vector). Regarding the storage format, the ELL format has shorter runtimes, only for *inline_1* and *audikw_1* the CSR format is the faster one. These latter two matrices have nearly twice the number of entries per row and here the CSR format outperforms the ELL format. Published with kind permission of ©Fraunhofer SCAI 2016. All Rights Reserved

introduce a negligible performance loss. Furthermore, this also proves that in this case our (memory) model works successfully.

On a single GPU results for the CSR format also are nearly the same with a tiny benefit in favor of LAMA. This is not a big surprise as both packages rely on the CSR sparse matrix-vector multiplication provided by the cuSPARSE library, which dominates the runtime (about 80%). For the *axpy* and dot operations, LAMA calls cuBLAS routines while PETSc exploits implementations using the Thrust library. Analysis by the NVidia Visual Profiler shows a small advantage for LAMA, here.

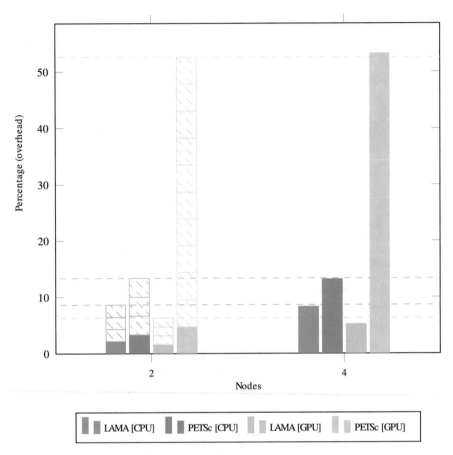

Fig. 8 Overhead for two/four node weak scaling LAMA—PETSc (CPU—CSR, GPU—ELL): Executed on 3D 27-point Poisson matrices with 10 million unknowns per node. Small overhead (under 5%) for both, PETSc and LAMA, on the CPU and GPU for scaling on two nodes. On four nodes, the overhead increases on CPU by a factor of 4 for both libraries. On GPU LAMAs overhead increases by less than factor 4 to 5%, while PETSc loses significant performance by more than factor 11 to 50%. Published with kind permission of ©Fraunhofer SCAI 2016. All Rights Reserved

When using the ELL format on a **single GPU**, there is currently a small advantage for PETSc in terms of performance. PETSc uses the corresponding kernel of the cuSPARSE library while LAMA uses an own CUDA kernel. However, the LAMA kernel provides the possibility to bind the right-hand-side vector in the matrix-vector multiplication to the texture cache of the GPU that can improve the data locality. On Nvidia's Kepler GPU architecture the use of the texture cache increases the performance slightly except for one test case. Nevertheless, the cuSPARSE kernel improved to previous versions and now outperforms the LAMA kernel.

Since the LAMA GPU-Kernels are also tuned for **multi GPU** (via MPI) usage, it is worthwhile to have a look at performance results in that context. We have extended the previous use case to a weak scaling one for both libraries (Fig. 8). Note that LAMA supports asynchronous execution and therefore the overlapping of communication and computation. We have not seen something similar in PETSc. Furthermore, halo data is treated differently in PETSc. In this case we observe a much better scalability for LAMA in direct comparison with PETSc. The communication overhead has more influence as on the CPU. This has two reasons: In our setup the GPU is already much faster than the CPU. Furthermore, the overhead increases due to the down- and upload of halo exchange data from and to the GPU. For a more detailed view on handling the halo data in LAMA and further results, see [4].

4 Summary

In this paper we presented the design, architecture, and performance analysis of LAMA. We showed that with the abstraction of the memory model and compute kernels, and the introduction of asynchronous execution and communication, it is possible to use LAMA to develop algorithms independently of the underlying hardware architecture. The central concept is the heterogeneous array that allows automatic up- and downloading of data, guarantees consistent use of multiple incarnations, and avoids invalid memory accesses.

That approach to our knowledge is novel and currently unique. It completely differs from our competitors' implementations.

Our approach speeds up the development process since communication between heterogeneous hardware parts and across node boundaries has already been encapsulated in our library. At the same time this approach also reduces the risk of programming errors.

We showed that such a design introduces an overhead in the computation. Compared to our strongest competitor PETSc, this overhead is fortunately negligible (CPU) or at least significantly small (single GPU). Moreover, the extra code for managing memory works in our favor when having a look at the multi-GPU results.

Our design introduces a kernel management composed of a registry with static registration at runtime. Employing KernelTraits in this case provides high flexibility and ensures already at compile time that the right methods are being used.

In LAMA we implement techniques for concealing necessary (and possibly time consuming) data transfers by making use of asynchronous execution. This is always advisable when working with accelerators with limited bandwidth interconnects. Furthermore, the "separation of concerns" design of our tasking library is implemented for every computation kernel. The library layer for inter-node parallelization handles the domain decomposition and communication. In addition, the underlying communication libraries are exchangeable and, hence, a developer can easily extend LAMA to support new hardware architectures.

The performance comparison with PETSc has shown that we achieve similar single node performance. The results of the performance comparison with single architecture implementations as well as PETSc underline that achieving good performance in general is not conflicting with a design focusing on portability and extensibility.

Currently, we are working on incorporating performance models for improving and adjusting the graph partitioning at runtime [10]. Application aware Co-scheduling of LAMA with other applications (or application parts) is another topic we currently investigate in the FAST project [2]. In addition, we continuously increase the number of available iterative solvers and pre-conditioners.

LAMA can be tested at http://libama.scai.fraunhofer.de:8080/lamaui/.

Acknowledgements This work was funded by the BMBF (German Federal Ministry of Education and Research) through the projects MACH, FAST, and WAVE.

Appendix

Test Environment

See Fig. 9 and Table 1.

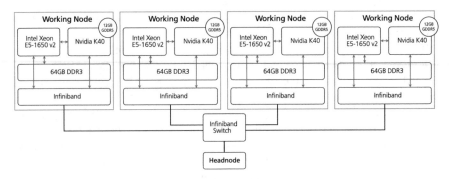

Fig. 9 Setup of the test system: Four node hybrid HPC system. Each node consists of an Intel Xeon® E5-1650v2 with 64GB RAM and a Nvidia© K40 GPU (12 GB GDDR 5 memory) as accelerator which is connected via PCI Express bus (15,754 MB/s). The nodes are connected over InfiniBand. Published with kind permission of ©Fraunhofer SCAI 2016. All Rights Reserved

Table 1 Used compilers and libraries for building LAMA and PETSc (both: development version of November 2015)

Category	Name	Version
Compiler	gcc	4.9.1
	g++	4.9.1
	nvcc	7.0
Support	boost	1.58.0
Communication	OpenMPI	1.10.1
BLAS	MKL	11.1.2
	cuBLAS	7.0.28
Sparse-matrix	cuSPARSE	7.0.28

Test Matrices

See Table 2 and Fig. 10.

Table 2 Representative set of large test matrices from the University of Florida sparse matrix collection [8] (sorted by number of total nonzeros): memory usage in MB for double precision

Name	Dimension		Nonzeros			Memory usage	
	Rows	Columns	Total	$\frac{\varnothing}{\text{row}}$	$\frac{\text{max}}{\text{row}}$	CSR	ELL
Fault_639	638, 802	638, 802	28,614,564	44	318	332.34	2327.18
inline_1	503,712	503,712	36,816,342	73	843	425.17	4861.42
Emilia_923	923,136	923,136	41,005,206	44	57	476.31	605.70
ldoor	952,203	952,203	46,522,475	48	77	539.67	842.71
Serena	1,391,349	1,391,349	64,531,701	46	249	749.12	3970.07
Audikw_1	943,695	943,695	77,651,847	82	345	895.86	3729.51

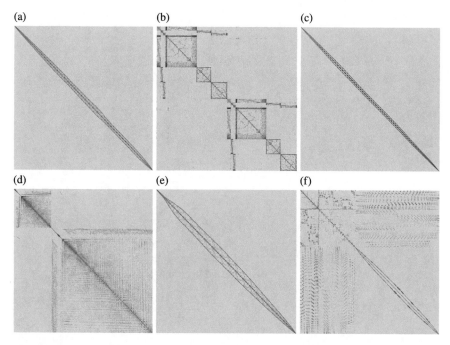

Fig. 10 Used test matrices. Published with kind permission of ©Fraunhofer SCAI 2016. All Rights Reserved. (**a**) fault_639. (**b**) inline_1. (**c**) Emilia_923. (**d**) ldoor. (**e**) Serena. (**f**) audikw_1

References

1. S. Balay, S. Abhyankar, M.F. Adams et al., PETSc Web page (2015). http://www.mcs.anl.gov/petsc
2. FaST Consortium, Find a Suitable Topology for Exascale Applications (FaST) (2015). https://www.fast-project.de/
3. G. Karypis, V. Kumar, A fast and highly quality multilevel scheme for partitioning irregular graphs. SIAM J. Sci. Comp. **20**, 359–392 (1999)
4. J. Kraus, M. Förster, T. Brandes, T. Soddemann, Using LAMA for efficient AMG on hybrid clusters. Comput. Sci. Res. Dev. **28**, 211–220 (2013)
5. Y. Saad, *Iterative Methods for Sparse Linear Systems*, 2nd edn. (Society for Industrial and Applied Mathematics, Philadelphia, 2003)
6. E. Strohmaier, J. Dongarra, H. Simon, M. Meuer, Top500, the list, November 2008 (2008). http://www.top500.org/lists/2008/11/
7. E. Strohmaier, J. Dongarra, H. Simon, M. Meuer, Top500, the list, November 2015 (2015). http://www.top500.org/lists/2015/11/
8. University of Florida, *The University of Florida Sparse Matrix Collection* (2015). https://www.cise.ufl.edu/research/sparse/matrices/
9. T. Veldhuizen, Expression templates. C++ Report **7**, 26–31 (1995)
10. WAVE Consortium, Eine portable HPC-Toolbox zur Simulation und Inversion von Wellenfeldern (WAVE) (2016). https://www.wave-project.de/

ModelCompare

Jochen Garcke, Mandar Pathare, and Nikhil Prabakaran

1 Introduction

Numerical simulations have been an essential tool in the virtual product development for many years now. Here, an approximate representation of a product by a finite element (FE) model is employed and, based on a mathematical model of the physical process, a numerical simulation is performed. For example in the automotive industry, the research and development process consists of generating multiple variants of a car CAE-model representing different configurations [5]. Furthermore, different operating conditions of the car and different situations, so-called load cases, are investigated, multiplying the amount of data once more. Post-processing software tools are readily available to display the 3D geometrical information of such a model and the results of the numerical simulation.

The complex structure of the data and its sheer size, the required 3D visualization of the geometry and the needed inspection of the associated design variables of each configuration prohibit a detailed comparative analysis by hand. For instance, in the automotive industry, different versions of a FE model are generated by morphing the geometry (mesh), re-meshing, altering the material data, thickness data and so on. Several tools, that are available in the market to record the changes made over time, are often time consuming and cumbersome to work with.

J. Garcke (✉)

Fraunhofer Institute for Algorithms and Scientific Computing SCAI, Schloss Birlinghoven, 53757 Sankt Augustin, Germany

Institute for Numerical Simulation, Rheinische Friedrich-Wilhelms-Universität Bonn, Wegelerstr. 6, 53115 Bonn, Germany
e-mail: jochen.garcke@scai.fraunhofer.de

M. Pathare • N. Prabakaran
Fraunhofer Institute for Algorithms and Scientific Computing SCAI, Schloss Birlinghoven, 53757 Sankt Augustin, Germany

© Springer International Publishing AG 2017
M. Griebel et al. (eds.), *Scientific Computing and Algorithms in Industrial Simulations*, DOI 10.1007/978-3-319-62458-7_10

However, our ModelCompare tool automatically determines the differences in the configurations and provides an easy and interactive access to the difference data and its visualization.

2 Development History

Fraunhofer SCAI has a long history of data analysis and data processing in the automotive industry, which for example resulted in the spin-off SIDACT[1] with its software products FEMZIP (for data compression) and DiffCrash (for an analysis of the robustness of crash simulation models). In the course of several projects in the past years (FEMMINER, SIMDATA-NL) and ongoing ones (VAVID), which were and are supported by the German Bundesministerium für Bildung und Forschung (BMBF), new data analysis techniques were studied and developed for the investigation of bundles of large scale numerical simulation data [1–4]. The efficient and easy identification of changes in the configuration of a numerical simulation was, and is, needed for the more advanced data analysis procedures developed in these projects. But as it turned out, it is on its own already a technology which is of use for the engineer in the automotive industry.

3 Capabilities

The ModelCompare software developed by SCAI supports the engineer in ana- lyzing the configurations obtained from the pre-processing step in the product development chain, where these configurations would then be used for the actual numerical simulation. Already at this step, automated processing scripts, meshing, parametrized localized plate thickness changes, or spotweld placements can result in a large number of changes. ModelCompare assists the user by providing elegant visuals of the differences in two similarly discretized FE models while seamlessly integrating with the specialized visualization software Animator.[2] During a product development cycle, several design modifications are made over a period of time by one or many users. Although several PLM tools are available in the market to record the design changes made over time, a quick interactive view of all the changes made between two design setups, in particular including the effect of parametrized changes within the post-processor, is so far not available.

Roughly speaking, based on previous designs, a concept design is first set that will be subsequently refined in several phases of the product design by changing the geometry, material data and other design related parameters in order to fulfill functional or regulatory constraints. The steps involved in the pre-processing include

---

[1] www.sidact.com.

[2] Animator4 from GNS mbH, gns-mbh.com/animator.html.

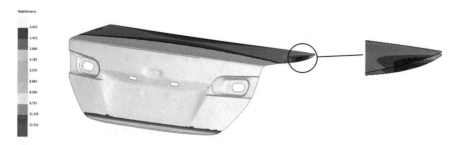

Fig. 1 The bonet of a Toyota Camry is morphed slightly at three positions to fit new design requirements, as identified by ModelCompare. Published with kind permission of ©Fraunhofer SCAI 2016. All Rights Reserved

geometry decomposition, discretization by a FE model, setting up connecting elements such as rigid body elements (RBEs) and spotwelds, defining boundary conditions as well as material parameters. New input models may be generated by varying one or many of the steps involved in the pre-processing. Note that this step is often given to engineering companies by the OEMs in the development process, further increasing the need for an easy investigation tool for determining the changes in configuration.

The ModelCompare software analyzes two models to detect changes in geometry and in attributes such as material property and part thickness. In the following we give examples for the different type of changes, illustrated on a model for the Toyota Camry from the Center for Collision Safety and Analysis (CCSA) repository.[3]

3.1 Detection of Geometry Changes

Geometry changes, or changes in the mesh, occur primarily due to morphing or remeshing of certain patches or surfaces in a part. ModelCompare identifies the changes made in the corresponding part(s) of the other model, based on the geometry (mesh) of the parts, and is independent of meta data of the parts such as their identifier or name. Geometrically changed parts have a one-to-one correspondence, i.e. for a part in one model exactly one part is found in the other model. Examples of geometry changes performed in the front of a car are shown in Figs. 1 and 2.

Note that some parts, which have the same mesh configuration in both models, might have one or a few new or missing finite elements. In physical terms, an element has material in the one model, while it has no material in the other one, i.e. it is a hole. The ModelCompare tool also detects such situations and finds the position of such elements.

---

[3]Finite Element Model Archive, www.ccsa.gmu.edu/models/2012-toyota-camry/.

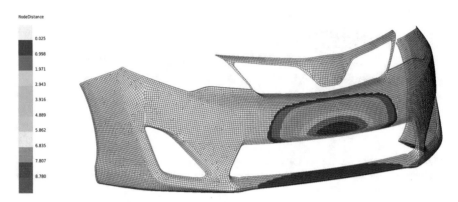

Fig. 2 The bumper of a Toyota Camry is morphed to fit new design requirements, as recognized by ModelCompare. Published with kind permission of ©Fraunhofer SCAI 2016. All Rights Reserved

Fig. 3 The front door assembly of a Toyota Camry model, which was described as a single part in one model, was found to be described as a group of sub parts. Published with kind permission of ©Fraunhofer SCAI 2016. All Rights Reserved

3.2 Detection of MultiParts

A part in one model can be split into many parts in the other model. Here the parts have one-to-many correspondence, i.e. one part in one model corresponds to a combination of some parts in the other model. This is illustrated in Fig. 3. Note that establishing such a correspondence between a part and parts that are merged together in the other model is based on certain assumptions, which can be problematic, i.e. bigger changes can result in misclassifications. Finally, parts which appear only in one of the two models are also identified and categorized accordingly.

For both, the geometry changes and the 'MultiParts' changes, the tool can also give a measure of the uncertainty of the matching, which becomes relevant for larger changes.

3.3 Spotwelds and Rigid Body Elements

Spotwelds and rigid body elements (RBEs) are also diagnosed for changes in their configuration, different positions or connections. In Fig. 4 we show a change in the spotwelds, while Fig. 5 shows corresponding RBEs from the two models that have different so-called master node positions. ModelCompare determines such differences automatically, thus giving the engineer important insight about changes

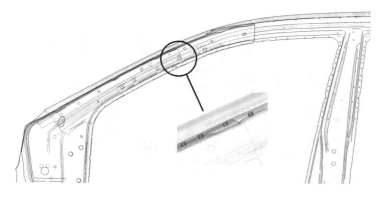

Fig. 4 Spotwelds on the A-Pillar which have no direct correspondence in the other model. Published with kind permission of ©Fraunhofer SCAI 2016. All Rights Reserved

Fig. 5 RBEs from the two models with different master node position. Published with kind permission of ©Fraunhofer SCAI 2016. All Rights Reserved

(a)   (b)

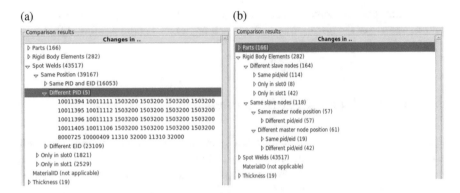

Fig. 6 Changes to spotwelds and rigid body elements as shown in the plugin GUI. Published with kind permission of ©Fraunhofer SCAI 2016. All Rights Reserved. (**a**) Differences in the spotweld configuration. (**b**) Differences in the RBE configuration

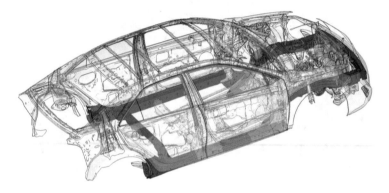

Fig. 7 All parts with thickness changes can be displayed for visual identification. Published with kind permission of ©Fraunhofer SCAI 2016. All Rights Reserved

in the model. All these differences are organized and presented in several categories, as can be seen in Fig. 6.

3.4 Detection of Material-ID and Thickness Changes

In addition to changes in geometry or position, differences in attributes such as material properties and thickness are also identified. While material properties are per part, the thickness can be given per part or per element (described by a parametrized local thickness). Besides the information per part, it is also helpful to show all parts with changes in the full car, with the remaining parts greyed out, as is shown in Fig. 7.

4 Outlook

Nowadays, engineers often analyze and compare different numerical simulation results using their own engineering knowledge. This is generally limited to the simultaneous analysis of only a very few full simulation results at a time or of simplified scalar quantities of interest, which describe the full complexity of the simulation results only in a limited fashion. Therefore, there is need for a more efficient product development process which overcomes the current limitations. For example, the use of post processing quantities, that are single scalar quantities or vectors, does not allow an in-depth analysis of 3D deformations. Indeed, an efficient and detailed analysis of this type for hundreds of design changes is nowadays a challenge in industrial practice.

The described ModelCompare plugin allows the comparison of two input configurations for a numerical simulation. A natural extension is such a comparison, or data organisation, over a bundle of configurations, which would allow an easy overview of the design changes made by engineers, alone or in a group.

Besides such an investigation of the input configuration, the comparison of multiple simulation data sets using the full output data of the numerical simulation, or large parts of it, is currently being investigated and developed, see e.g. [2–4]. It is envisioned that the combination of data analysis for the results of a numerical simulation and the data organization and analysis of the input configurations will provide essential software components for the virtual product development process, both enhancing and simplifying the work for an engineer.

References

1. B. Bohn, J. Garcke, R. Iza Teran et al., Analysis of car crash simulation data with nonlinear machine learning methods, in *Proceedings of the ICCS 2013*. Procedia Computer Science, vol. 18 (Elsevier, Amsterdam, 2013), pp. 621–630
2. J. Garcke, R. Iza-Teran, Machine learning approaches for repositories of numerical simulation results, in *10th European LS-DYNA Conference. Proceedings* CD-ROM: 15–17 June 2015, Würzburg, DYNAmore GmbH, Stuttgart (2015)
3. R. Iza Teran, Enabling the analysis of finite element simulation bundles. Int. J. Uncertain. Quantif. **4**, 95–110 (2014)
4. R. Iza-Teran, J. Garcke, Data analytics for simulation repositories in industry, in *44. Jahrestagung der Gesellschaft für Informatik, Informatik 2014*, Big Data - Komplexität meistern, 22.-26. September 2014 in Stuttgart, Deutschland, ed. by E. Plödereder, L. Grunske, E. Schneider, D. Ull (2014), pp. 161–167
5. M. Meywerk, *CAE-Methoden in der Fahrzeugtechnik* (Springer, Berlin, 2007)

Rapid Enriched Simulation Application Development with PUMA

Marc Alexander Schweitzer and Albert Ziegenhagel

1 Introduction

Simulation application development typically requires a lot of human effort to achieve the required overall efficiency, especially in a performance-critical industrial setting. This issue is becoming ever more challenging when neither the mathematical model nor the employed numerical scheme are fixed a priori. In many applications, engineers need to make substantial changes to the mathematical model over time to account for additional physics that need to be captured in the simulation to provide realistic results. Moreover, mathematicians constantly devise novel discretization techniques and numerical algorithms, based on different data structures etc., to construct more efficient approximation schemes. Finally, the hardware of high-performance computing equipment is changing drastically over very short periods of time, so that any implementation optimized for a specific platform becomes irrelevant fast. Thus, the timely evaluation of a new mathematical model using a novel specialized approximation technique in an industrial application context is essentially not feasible at the moment.

With the PUMA framework we shrink this gap and allow modelling engineers to quickly implement various mathematical models, and at the same time to improve the approximation properties of the employed discretization techniques in a stable

M.A. Schweitzer (✉)
Fraunhofer Institute for Algorithms and Scientific Computing SCAI, Schloss Birlinghoven, 53757 Sankt Augustin, Germany

Institute for Numerical Simulation, Rheinische Friedrich-Wilhelms-Universität Bonn, Wegelerstr. 6, 53115 Bonn, Germany
e-mail: marc.alexander.schweitzer@scai.fraunhofer.de

A. Ziegenhagel
Fraunhofer Institute for Algorithms and Scientific Computing SCAI, Schloss Birlinghoven, 53757 Sankt Augustin, Germany

© Springer International Publishing AG 2017
M. Griebel et al. (eds.), *Scientific Computing and Algorithms in Industrial Simulations*, DOI 10.1007/978-3-319-62458-7_11

fashion for each model individually. To this end, PUMA is based on the partition of unity method (PUM) [5, 6, 24, 25] which is a generalization of the classical finite element method (FEM), the current workhorse of industrial simulation. It allows for the use of problem-dependent approximation functions to attain higher order and fast convergence independent of the regularity of the problem at hand. To enable the user to quickly setup a new simulation application or mathematical model, PUMA provides its own expression language, see also e.g. [1, 8, 9, 17, 20, 21], which is then compiled into efficient C++ code by PUMA's general expression compiler (GECO).

In this paper, we give an introduction into PUMA's capabilities and demonstrate its ease of use via some sample applications from hydraulic fracturing and large deformation dynamics. We begin with a short introduction to the PUM which is the mathematical foundation of PUMA in the following section before we give an overview of the design of the PUMA software framework in Sect. 3. In Sect. 4 we present some example applications showcasing the straightforward realization of stable enriched approximations with PUMA. Moreover, we show that very accurate approximations can be attained simply by providing some problem-dependent or physics-based enrichment functions.

2   Partition of Unity Methods

The partition of unity method (PUM) was introduced in [5, 6] as an approach to generalize the classical finite element method (FEM), see also [4]. The abstract ingredients which make up a PUM space

$$V^{\mathrm{PU}} := \sum_{i=1}^{N} \varphi_i V_i = \mathrm{span}\langle \varphi_i \vartheta_i^m \rangle \tag{1}$$

are a partition of unity (PU) $\{\varphi_i : i = 1, \ldots, N\}$ and a collection of local approximation spaces $V_i := V_i(\omega_i) := \mathrm{span}\langle \vartheta_i^m \rangle_{m=1}^{d_{V_i}}$ defined on the patches $\omega_i := \mathrm{supp}(\varphi_i)$ for $i = 1, \ldots, N$. Thus, the shape functions of a PUM space are simply defined as the products of the PU functions φ_i and the local approximation functions ϑ_i^m. The PU functions provide the locality and global regularity of the product functions $\varphi_i \vartheta_i^m$ whereas the functions ϑ_i^m equip V^{PU} with its approximation power [6, 31]. Note that there are no constraints imposed on the choice of the local spaces V_i, i.e., they are completely independent of each other, which makes the PUM an extremely versatile and highly efficient approximation technique.

Unlike in the classical finite element method (FEM), where the shape functions are piecewise polynomials, the PUM allows for the use of problem-dependent local approximation spaces

$$V_i := \mathcal{P}_i + \mathcal{E}_i = \mathrm{span}\langle \psi_i^t, \eta_i^s \rangle = \mathrm{span}\langle \vartheta_i^m \rangle \tag{2}$$

which are usually comprised of a space of polynomials $\mathcal{P}_i$ and a so-called enrichment space $\mathcal{E}_i$. The role of these enrichment spaces is to resolve special local behavior of the solution, e.g. discontinuities or singularities, to obtain a convergence behavior that is independent of the regularity of the sought solution. While this improved approximation property holds in principle for any PUM, the stability of the respective PUM is strongly dependent on the concrete choice of the employed PU [11, 18, 30].

Choosing classical piecewise linear finite element (FE) functions for the PU, we obtain the so-called extended or generalized finite element method (XFEM/GFEM) [7, 22, 35, 36]. Even though this realization of a PUM is probably the most widely available approach it imposes strong limitations on the allowable local approximation spaces to attain the stability [3, 16, 38] of the overall method which obviously has an adverse effect on the versatility of the approach. The particle-partition of unity method (PPUM) developed in [12–15, 24], however, employs a meshfree approach to the construction of the employed PU and in particular advocates the use of a so-called flat-top PU. Due to this construction, the stability of the global shape functions [29] can be ensured for arbitrary enrichments and the PPUM thereby allows to utilize the versatility of the PUM in a stable way to a much greater extent than the XFEM/GFEM [3, 16, 38]. Thus, we pursue the latter approach in PUMA. Let us quickly summarize the respective construction and ingredients in the following.

The flat-top PU functions φ_i employed in PUMA are non-negative and constructed by Shepard's method [34] subordinate to a cover consisting of tensor-product patches ω_i, which are obtained from an octree construction, see [24, 25] for details. The flat-top property of the PU ensures the stability of the product functions $\varphi_i \vartheta_i^m$ [29] for arbitrary enrichment functions [11, 28, 30]. Due to the non-negativity of the PU functions there is moreover a variational mass lumping scheme [31] which is also applicable to arbitrary local approximation spaces. Within PUMA essential boundary conditions can be applied in a conforming fashion [27] as well as via a non-conforming approach via Nitsche's method [15]. Furthermore, PUMA comes with efficient multilevel solvers [14], subdomain-based error estimators and classical adaptive h-, p-, and hp-refinement [26] which can be employed in conjunction with problem-dependent enrichments.

3 PUMA Framework Design

The PUMA framework, see Fig. 1, employs a multi-layered architecture to hide the internal methodological and implementational complexity from the user. The computational core of the framework is bundled in the PaUnT module which is implemented in ISO-standard C++ to provide platform independence, computational efficiency, and the possibility to interact with many scientific third party libraries. The interface level separates the details of the computational core from the user. Currently, PUMA provides a Python interface consisting of a collection of thin wrappers which provide easy access to the higher level functionality of PaUnT. On the application level, the user implements the simulation application with the

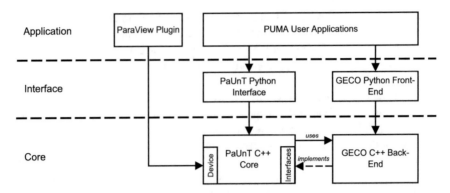

Fig. 1 Components of the multi-layered architecture of the PUMA framework. Published with kind permission of ©Fraunhofer SCAI 2016. All Rights Reserved

help of these Python interfaces which essentially allow to control all aspects of the method but do not require the user to be aware of all details of the PUM or its implementation within the computational core of PUMA.

Let us now take a quick look at each of PUMA's software components, what their specific tasks are, how they are implemented and how they interact with each other:

PaUnT C++ Core: This is the heart of PUMA. It includes the implementation of all the main ingredients of the PUM e.g. the stable construction of function spaces with enrichments, adaptive refinement, handling of geometry, numerical integration for the assembly of the linear systems as well as PUMA's (linear) algebra backend. The core itself is written in modern C++, utilizing software engineering paradigms like object oriented and generic programming. The highly modularized code is easily extendable by implementing well posed class interfaces for key components throughout the system. Moreover, all data structures and components of the computational core are implemented for distributed parallel computing via MPI and provide automatic dynamic load-balancing schemes.

PaUnT Python Interface: This is mostly a thin wrapper on top of the higher level classes and functions of the underlying PaUnT core. Additionally, it encompasses some general and domain-dependent toolkits that provide high level abstraction and convenience functions such as simple coordinate transformations and e.g. various non-linear material models in continuum mechanics.

GECO C++ Back-End: This is the key component of PUMA's user interface. GECO (Generic Expression COmpiler) is an optimizing expression compiler operating on an abstract intermediate representation (IR) for mathematical expressions. It includes code generators for PaUnT's weak formulations, enrichments and general scalar/vector field expressions that generate efficient C++ code and implement the respective class interfaces provided by the PaUnT core. In a second pass, a C++ compiler is invoked that translates the generated source code into shared libraries which are then consumed by PUMA during the run time of a simulation.

GECO Python Front-End: This component allows the user to specify mathematical expressions via PUMA's own expression language in a notation that is very close to typical mathematical formulations, see also [1, 8, 9, 17]. By utilizing Python operator overloading and additional Python functions this component first stores the user provided expressions in an abstract syntax tree that is then translated into the IR for the GECO Back-End.

ParaView Plugin: This component integrates into the very powerful visualization and post-processing tool ParaView [2]. It allows ParaView to access PUMA simulation results with the full resolution of the underlying function space (including user defined arbitrary enrichments) by making the PaUnT core directly available in ParaView. Therefore, it enables the use of a highly compact and memory-efficient data format to store PUMA's simulation results on disk for later post-processing.

4   Application Examples

In this section we present some examples showing the ease of use of PUMA as well as the versatility and improved approximation properties of the PUM in general.

Example 1 (Fracture Mechanics)   Consider the fracture mechanics problem

$$
\begin{aligned}
- \mathbf{div}\, \sigma(\mathbf{u}) &= 0 \quad \text{in } \Omega = (-1, 1)^2, \\
\mathbf{u} &= \mathbf{g}_D \text{ on } \Gamma_D, \\
\sigma(\mathbf{u}) \cdot \mathbf{n} &= \mathbf{g}_N \text{ on } \Gamma_N = \partial\Omega \setminus \Gamma_D, \\
\sigma(\mathbf{u}) \cdot n &= 0 \quad \text{on } C \subset \Omega
\end{aligned}
\tag{3}
$$

in two dimensions with $C = \{(x, y) \in \Omega \,|\, y = 0, x \in (-1, 0)\}$ in a homogeneous and isotropic linearly elastic material; i.e., the stress is given by $\sigma(\mathbf{u}) := C\varepsilon(\mathbf{u}) = 2\mu\varepsilon(\mathbf{u}) + \lambda\,\text{trace}(\varepsilon(\mathbf{u}))\mathbb{I}$ with the strain $\varepsilon(u) := \frac{1}{2}(\nabla\mathbf{u} + (\nabla\mathbf{u})^T)$. The point $T = (0, 0)$ is referred to as the crack tip.

It is well known that the approximation of (3) poses several challenges for classical FEM which must be addressed by adaptive mesh-refinement: The solution u exhibits a discontinuity across the crack C and is singular in the crack tip T. In the PUM, however, these issues are rather simple to resolve by the use of available appropriate enrichment functions which are obtained from the asymptotic expansion of u in the vicinity of the crack tip T, see e.g. [37]. To this end, a PUM patch ω_i that is completely cut by the crack C, i.e. that does not contain the crack tip T, is enriched with the help of the discontinuous Heaviside function

$$
H_{\pm}^C(x) = \begin{cases} 1 & \text{if } x \cdot n_C > 0 \\ -1 & \text{else} \end{cases},
\tag{4}
$$

where n_C denotes the normal to the crack C, so that

$$\mathcal{E}_i := H_\pm^C \mathcal{P}^{p_i} \quad \text{and} \quad V_i := \mathcal{P}^{p_i} + \mathcal{E}_i \tag{5}$$

is employed on these patches. If the patch ω_i contains the crack tip T, then the patch is enriched by the respective space of singular tip functions

$$\mathcal{E}_{\text{tip}} := \text{span}\langle \sqrt{r}\cos\frac{\theta}{2}, \sqrt{r}\sin\frac{\theta}{2}, \sqrt{r}\sin\theta\sin\frac{\theta}{2}, \sqrt{r}\sin\theta\cos\frac{\theta}{2}\rangle \tag{6}$$

given in local polar coordinates with respect to the tip T to resolve the singularity of the solution u at T. Moreover, we enrich the patches in the vicinity of the crack tip T by (6) to ensure an optimal convergence behavior of the global overall scheme with simple uniform h-refinement, i.e., if $\text{dist}(T, \omega_i) < \rho_\mathcal{E}$ for some fixed $\rho_\mathcal{E} > 0$, we use

$$\mathcal{E}_i := \mathcal{E}_{\text{tip}}|_{\omega_i} \quad \text{and} \quad V_i := \mathcal{P}^{p_i} + \mathcal{E}_i. \tag{7}$$

It is however well-known [29, 30] that using this enrichment approach directly leads to ill-conditioned or singular stiffness matrices. Thus, PUMA provides an automatic transformation or preconditioner [29] to allow for a stable discretization regardless of the enrichment scheme and the enrichment functions employed. The Python script given in Algorithm 1 shows how this stable enriched approximation of (3) can be set up easily within PUMA.

Moreover, PUMA has a built-in multilevel solver [14, 28] applicable to arbitrarily enriched discretizations that allows for the efficient solution of the resulting stiffness matrices which is directly accessible with some minor modifications (see Algorithm 2) to the script given in Algorithm 1. The obtained convergence history for a conjugate gradient solver preconditioned by a V(2,2)-multilevel cycle is depicted in the plots given in Fig. 2. From these plots we can clearly observe the anticipated optimal convergence behavior of PUMA's multilevel solver for the enriched discretization considered. All components of the PUMA framework are furthermore fully parallel and the scalability of PUMA is close to optimal, see Fig. 2.

After the approximate solution is obtained efficiently, it must be available for domain-dependent post-processing purposes. To this end, PUMA provides a plugin to ParaView [2] which allows to load PUMA's native PNT file format and gives ParaView access to the enriched discretization space V^{PU} so that all of ParaView's filters, etc. can be utilized in the post-processing of the computed data, compare Fig. 3.

The usual quantities of interest in linear elastic fracture mechanics are the stress intensity factors (SIFs) K^I and K^{II} which are computed directly from the approximate solution $\mathbf{u}^{\text{PU}}$ via some extraction technique such as the J-integral [10, 23] or the contour integral method (CIM) [37] and not via post-processing tools such as ParaView. Thus, PUMA's solid mechanics toolbox comes with various extraction techniques. In the following, we compute the approximate SIFs K_I^{PU} and

---

Algorithm 1: PUMA Python script for Example 1

```
# Import packages
from puma import *
# Create domain and boundary segments and crack
omega = RectangleDomain(Point(-1.,-1.), Point(1.,1.))
gamma_D = PhysicalGroupSubDomain(omega, "Right", "Bottom")
gamma_N = PhysicalGroupSubDomain(omega, "Top")
crack = CrackPolyline(omega, Point(-1.1,0.), Point(0.,0.))
# Set material properties
material = LinearElasticMaterialModel(omega, E=1, nu=0.3)
# Create function space
V = PumSpace(omega, level=5, polynomial_degree=1, polynomial_type=
    ↪ 'Legendre Tensor Product')
# Create enrichments and enrich function space
x = Position()
T = Translation(-crack.second_tip()) & Rotation(crack.
    ↪ second_tip_direction())
near_crack_tip = distance_point_box(crack.second_tip(),
    ↪ PatchStretchedDomain()) < 0.5
V.enrich(sintheta2(T(x)), EnrichmentApplicationType.additive,
    ↪ near_crack_tip)
V.enrich(costheta2(T(x)), EnrichmentApplicationType.additive,
    ↪ near_crack_tip)
V.enrich(sintheta2sintheta(T(x)), EnrichmentApplicationType.
    ↪ additive, near_crack_tip)
V.enrich(costheta2sintheta(T(x)), EnrichmentApplicationType.
    ↪ additive, near_crack_tip)
at_crack = intersects_box_line(PatchStretchedDomain(), crack.
    ↪ first_tip(), crack.second_tip()) & Not(within_point_box(
    ↪ crack.second_tip(), PatchStretchedDomain()))
V.enrich(heaviside(T(x)), EnrichmentApplicationType.multiplicative,
    ↪  at_crack)
# Create transformation to stable basis
u = VectorTrialFunction(V)
v = VectorTestFunction(V)
st = create_stable_transformation((dot(u,v)+inner(grad(u),grad(v)))
    ↪ *dx)
# Define boundary conditions
g_D = (0,0)
bcs = [DirichletBoundaryCondition(V, gamma_D, g_D)]
# Define loading conditions
f = (0,-0.1)
n = Normal()
g_N = 0.1 * n
# Define variational problem
u = VectorTrialFunction(V)
v = VectorTestFunction(V)
a = material.second_variation_of_strain_energy(u,v)
L = dot(f,v)*dx + dot(g_N,v)*ds(gamma_N)
# Compute solution
u = VectorFunction(V)
solve(a == L, u, bcs, st)
# Output solution
write_to_pnt("crack_solution.pnt", (u, "u"), material)
```

Algorithm 2: PUMA Python interface to access the multilevel solver

```
# Create function space
coarsest_level = 2
finest_level   = 6
V = MultilevelPumSpace(omega, level=finest_level, polynomial_degree=1,
    ↪ polynomial_type='Legendre Tensor Product')

.

.

# Compute solution
u = VectorFunction(V)
solver_parameters = {
                      'linear_solver' :
                      {
                          'type' : 'cg',
                          'tolerance' : 1e-15,
                          'preconditioner' :
                          {
                              'type' : 'multilevel',
                              'quiet' : 'False',
                              'coarsest_level' : coarsest_level,
                              'pre_smoothing_steps' : 2,
                              'post_smoothing_steps' : 2,
                              'coarse_level_solver' : {'type' : 'lu'}
                          }
                      }
                    }
solve(a == L, u, bcs, st, solver_parameters=solver_parameters)
```

K_{II}^{PU} obtained from $\mathbf{u}^{\mathrm{PU}}$ via the CIM for several extraction radii to validate the convergence properties of our PUM scheme. To this end, we choose the boundary data in (3) such that the solution to (3) is given by

$$\mathbf{u} = \begin{pmatrix} u_x \\ u_y \end{pmatrix} = \frac{\sqrt{r}}{2\mu\sqrt{2\pi}}\left(K_I\mathbf{\Psi}_1^I(\theta) + r\mathbf{\Psi}_2^I(\theta) + K_{II}\mathbf{\Psi}_1^{II}(\theta) + r\mathbf{\Psi}_2^{II}(\theta)\right) \qquad (8)$$

with $K^I = K^{II} = 1$ and

$$\mathbf{\Psi}_i^I(\theta) := \begin{pmatrix} (\kappa - Q_i^I(\lambda_i^I + 1))\cos(\lambda_i^I\theta) - \lambda_i^I\cos((\lambda_i^I - 2)\theta) \\ (\kappa + Q_i^I(\lambda_i^I + 1))\sin(\lambda_i^I\theta) + \lambda_i^I\sin((\lambda_i^I - 2)\theta) \end{pmatrix},$$

$$\mathbf{\Psi}_i^{II}(\theta) := \begin{pmatrix} (\kappa - Q_i^{II}(\lambda_i^{II} + 1))\sin(\lambda_i^{II}\theta) - \lambda_i^{II}\sin((\lambda_i^{II} - 2)\theta) \\ -(\kappa + Q_i^{II}(\lambda_i^{II} + 1))\cos(\lambda_i^{II}\theta) - \lambda_i^{II}\cos((\lambda_i^{II} - 2)\theta) \end{pmatrix},$$

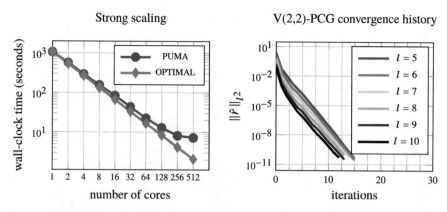

Fig. 2 Strong parallel scaling of PUMA, including discretization and multilevel solution for a scalar PDE problem using DOF= 3,145,728 only. Convergence history of the residual $\|\hat{r}\|_{l^2}$ for a conjugate gradient solver preconditioned with a V(2,2) cycle of PUMA's built-in multilevel iteration up to level $l = 10$, i.e., a total number of stable basis functions of DOF=6,602,594 including 311,138 non-polynomial enrichment functions. Published with kind permission of ©Fraunhofer SCAI 2016. All Rights Reserved

where $\lambda_i^I = \lambda_i^{II} = \frac{1}{2}, \frac{3}{2}, 2, \frac{5}{2}, \ldots$ and

$$Q_i^I := -\frac{\lambda_i^I - 1}{\lambda_i^I + 1} \frac{\sin((\lambda_i^I - 1)\pi)}{\sin((\lambda_i^I + 1)\pi)}, \quad Q_i^{II} := -\frac{\sin((\lambda_i^{II} - 1)\pi)}{\sin((\lambda_i^{II} + 1)\pi)},$$

and $\kappa = 3 - \frac{2\lambda}{\lambda+\mu}$ denotes the material's Kolosov constant. The relative errors $e_I := |K_I^{PU} - K_I|/K_I$ and $e_{II} := |K_{II}^{PU} - K_{II}|/K_{II}$ associated with the extracted approximate SIFs K_I^{PU} and K_{II}^{PU}, respectively, are given in the plots depicted in Fig. 4. Here, we employed four different extraction radii $\rho_E^0 = \frac{4}{3}\rho_E$, $\rho_E^1 = \rho_E$, $\rho_E^2 = \frac{1}{2}\rho_E$, $\rho_E^3 = \frac{1}{4}\rho_E$ to demonstrate the superior convergence behavior due to enrichment. From these plots and measured convergence rates τ given in Table 1 we can clearly observe that the extracted SIFs converge with the optimal rate of $\tau \approx 0.5$ due to enrichment for the larger extraction radii ρ_E^0 and ρ_E^1, i.e. far away from the crack tip singularity where SIFs are usually extracted in classical FEM.[1] For the smaller extraction radii ρ_E^2 and ρ_E^3, however, we find a much better convergence behavior with $\tau \approx 1$, i.e. a local super-convergence, for our PUM [29] due to the use of many enrichment functions within a large zone around the crack tip. Thus, in PUMA we can compute the quantities of interest with higher order convergence and much less degrees of freedom than in classical FEM without compromising the stability of the discretization. The contour plot of the dimension of the local enriched approximation spaces $\dim(V_i)$ depicted in Fig. 5 (right) shows the effect of PUMA's

---

[1]Note that only for the largest extraction radius the SIFs are computed from local approximation spaces without any enrichments.

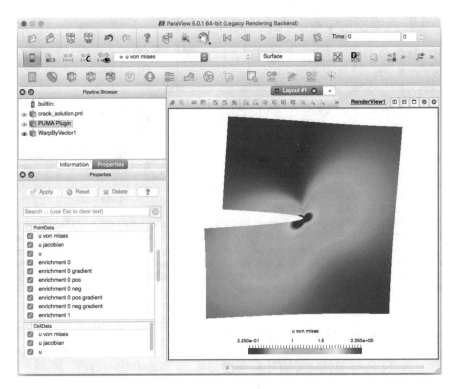

Fig. 3 Post-processing of an enriched PUMA approximation via ParaView. PUMA provides a plugin for ParaView so that PUMA's native PNT file format can be read and any enriched discretization space becomes available in ParaView. Thus, the enriched approximation can be evaluated at an arbitrary location and all ParaView filters, etc. can be utilized. Published with kind permission of ©Fraunhofer SCAI 2016. All Rights Reserved

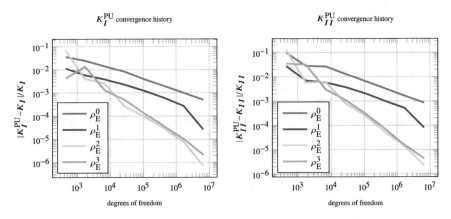

Fig. 4 Convergence history of K_I^{PU} and K_{II}^{PU} for the different extraction radii ρ_E^i with $i = 0, 1, 2, 3$. Published with kind permission of ©Fraunhofer SCAI 2016. All Rights Reserved

Table 1 Convergence history of K_I^{PU} and K_{II}^{PU}

D_{OF}	τ_I^0	τ_{II}^0	τ_I^1	τ_{II}^1	τ_I^2	τ_{II}^2	τ_I^3	τ_{II}^3
1748	0.3	0.9	0.5	1.0	2.1	2.3	−0.9	0.1
6836	0.4	0.0	0.3	0.1	0.3	0.0	1.7	1.7
26,996	0.4	0.5	0.4	0.3	1.6	1.5	0.7	0.7
107,212	0.6	0.5	0.5	0.4	0.8	0.8	1.0	0.9
422,300	0.5	0.5	0.5	0.5	0.8	1.1	0.9	1.0
1,676,884	0.5	0.5	0.6	0.5	1.0	1.0	0.9	1.1
6,602,594	0.5	0.5	1.7	1.4	1.7	1.3	1.2	0.9

Given are the number of degrees of freedom D_{OF} and the respective convergence rates $\tau^i := \log\left(\frac{e_l^i}{e_{l-1}^i}\right)/\log\left(\frac{D_{OF_l}}{D_{OF_{l-1}}}\right)$ for the different extraction radii ρ_E^i with $i = 0, 1, 2, 3$ where $e^i := |K^{PU,i} - K|/K$

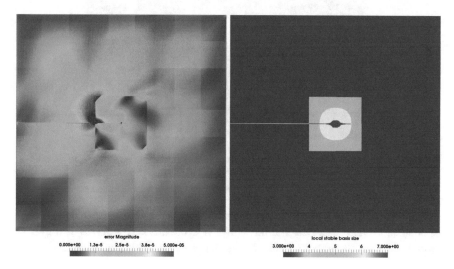

error Magnitude

0.000e+00   1.3e-5   2.5e-5   3.8e-5   5.000e-05

local stable basis size

3.000e+00   4   5   6   7.000e+00

Fig. 5 Contour plot of the magnitude of the error $\|\mathbf{u}(x) - \mathbf{u}_J^{PU}(x)\|_{\mathbb{R}^2}$ (*left*) and the distribution of the enrichment degrees of freedom (*right*) on level $J = 10$ ($\dim(V_i) \geq 7$ (*red*), $\dim(V_i) = 6$ (*yellow*), $\dim(V_i) = 3$ (*blue*)). Here, we can clearly observe the elimination of certain enrichment functions away from the crack tip to ensure stability. Published with kind permission of ©Fraunhofer SCAI 2016. All Rights Reserved

stability transformation which eliminates linearly dependent enrichment functions and yields a well-conditioned stiffness matrix while retaining the improved local approximation power due to the enrichment, compare Fig. 5 (left).

Thus, in a quasi-static crack growth application, as depicted in Fig. 6, PUMA requires only a very small number of degrees of freedom to provide high fidelity results. Note also that even though there is no closed form asymptotic expansion in three space dimensions for (3) we can still construct appropriate enrichments in three dimensions based on the two dimensional expansion, compare Fig. 7.

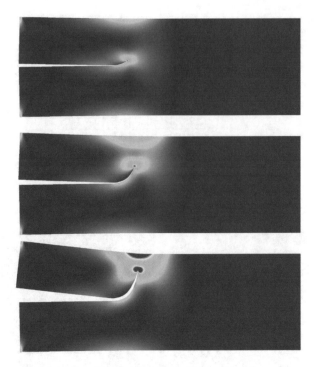

Fig. 6 Contour plots of the von Mises stress for several snapshots of a crack growth simulation. Published with kind permission of ©Fraunhofer SCAI 2016. All Rights Reserved

Moreover, PUMA's block-operations allow the user to implement the treatment of coupled systems of PDEs and multiphysics problems easily. As an example, let us consider the following hydraulic fracturing problem [19]: Find (u, p) such that

$$
\langle \sigma(u), \epsilon(v) \rangle_\Omega + \langle \mathrm{cod}(\dot{u}), q \rangle_{\Gamma_f} - \langle \sigma_0 \cdot n, v \rangle_{\partial\Omega} - \langle p, v \rangle_{\Gamma_f} = 0
$$
$$
\frac{1}{\mu_f} \langle \mathrm{cod}(u)^3 \nabla p, \nabla q \rangle_{\Gamma_f} - \langle \frac{Q_0}{2} \delta_0, q \rangle_{\Gamma_f} = 0 \tag{9}
$$

where Γ_f denotes the section of the crack C that is filled with fluid and the crack opening displacement $\mathrm{cod}(u) := (u_+ - u_-) \cdot n_C$ denotes the jump across the crack. Due to the injection of fluid, both Γ_f and C are growing over time and must also be tracked during the simulation. The respective formulation of this nonlinear coupled problem (9) in PUMA is sketched in Algorithm 3. Some snapshots of the computed pressure and crack opening displacement at several time-steps are depicted on the respective deformed crack configuration in Figs. 8 and 9. Again, we observe a high fidelity of the approximation due to the enrichment, i.e., we can use a much smaller number of degrees of freedom in PUMA than in classical FEM to attain the same accuracy. Thus, the run time for this time-dependent nonlinear problem is substantially reduced.

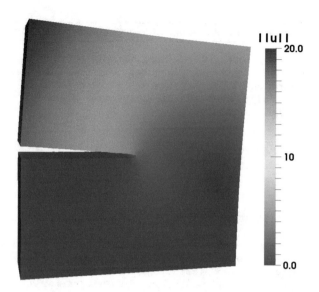

Fig. 7 Contour plots of the approximate displacements for a through the thickness crack in three space dimensions. Published with kind permission of ©Fraunhofer SCAI 2016. All Rights Reserved

Example 2 (Explicit Dynamics) Explicit time-stepping techniques are often necessary in the simulation of highly dynamic processes. The equations of elastodynamics

$$
\begin{aligned}
\rho\ddot{u} - \mathbf{div}\,\sigma(u) &= 0 \text{ in } \Omega \times (0, T), & u(\cdot, 0) &= u_0 \text{ in } \Omega, \\
u(x, t) &= 0 \text{ for } x \in \Gamma_D, t > 0, & \dot{u}(\cdot, 0) &= v_0 \text{ in } \Omega, \\
\sigma(\mathbf{u}) \cdot n &= 0 \text{ for } x \in \partial\Omega \setminus \Gamma_D, t > 0,
\end{aligned}
\tag{10}
$$

for instance, are often discretized by a central difference scheme

$$
\bar{M}\tilde{a}_n = K\tilde{u}_n, \quad \tilde{v}_{n+1/2} = \delta t_{n+1/2}\tilde{a}_n + \tilde{v}_{n-1/2}, \quad \tilde{u}_{n+1} = \delta t_n\tilde{v}_{n+1/2} + \tilde{u}_n, \tag{11}
$$

where $\tilde{u}_{n+1}$ refers to the coefficient-vector of the discrete displacement at time t_{n+1}, $\tilde{v}_{n+1/2}$ is the velocity coefficient-vector at time $t_{n+1/2} = \frac{1}{2}(t_{n+1} + t_n)$, and $\tilde{a}_n$ encodes the acceleration at time t_n. Moreover, $\bar{M}$ denotes either the consistent mass matrix M or an approximation to the mass matrix, a so-called lumped mass matrix, which allows for simple direct inversion. Note that the construction of an appropriate lumped mass matrix is a non-trivial task in classical FEM for higher order elements. Somewhat surprisingly, the PUM allows for the construction of a high-quality lumped mass matrix for arbitrary local approximation spaces, e.g. higher order polynomials with arbitrary enrichments [31], which is directly available in PUMA. To demonstrate the effectiveness of this general lumping approach, we first consider (10) for a Saint Venant-Kirchhoff hyperelastic material with $E = 10^5$ and $v = 0.3$ on $\Omega = (-8, 8) \times (-1, 1) \times (-1, 1)$. We discretize

---

Algorithm 3: PUMA problem formulation for hydraulic fracture problem (9)

```
V = PumSpace(omega_rock, level=displacement_refinement_level,
    ↪ polynomial_degree=displacement_polynomial_degree)
P = PumSpace(crack_fluid, level=pressure_refinement_level,
    ↪ polynomial_degree=pressure_polynomial_degree)
W = ProductFunctionSpace(V, P)
V = W[0]
P = W[1]
u = VectorFunction(V)
p = ScalarFunction(P)
w = (u,p)
.
.
.
R_u = material.first_variation_of_strain_energy(v,u) - L_u -
    ↪ L_u_star
R_p = delta_t*inner(conductivity_u(u)*prime(p), prime(q))*ds(
    ↪ gamma_p) + cod(u)*q*ds(gamma_p) - cod(u_last)*q*ds(gamma_p)
    ↪ - delta_t*Q_0*q*dp(point_0)
J_u_du = material.second_variation_of_strain_energy(deltau,v,None)
J_u_dp = -inner(deltap,cod(v))*ds(gamma_p)
J_p_du = delta_t*inner(Dconductivity_u(u)*cod(deltau)*prime(p),
    ↪ prime(q))*ds(gamma_p) + cod(deltau)*q*ds(gamma_p)
J_p_dp = delta_t*inner(conductivity_u(u)*prime(deltap), prime(q))*
    ↪ ds(gamma_p)
R = (R_u, R_P)
J = ((J_u_du, J_u_dp),
     (J_p_du, J_p_dp))
.
.
.
while time < end_time:
    .
    .
    solve(R == 0, w, bcs, sts, J)
    .
    .
```

---

the problem with cubic Legendre polynomials with $v_0 = 0$ and choose the initial displacement u_0 as the solution to the respective nonlinear equilibrium problem

$$
\begin{aligned}
-\operatorname{\mathbf{div}}\sigma(\mathbf{u}) &= 0 \quad \text{in } \Omega, \\
\mathbf{u} &= 0 \quad \text{on } \Gamma_D, \\
\mathbf{u} &= \mathbf{g}_T \text{ on } \Gamma_T, \\
\sigma(\mathbf{u})\cdot n &= 0 \quad \text{on } \partial\Omega \setminus (\Gamma_D \cup \Gamma_T)
\end{aligned}
\tag{12}
$$

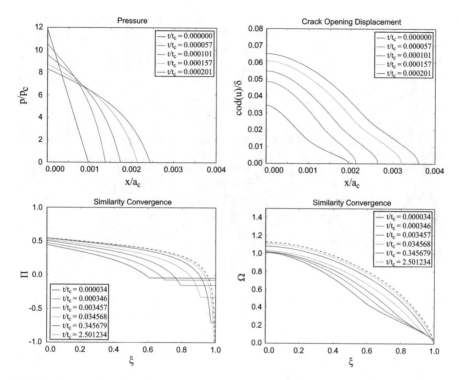

Fig. 8 Computed profiles for the pressure p, the crack opening displacements $\mathrm{cod}(u)$, and the respective dimensionless similarity variables $\Pi(\xi, \tau) := \tau^{1/3} (p/p_c - \sigma_0)$ and $\Omega(\xi, \tau) := \tau^{-1/3} \mathrm{cod}(u)/\delta$ where $\tau := V_0/a_c\delta + t/t_c$ and $\xi := xt_c/a$ denote similarity variables for the time t and the position x with a_c, p_c, t_c, δ depending on the initial position of the fluid tip, the injection rate Q_0, and the material parameters μ_f, E, ν, cmp. [19]. Published with kind permission of ©Fraunhofer SCAI 2016. All Rights Reserved

with $c = (c_0, c_1, c_2)^T = (0, 0, 0)^T \in \mathbb{R}^3$ and $\theta = 8\pi/3$ in

$$
\mathbf{g}_T(x) = \begin{pmatrix} 0 \\ c_1 + (x_1 - c_1)\cos(\theta) - (x_2 - c_2)\sin(\theta) - x_1 \\ c_2 + (x_1 - c_1)\sin(\theta) + (x_2 - c_2)\cos(\theta) - x_2 \end{pmatrix}
$$

and the boundary segments

$$
\Gamma_D := \{x = (x_0, x_1, x_2) \in \partial\Omega : x_0 = -8\}, \quad \Gamma_T := \{x = (x_0, x_1, x_2) \in \partial\Omega : x_0 = 8\}.
$$

The plots depicted in Fig. 10 show a few snapshots of this highly dynamic large deformation simulation with a higher order PUM in space. We can clearly observe that the propagation of the waves is captured in a stable fashion by our lumping scheme for this higher order discretization. Yet, the overall run time for this

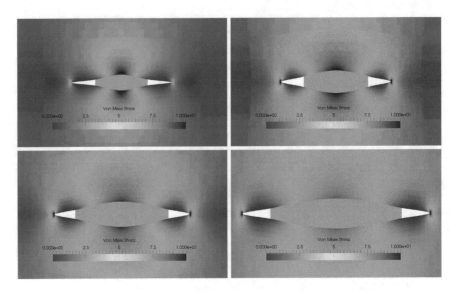

Fig. 9 Some snapshots of the computed fluid propagation and respective crack growth obtained for (9). Published with kind permission of ©Fraunhofer SCAI 2016. All Rights Reserved

simulation with a higher order space discretization is substantially reduced due to PUMA's general mass lumping scheme.

We moreover give some snapshots of a simulation of a slit membrane problem to show that PUMA's lumping technique works equally well even for enriched higher order discretizations. To this end, we consider the scalar wave equation

$$\ddot{u} - \Delta u = 0 \text{ in } \Omega \times (0, T), \qquad u(\cdot, 0) = \gamma \text{ in } \Omega,$$
$$u(x, t) = 0 \text{ for } x \in \partial\Omega, t > 0, \ \dot{u}(\cdot, 0) = 0 \text{ in } \Omega, \tag{13}$$

on a slit domain $\Omega = (-1, 1) \setminus C$ with $C = \{(x, y) : x \in [-1, 0], y = 0\}$ with an initial velocity

$$\gamma(x) = \exp\left(\frac{-\|x - c\|^2}{\sigma^2}\right) \quad \text{with } c = (-0.5, -0.5)^T, \sigma = 0.25. \tag{14}$$

We discretize the problem with our enriched PUM using cubic Legendre polynomials as local approximation spaces V_i throughout the domain and the enriched spaces (5) at the crack C and the spaces (7) in the vicinity of the crack tip $(0, 0)^T$. Thus, in this example the lumping scheme has to cope with higher order basis functions, discontinuous basis functions and singular basis functions. From the obtained numerical results depicted in Fig. 11 we can clearly observe the stability of the PUMA simulation. Moreover, the results of [31] show the superior efficiency of the lumped discretization compared with the consistent approach for arbitrary local approximation spaces e.g. by allowing for a larger critical time-step size.

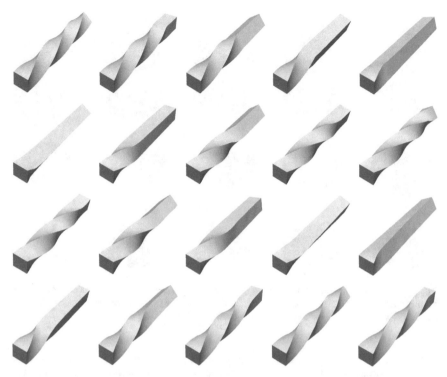

Fig. 10 Deformed material configuration at time steps $n = 200i$ for $i = 0, 1, \ldots, 19$ with constant $\delta t = 10^{-4}$. The bar undergoes a full cycle in deformation in roughly 3600 timesteps, i.e. the respective wave has a period of approximately 0.36. Published with kind permission of ©Fraunhofer SCAI 2016. All Rights Reserved

---

Algorithm 4: PUMA Python interface to access explicit time stepping schemes

---

```
deltau = VectorTrialFunction(V, with_phi=False)
v = VectorTestFunction(V)
lumped_mass_bilinear_form = inner(deltau, v)*dx
time.assign(0)
internal_forces_linear_form = material.
    ↪ first_variation_of_strain_energy(v, u)
internal_forces_bilinear_form = material.
    ↪ second_variation_of_strain_energy(deltau, v, u)
integrator = CentralDifferenceIntegrator(lumped_mass_bilinear_form,
    ↪ internal_forces_linear_form, internal_forces_bilinear_form,
    ↪ external_forces_linear_form, u, udot, uddot, bcs, g, gdot, st)
time_step = 0
end_time = time.value() + 10
while time.value() < end_time:
    integrator.step(time.value(), deltat, time_step)
    time.assign(time.value() + deltat)
    time_step += 1
```

Fig. 11 Deformed material configuration (exaggerated by a factor of 100 and color coded by the absolute value of the velocity) $n = 100 + 200i$ for $i = 0, 1, \ldots, 19$ with constant $\delta t = 10^{-3}$. Published with kind permission of ©Fraunhofer SCAI 2016. All Rights Reserved

Altogether, PUMA's general mass lumping scheme reduces the overall run time for this simulation with an enriched higher order space discretization dramatically.

5   Concluding Remarks

In this paper we presented the basic building blocks of the PUMA software framework. PUMA is designed to allow for the rapid development of simulation applications using generalized finite element techniques based on the partition of unity method and thus can directly utilize user insight, domain-specific information and physics-based basis functions to reduce the computational cost substantially. We demonstrated some of PUMA's capabilities and high fidelity approximation properties in selected applications. However, PUMA has many more uses. For instance, it can be utilized to embed its improved approximation qualities locally into any commercial FEM package [33] and it also allows for the incorporation of numerically computed enrichment functions, e.g., multiscale enrichment functions [32]. Overall, PUMA allows modelling engineers to quickly implement various mathematical models, and at the same time to improve the approximation properties of the employed discretization techniques in a stable fashion for each model individually and thereby enables the rapid evaluation of novel models in an industrial application setting.

References

1. M.S. Alnæs, J. Blechta, J. Hake et al., The FEniCS project version 1.5. Arch. Numer. Softw. **3**, 9–23 (2015)
2. U. Ayachit, *The ParaView Guide: A Parallel Visualization Application* (Kitware, New York, 2015)
3. I. Babuška, U. Banerjee, Stable generalized finite element method (SGFEM). Comput. Methods Appl. Mech. Eng. **201–204**, 91–111 (2012)
4. I. Babuška, G. Caloz, J.E. Osborn, Special finite element methods for a class of second order elliptic problems with rough coefficients. SIAM J. Numer. Anal. **31**, 945–981 (1994)
5. I. Babuška, J.M. Melenk, The partition of unity finite element method: basic theory and applications. Comput. Methods Appl. Mech. Eng. **139**, 289–314 (1996). Special Issue on Meshless Methods
6. I. Babuška, J.M. Melenk, The partition of unity method. Int. J. Numer. Methods Eng. **40**, 727–758 (1997)
7. T. Belytschko, T. Black, Elastic crack growth in finite elements with minimal remeshing. Int. J. Numer. Methods Eng. **45**, 601–620 (1999)
8. G.L. Bernstein, F. Kjolstad, Why new programming languages for simulation? ACM Trans. Graph. **35**, 20e:1–20e:3 (2016)
9. G.L. Bernstein, C. Shah, C. Lemire et al., Ebb: a DSL for physical simulation on CPUs and GPUs. ACM Trans. Graph. **35**, 21:1–21:12 (2016)
10. G.P. Cherepanov, The propagation of cracks in a continuous medium. J. Appl. Math. Mech. **31**, 503–512 (1967)
11. T.-P. Fries, T. Belytschko, The extended/generalized finite element method: an overview of the method and its applications. Int. J. Numer. Methods Eng. **84**, 253–304 (2010)
12. M. Griebel, M.A. Schweitzer, A particle-partition of unity method for the solution of elliptic, parabolic and hyperbolic PDE. SIAM J. Sci. Comput. **22**, 853–890 (2000)
13. M. Griebel, M.A. Schweitzer, A particle-partition of unity method—Part II: efficient cover construction and reliable integration. SIAM J. Sci. Comput. **23**, 1655–1682 (2002)
14. M. Griebel, M.A. Schweitzer, A particle-partition of unity method—Part III: a multilevel solver. SIAM J. Sci. Comput. **24**, 377–409 (2002)
15. M. Griebel, M.A. Schweitzer, A particle-partition of unity method—Part V: boundary conditions, in *Geometric Analysis and Nonlinear Partial Differential Equations*, ed. by S. Hildebrandt, H. Karcher (Springer, Berlin, 2002), pp. 517–540
16. V. Gupta, C. Duarte, I. Babuška, U. Banerjee, Stable GFEM (SGFEM): improved conditioning and accuracy of GFEM/XFEM for three-dimensional fracture mechanics. Comput. Methods Appl. Mech. Eng. **289**, 355–386 (2015)
17. F. Hecht, New development in FreeFem++. J. Numer. Math. **20**, 251–265 (2012)
18. A. Huerta, T. Belytschko, T. Fernández-Méndez, T. Rabczuk, *Meshfree Methods*. Encyclopedia of Computational Mechanics, vol. 1, chap. 10 (Wiley, New York, 2004), pp. 279–309
19. M.J. Hunsweck, Y. Shen, A.J. Lew, A finite element approach to the simulation of hydraulic fractures with lag. Int. J. Numer. Anal. Methods Geomech. **37**, 993–1015 (2013)
20. F. Kjolstad, S. Kamil, J. Ragan-Kelley et al., Simit: a language for physical simulation. ACM Trans. Graph. **35** 20:1–20:21 (2016)
21. A. Logg, K.-A. Mardal, G.N. Wells et al., *Automated Solution of Differential Equations by the Finite Element Method* (Springer, Berlin, 2012)
22. N. Moës, J. Dolbow, T. Belytschko, A finite element method for crack growth without remeshing. Int. J. Numer. Methods Eng. **46**, 131–150 (1999)
23. J.R. Rice, A path independent integral and the approximate analysis of strain concentration by notches and cracks. J. Appl. Mech. **35**, 379–386 (1968)
24. M.A. Schweitzer, *A Parallel Multilevel Partition of Unity Method for Elliptic Partial Differential Equations*. Lecture Notes in Computational Science and Engineering, vol. 29 (Springer, Berlin, 2003)

25. M.A. Schweitzer, *Meshfree and Generalized Finite Element Methods*. Habilitation, Institute for Numerical Simulation, University of Bonn (2008)
26. M.A. Schweitzer, An adaptive hp-version of the multilevel particle–partition of unity method. Comput. Methods Appl. Mech. Eng. **198**, 1260–1272 (2009)
27. M.A. Schweitzer, An algebraic treatment of essential boundary conditions in the particle–partition of unity method. SIAM J. Sci. Comput. **31**, 1581–1602 (2009)
28. M.A. Schweitzer, Multilevel particle–partition of unity method. Numer. Math. **118**, 307–328 (2011)
29. M.A. Schweitzer, Stable enrichment and local preconditioning in the particle–partition of unity method. Numer. Math. **118**, 137–170 (2011)
30. M.A. Schweitzer, Generalizations of the finite element method. Cent. Eur. J. Math. **10**, 3–24 (2012)
31. M.A. Schweitzer, Variational mass lumping in the partition of unity method. SIAM J. Sci. Comput. **35**, A1073–A1097 (2013)
32. M.A. Schweitzer, S. Wu, Evaluation of local multiscale approximation spaces for partition of unity methods, in *Meshfree Methods for Partial Differential Equations VIII*, ed. by M. Griebel, M.A. Schweitzer. Lecture Notes in Science and Engineering, vol. 115 (2017), pp. 163–194
33. M.A. Schweitzer, A. Ziegenhagel, Embedding enriched partition of unity approximations in finite element simulations, in *Meshfree Methods for Partial Differential Equations VIII*, ed. by M. Griebel, M.A. Schweitzer. Lecture Notes in Science and Engineering, vol. 115 (2017), pp. 195–204
34. D. Shepard, A two-dimensional interpolation function for irregularly spaced data, in *Proceedings 1968 ACM National Conference* (ACM, New York, 1968), pp. 517–524
35. T. Strouboulis, I. Babuška, K. Copps, The design and analysis of the generalized finite element method. Comput. Methods Appl. Mech. Eng. **181**, 43–69 (2000)
36. T. Strouboulis, K. Copps, I. Babuška, The generalized finite element method. Comput. Methods Appl. Mech. Eng. **190**, 4081–4193 (2001)
37. B. Szabó, I. Babuška, *Finite Element Analysis* (Wiley, New York, 1991)
38. Q. Zhang, U. Banerjee, I. Babuška, Higher order stable generalized finite element method. Numer. Math. **128**, 1–29 (2014)

Part III
Applications and Show Cases

Applying CFD for the Design of an Air-Liquid Interface In-Vitro Testing Method for Inhalable Compounds

Carsten Brodbeck, Jan Knebel, Detlef Ritter, and Klaus Wolf

1 Introduction

Within the last decade the air-liquid interface (ALI) cell culture technology became the state-of-the-art method for in-vitro testing of airborne substances. Cells are cultured on microporous membranes and thereby get efficiently into contact with the test atmosphere apically while being supplied with nutrients and being humidified from the basal side of the membrane. Biological models like cell lines or primary cells from the human lung or complex ex-vivo models like precision cut lung slices (PCLS) can be applied. However, especially for the application of this technology on the testing of a broader range of airborne materials like droplet and particle aerosols, there is still no general scientific consensus on the question of the most suitable exposure design. Concepts including complex physical effects like electrostatic deposition of particles or thermophoresis have been introduced to address specific scientific and technical questions in this context. With the objective of improving the understanding and establishing a concept for a general application of the ALI culture technology for the testing of varying kinds of airborne materials, this study aimed, by the use of computational fluid dynamics (CFD), at (1) the characterization of the influence of the aerosol conduction system on the particle deposition, (2) an analysis of the behavior of a liquid supply system and (3) an investigation of a procedure for sampling laboratory generated test aerosols.

C. Brodbeck • K. Wolf (✉)
Fraunhofer Institute for Algorithms and Scientific Computing SCAI, Schloss Birlinghoven, Sankt Augustin 53757, Germany
e-mail: klaus.wolf@scai.fraunhofer.de

J. Knebel • D. Ritter
Fraunhofer Institute for Toxicology and Experimental Medicine ITEM, Nikolai-Fuchs-Straße 1, 30625 Hannover, Germany

© Springer International Publishing AG 2017
M. Griebel et al. (eds.), *Scientific Computing and Algorithms in Industrial Simulations*, DOI 10.1007/978-3-319-62458-7_12

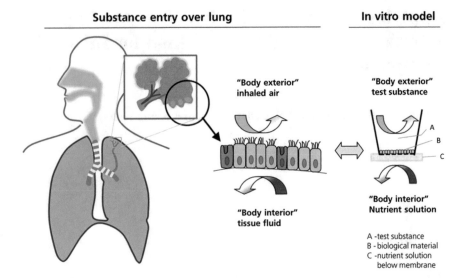

Fig. 1 In-vitro models such as air-liquid interface cultures replace the real conditions in a human lung [4]. Published with kind permission of ©Fraunhofer-Gesellschaft 2016. All Rights Reserved

2   In-Vitro Air-Liquid Interface

An in-vitro air-liquid interface culture reproduces typical human physical barrier systems, such as the intestinal mucosa or bronchial epithelial cells in the lung in a laboratory scaled setup (Fig. 1). A nutrient solution ensures that the applied cells are kept alive. The procedures to investigate substances with in-vitro systems are generally divided into three single components: The "cell system" (cell lines, primary cells or other biological models), the "exposure step" (where the concrete contact between the biological model and the test substance takes place) and the "read out" (the investigation of the biological effect of the test substance on the biological model by measurement of a cellular parameter such as "viability"). These three elements are fundamental for the establishment of any efficient and meaningful test system. The optimization of each individual step and the integration of air-liquid interface cultures into the overall process comprises the potential for a powerful test system with high integrity, efficiency and biological relevance.

In contrast to this aim, the exposure step is generally organized as an isolated step in usual experimental cell-based processes during testing of airborne materials. This leads to an unfavorable, time-consuming, expensive and cell-damaging handling of the biological test system. As one fundamental improvement amongst others of this common situation, the air-liquid interface P.R.I.T.® ExpoCube® applied in this project (Fig. 2) enables a completely integrated workflow based on an "all-in-one-plate" concept by the use of standard multiwell-plates throughout the whole experimental process.

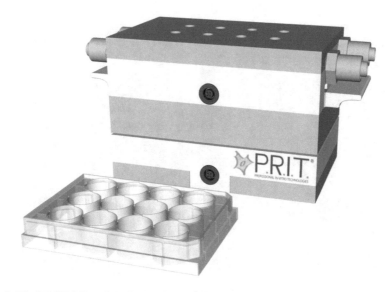

Fig. 2 The P.R.I.T.® ExpoCube® with a standard multiwell plate. Published with kind permission of ©Fraunhofer-Gesellschaft 2016. All Rights Reserved

3 Simulating the Aerosol Conduction System

The test aerosol is conducted to the air-liquid interface in order to bring it into contact with the human cells, as shown in Fig. 3. Briefly, it is transported through the supply pipe with a higher volume flow and conducted into the deposition areas using lower exposure flows. Exposure volume flows are limited in order to reduce physical stress during the exposure of the cells and to conserve their viability. As encountered in many biomedical applications, it is therefore necessary to branch off a defined amount of fluid. In this context, it is not only the aim to increase the branched off amount of aerosols to get a more obvious and faster aerosol deposition, but it is also desired to get a homogeneous distribution into the multiple branches and on the surfaces which are covered with the human cells. For the present design of the exposure device the following questions were stated:

1. Which design of the drain of the deposition device will cause a more homogeneous flow distribution?
2. How should the deposition device geometry be modified to enhance particle deposition on cell membranes?
3. Is there a possibility to enhance particle mass flow into the branches feeding the deposition devices without increasing the fluid flow rate in the branches?

Our numerical analyses started with an investigation focused on the shape of the nozzle which feeds the membrane and on the shape of the drain and their influences on the deposition of particles of certain sizes. For this purpose we extracted one

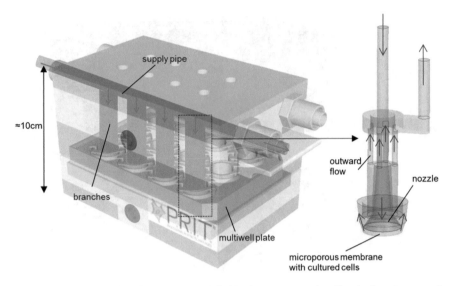

Fig. 3 P.R.I.T.® ExpoCube® (depicted on *left* side, inner construction disguised) and extracted CFD model of the deposition device (*right*) [4]. Published with kind permission of ©Fraunhofer-Gesellschaft 2016. All Rights Reserved

of the several deposition devices depicted as simulation model in Fig. 3. In the simulation model the fluid flow of air was assumed to be laminar and a steady state solver was applied. Due to the small portion of aerosols in the flow, the impact of the particles (Ø50 to 3000 nm as cumulative distribution function) on the carrier flow could be neglected, so that the two phases could be simulated consecutively. In all simulations with aerosols, the particles were modeled in a Lagrangian frame of reference with a spherical shape and gravity influence. Particles which came into contact with a solid wall were regarded as trapped to that wall. Particle injection points were placed evenly in the inflow cross-section and in some cases the inflow area was extended in upstream direction to account for the particle and flow history in the pipe system. The number of particle injection points was increased until the particle deposition behavior did not change any further.

Regarding the outflow geometry it could be seen that a change in the number and organization of drains smoothed the flow profile in the vicinity of the cell membrane in circumferential direction. The shape of the nozzle was diversely modified, but the simulations revealed that the original shape was nearly optimal concerning the particle deposition efficiency. Although thus only small geometrical changes were applied, the reliance in the functionality of the design was thereby increased. In Fig. 4 the results for two actually differing shapes are exemplarily compared. Although the velocity distribution is quite different, the deposition structure of the particles on the cell membrane appeared rather similar. These simulations were accomplished with STAR-CCM+® (Siemens).

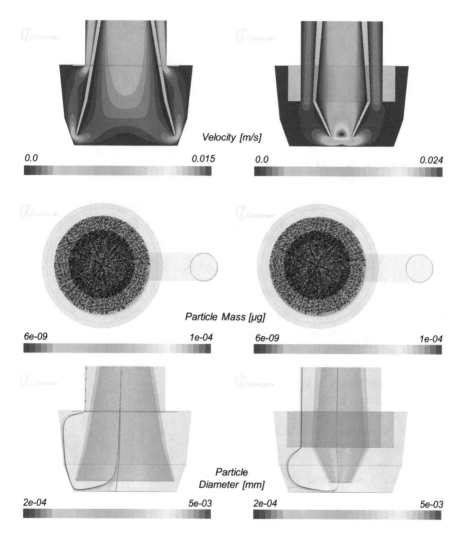

Fig. 4 Comparison of two exemplarily selected geometries, velocity distribution, particle deposition on cell membrane and particle tracks of different particle sizes for one injection point [4]. Published with kind permission of ©Fraunhofer-Gesellschaft 2016. All Rights Reserved

Due to the partially submicron size of the particles non-continuum effects like Brownian motion and a Stokes-Cunningham particle drag law (slip correction) were taken into account for further simulations. Both effects are provided inside the ANSYS Fluent flow solver where the Brownian motion is modeled as a Gaussian white noise process with a spectral intensity [2]. The Cunningham correction factor C_c was taken from [6]:

$$C_c = \begin{cases} 1 + \frac{2.52\lambda}{d_p} & : \quad d_p \geq 100\,\text{nm} \\ 1 + \frac{\lambda}{d_p}\left[2.34 + 1.05\exp\left(-0.39\frac{d_p}{\lambda}\right)\right] & : \quad d_p < 100\,\text{nm} \end{cases} \tag{1}$$

with
λ = mean free path of fluid (ambient air $\approx$ 68nm)
d_p = particle diameter

In order to increase the deposition rate further the effects of thermophoretic forces acting on particles in opposite direction of the temperature gradient were considered in the simulations according to [1]:

$$\mathbf{F} = -\frac{6\pi d_p \mu^2 C_s(K + C_t Kn)}{\rho(1 + 3C_m Kn)(1 + 2K + 2C_t Kn)}\frac{1}{m_p T}\nabla T \tag{2}$$

with
$\mathbf{F}$ = force vector due to thermophoresis
K = k/k_p
k = fluid thermal conductivity based on translational energy only = $(15/4)\mu R$
k_p = particle thermal conductivity
Kn = Knudsen number = $2\lambda/d_p$
λ = mean free path of fluid
μ = fluid viscosity
ρ = fluid density
m_p = particle mass
T = local fluid temperature
C_s = 1.17
C_t = 2.18
C_m = 1.14

To obtain a defined temperature distribution inside the deposition device which triggers thermophoresis a conjugate heat transfer model including solid parts was set up and the results were applied as approximated boundary conditions in further simulations where thermophoresis was considered. For particle sizes below 1 μm the above described effects revealed a significant influence on the particle behavior. We could confirm our previous improvements of the deposition device geometry simulating the different models in ANSYS Fluent (Fig. 5). For the purpose of a clearer evaluation, the cumulative distribution for the particle sizes was dropped and a representative number of particle sizes was simulated individually. In Fig. 5 it can be observed that particles smaller than 15 nm are not deposited; this effect showed

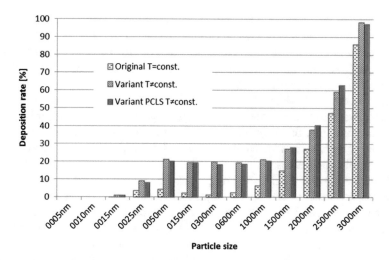

Fig. 5 Comparison of deposition rates for different device geometries and activated thermophoresis, additional control for thicker lung slices (PCLS) [4]. Published with kind permission of ©Fraunhofer-Gesellschaft 2016. All Rights Reserved

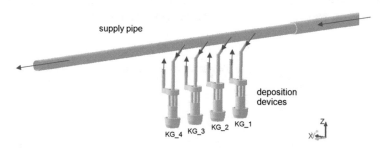

Fig. 6 ANSYS Fluent model of one row; supply pipe and four deposition devices with flow pattern [4]. Published with kind permission of ©Fraunhofer-Gesellschaft 2016. All Rights Reserved

up especially after the Brownian motion was included in the simulations. Tracks of these particles exhibited that the particles are hardly following the direction of the carrier flow. Even with all the non-continuum models included it is undetermined, if the flow models are able to cover the behavior of such extremely small particles.

In a next step, the geometry of the simulation model was extended to account for the supply pipe and the branches first of all without and then with the deposition devices (Fig. 6). Concerning the original geometry it could be observed both in experiments and simulations that the deposition rate decreases continuously regarding the sequence of the branches in flow direction. This behavior is due to a loss of particles in the supply pipe by reason of gravity and branched off particles.

In order to improve the system further it was necessary to realize the exact correlations between the flow pattern of the carrier fluid and the particle tracks. It was therefore inevitable to reveal the history of the particle tracks in regard to the branch they turn in. Therefore, a post-processing tool to plot the particle history in a colored manner using particle tracking files of ANSYS Fluent was developed. Figure 7 shows exemplarily the history of the particles of Ø1000 nm for the original and a modified geometry. Regarding the original geometry it is obvious that the branched off particles are layered and the amount is reduced concerning the subsequent branches. Many particles leave the supply pipe without having a chance to enter a branch. Other particles are lost on the pipe wall. To overcome this problem the idea came up to break up the layered order by implementing small appliances right before the branches to disturb the flow and thereby to drive more particles into branch feeding flow areas.

In order to disturb the flow several geometries, like bumps, nozzles etc. in different dimensions and positions were implemented and simulated. It was recognized that one definite geometry of the appliances within the pipe yielded the best particle branch flows concerning the absolute mass and the equal distribution. As visible in Fig. 7 for the modified pipe design this effect is due to the fact that a flow rotation (swirl) could be introduced in the pipe which pushes fresh particles in the vicinity of the pipe wall where the fluid is sucked into the branches. Since a variety of particle sizes had to be considered, a compromise concerning the magnitude of the rotational speed had to be found due to the fact that larger particles tend to be lost at the wall for stronger swirls. For a designated design with an increased particle

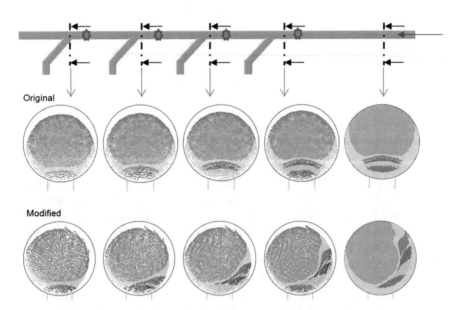

Fig. 7 History of particles in certain slices exemplary for a diameter of 1000 nm for the original and a modified geometry; *pink*=leave through supply pipe, *cyan*=lost on supply pipe, *red*=enter fourth branch, *blue*=enter third branch, *green*=enter second branch, *black*=enter first branch; asterisks indicate the position of appliances to disturb the flow [4]. Published with kind permission of ©Fraunhofer-Gesellschaft 2016. All Rights Reserved

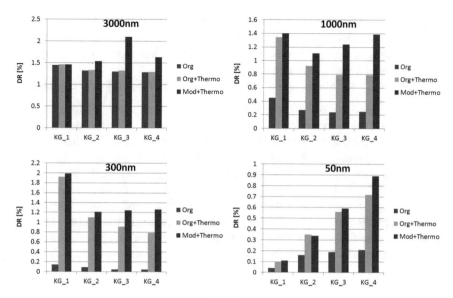

Fig. 8 Deposition rates (DR) for original design (Org), original design with thermophoresis (Org+Thermo) and modified design with thermophoresis (Mod+Thermo); the deposition rate is referred to the total particle mass injected [4]. Published with kind permission of ©Fraunhofer-Gesellschaft 2016. All Rights Reserved

mass flow into the branches, it was then verified if the deposition rate on the cell membrane measures up as well. Figure 8 shows the deposition rate related to the total injected particle mass for four different particles sizes. A significant increase of settled particles was calculated. This behavior is in a major part due to the effect of thermophoresis but it was further increased through the modifications in the supply pipe for several particle sizes. For larger particles the effect was smaller, but for medium and smaller sized particles ($d <= 1000\,\text{nm}$) the increase was immense. The improvements could be fully confirmed by experiments and are fundamental for further developments.

4 Simulating the Liquid Supply System

For long-term exposures the exposed cells can be supplied by perfusion of the nutrient solution in a special plate ("perfusion plate") through a ductwork which is then integrated in the P.R.I.T.® ExpoCube®. The main questions arising here are the following:

1. Is there an equal distribution of nutrient to all cell inserts?
2. What volume flow limit exists regarding a maximum pressure in order to avoid cell damage?

3. How fast can we flush the system, i.e. replace the fluid in the channels? Where are the areas which are problematic for flushing?
4. How should we design the channel cross-section to reduce the pressure loss and the pressure on the membranes, respectively ?
5. How should we design the channel geometry to avoid clogging by air bubbles?

To answer questions 1–4 we set up a model of the fluid system, where we took advantage of the symmetrical shape. Figure 9 shows a principle sketch of the geometry as half model with four cell inserts. For several volume flows and different cross-sectional areas and shapes the pressure on the cell membranes and the volume flows in the branches were determined by steady state simulations. The flushing behavior was modeled by applying transient simulations. For these simulations STAR-CCM+® was used. In the flushing simulations it was observed that some areas, as depicted in Fig. 10, exhibited problems with a proper removal of old liquid due to the specific shape of the channels and sharp corners in the cell membrane area. These areas were redesigned, e.g. the channel shape was modified. It is shown

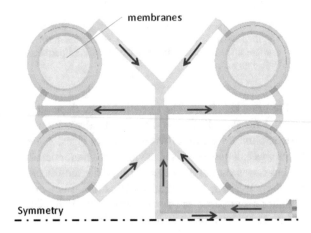

Fig. 9 Schematic representation of the geometry of the liquid system, half model due to symmetry, inflow (*blue*) and outflow (*red*) [4]. Published with kind permission of ©Fraunhofer-Gesellschaft 2016. All Rights Reserved

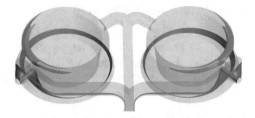

Fig. 10 Simulation result showing part of the liquid system after 3 min flushing, areas with remains of old fluid are visible [4]. Published with kind permission of ©Fraunhofer-Gesellschaft 2016. All Rights Reserved

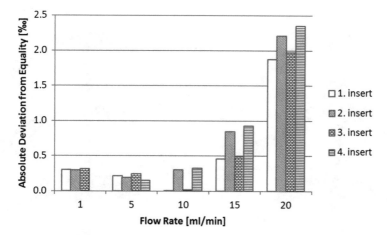

Fig. 11 Absolute deviation from equal supply of liquid to the four cell culture inserts through branched channel system simulated in the half model for various flow rates. Published with kind permission of ©Fraunhofer-Gesellschaft 2016. All Rights Reserved

below that this also improved the performance of the system when air bubbles were present. For the flushing time it is advantageous to reduce the overall fluid mass in the system by reducing the channel cross-sectional area, but on the other hand this generates higher pressure losses and thus leads to higher pressures on the cell membranes. Therefore, a compromise had to be made by defining intermediate values for the channel diameter. With respect to the distribution of the fluid to supply all cell membranes sufficiently, it became obvious that the original channel system was already chosen in a good manner. The deviation from uniformity ranged only from less than one per mill for low flow rates to a maximum of ca. two per mill for higher flow rates (Fig. 11). In this case the flow simulations were not further applied to improve the system. However, they strengthened the confidence in the carried out design.

4.1 Clogging in Liquid Channels

A well-known problem of channels with small cross-sectional areas is when unavoidable air bubbles are stuck in the system. This effect is called clogging [7] and emerges when small channels furcate and fluid forces through surface tensions reach the same order of magnitude as flow driving forces like pressure. It is observed that clogging occurs preferentially in channel contractions [8] and channel elbows [5]. When a bubble blocks a channel, this part of the system is then insufficiently supplied with nutrient solution. It is a current but rather makeshift procedure to actuate bubbles by clicking one's finger at the channel (if reachable) or bouncing the whole apparatus. In order to reduce the tendency for clogging, simulations with different channel geometries were conducted in ANSYS Fluent. The model size was

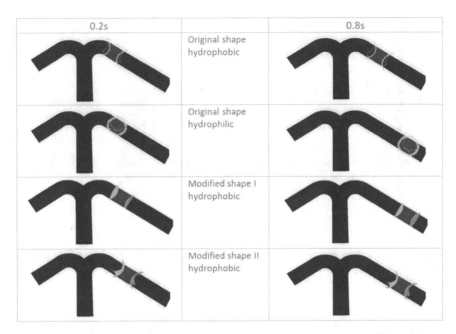

Fig. 12 Bubble position after t=0.2s and t=0.8s for different channel layouts which illustrates the movability of the bubble [4]. Published with kind permission of ©Fraunhofer-Gesellschaft 2016. All Rights Reserved

successively reduced due to the costly computing times of the transient multi-phase simulations.

To verify bubble behavior in a channel, a small model with different cross-sectional shapes was built up including a variation of the wetting behavior of the surface. For simplicity the contact angles were generally set to 45° for a hydrophobic and 135° for a hydrophilic surface. The flow is forked and a bubble is initially placed in one of the branches. In Fig. 12 the movability of the bubble is depicted for selective simulations and it is evident, that the modified channel layouts are better than the original cross-section shape with a hydrophobic surface. Also, due to the already experienced advantages for the flushing of the system, the modified cross-section for a hydrophobic surface was selected. One might achieve even better results combining the modified shape with a hydrophilic surface, but this would lead to higher production costs.

5   Simulating an Aerosol Sampling Box

The use of the P.R.I.T.® ExpoCube® allows the collection of test atmospheres from almost any source, as long as it is under ambient pressure. In an actual cooperation with a company from the cosmetics industry the possible toxic effects of a hair

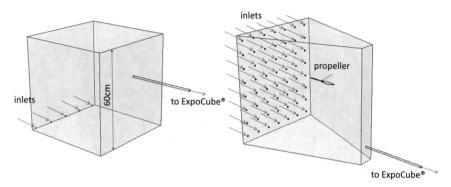

Fig. 13 CFD models of the original AE-box (*left*) and the modified design (right). Published with kind permission of ©Fraunhofer-Gesellschaft 2016. All Rights Reserved

care product were investigated. In order to reproduce real conditions the aerosol samples have to be generated and transferred into the P.R.I.T.® ExpoCube®. For this purpose an aerosol box (AE-box) was developed in a glovebox design (Fig. 13). The laboratory test should cover the realistic customer and workplace relevant procedure, including as far as possible the full size range of emanating particles. During experimental applications it became obvious that the amount of extracted larger particles for the present AE-box was insufficient and the desired range of sizes could not be covered. In order to reveal the reasons for this particle behavior, a CFD model was built up and particles with a diameter of $0.1–10\,\mu m$ (inhalable particle sizes, human lung) were simulated. In accordance with the aerosol simulation models of the conduction system (Sect. 2), Brownian motion and a Cunningham correction were modeled here as well. However, the flow was not assumed to be laminar but a standard $k - \epsilon$ turbulence model was applied. Several volume flows from 200 to 1000 ml/min were investigated and the extracted amount of particles and their tracks were analyzed. The simulation results clearly showed that the original AE-box is not able to catch particles with larger sizes as these particles tend to fall almost directly to the bottom wall (Fig. 14 left side). The first approach to overcome this deficiency comprised an increase of the volume flow. However, since the concentration of the test aerosols within the box decreases with the increase of the volume flow through the box, the latter one had to be limited to preserve a relevant aerosol concentration for sampling and testing. Due to this fact several additional modifications were applied (Fig. 13 right side):

- The box was tapered in the outlet region to get higher flow velocities respectively a higher particle drag.
- More inlets were placed on the sidewall to reduce extensive areas of resting fluid in the box.
- A propeller was inserted to prevent larger particles from directly falling to the ground.
- The outlet was moved downwards to account for the influence of gravity.

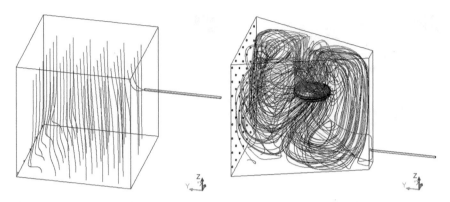

Fig. 14 Exemplary tracks for particles with ∅4 µm for the original and modified sampling box. Published with kind permission of ©Fraunhofer-Gesellschaft 2016. All Rights Reserved

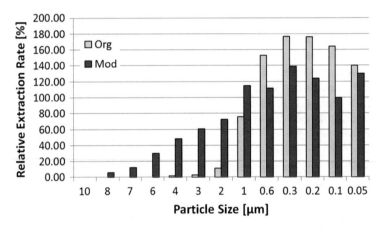

Fig. 15 Comparing the relative extraction rates of the original (Org) and modified (Mod) AE-box. Published with kind permission of ©Fraunhofer-Gesellschaft 2016. All Rights Reserved

The propeller was modeled in a moving reference frame with a frozen rotor approach [3]. The propeller speed, position and blow direction were varied to disclose dependencies regarding the amount of extracted particles.

The resulting complex particle tracks for an exemplary particle size are depicted in Fig. 14 (right side). Due to the complex propelled particle movement smaller particles are lost to a greater extent than for the original box design. However, the possibility of sampling larger particles from the box was introduced through this constellation. This is quite obvious in Fig. 15 where the extracted amounts are related to the average extracted amount in the respirable range of particle sizes; meaning particles which enter the deeper lung of humans (alveolar areas) with a particle diameter of 0.1–4 µm.

6 Conclusions

Methods of computational fluid dynamics were applied to improve an air-liquid interface technology for a compact in-vitro testing device (P.R.I.T.® ExpoCube®). Concerning the aerosol conduction and deposition system the simulation results gave a meaningful insight into the aerosol particle behavior and provided ways to enhance the amount of deposited particles on the cell cultures through geometrical and physical modifications. For that purpose, on the one hand the shape of the deposition device was modified to settle more particles and on the other hand the supply pipe was equipped with appliances to enhance the particle mass flow into the feeding branches and thus to the cell membranes. Additionally, thermophoresis effects were introduced to further increase the efficiency. The enhanced constellation was confirmed in experiments. Concerning the liquid nutrient supply system, which can be compactly included in the device, the simulations helped to improve the performance by the means of modifications of the channel geometry. Attention was turned to the ability to uniformly supply the cell inserts, to flush the system effectively without harming the cells and to improve the capability to overcome blockage by clogging effects of air bubbles. For the purpose of generating custom-designed particle samples, an aerosol extraction box was reworked resulting in a broader spectrum of sampled particle sizes.

Acknowledgements The authors would like to acknowledge the financial support supplied by Fraunhofer in the framework of the 4D program.

References

1. ANSYS Fluent Inc., *ANSYS Fluent Theory Guide—16.2.1.5. Thermophoretic Force*. Release 15.0
2. ANSYS Fluent Inc., *ANSYS Fluent Theory Guide—16.2.1.6. Brownian Force*. Release 15.0
3. ANSYS Fluent Inc., *ANSYS Fluent Theory Guide—2.3.1. The Multiple Reference Frame Model*. Release 15.0
4. C. Brodbeck, D. Ritter, J. Knebel, Improvements of an air-liquid interface in-vitro testing method for inhalable compounds using CFD-methods, in *A World of Engineering Simulation. Proceedings: NAFEMS World Congress 2015, 21–24 June 2015, San Diego, California, USA, National Agency for Finite Element Methods and Standards (NAFEMS)*, (2015)
5. T. Cubaud, C. Ho, Transport of bubbles in square microchannels. Phys. Fluids **16**, 4575–4585 (2004)
6. W. Hinds, *Aerosol Technology, Properties, Behavior and Measurement of Airborne Particles*. Dover Civil and Mechanical Engineering Series (John Wiley & Sons, Inc., New York, 1999)
7. M.J. Jensen, Bubbles in Microchannels, master's thesis, Technical University of Denmark, 2002
8. M.J. Jensen, G. Goranovic, H. Bruus, The clogging pressure of bubbles in hydrophilic microchannel contractions. J. Micromech. Microeng. **14**, 876–883 (2004)

A Mapping Procedure for the Computation of Flow-Induced Vibrations in Turbomachinery

Nadja Wirth and André Oeckerath

1 Introduction

Turbomachinery is deployed in a variety of industrial systems. Flow-induced vibrations can lead to a high noise emission and to blade fatigue which can endanger the integrity of the whole system. The blade excitation is caused by pressure fluctuations in the flow field generated by interactions between rotating and stationary blade rows [7]. Design optimization for the reduction of product size and weight leads inter alia to a reduction of the distance between rotor blades and stationary guide vanes. This increases the unsteady interactions and thereby the excitation of the already highly loaded blading in the whole flow channel.

Numerical simulations of those excitation forces and vibration responses lead to time- and cost-savings in the prototyping and testing of products. However, a classical transient simulation of the steady state flow conditions can be computationally highly expensive. A faster simulation approach—the harmonic CFD (Computational Fluid Dynamics) methods—and its application in a new mapping process to structural analyses is the topic of this paper.

Harmonic CFD simulations calculate the steady periodic flow field by frequency domain methods. The pressure excitations are expressed by complex fields which correspond to oscillations of a certain frequency, amplitude and phase lag. Superposing those oscillations leads to the transient pressure field at steady state conditions.

In order to compute the resulting blade vibrations, the complex pressure excitations are mapped to the structural FE (Finite Element) model and a frequency response analysis is performed.

N. Wirth (✉) • A. Oeckerath
Fraunhofer Institute for Algorithms and Scientific Computing SCAI, Schloss Birlinghoven, 53757 Sankt Augustin, Germany
e-mail: nadja.wirth@scai.fraunhofer.de

© Springer International Publishing AG 2017
M. Griebel et al. (eds.), *Scientific Computing and Algorithms in Industrial Simulations*, DOI 10.1007/978-3-319-62458-7_13

The mapping algorithm is able to transfer data between different periodic sections of a cyclic symmetric geometry, which is often applied in turbomachinery simulations. This mapping feature reduces modeling and simulation effort by using the periodicity information of the data.

The procedure is applied to a rotor in a two-stage high pressure axial turbine. Blade vibration responses to the pressure fluctuations are calculated and hot spots with high fatigue probability are located.

This article is based on the authors' publication [15], where the mapping procedure has been introduced for single-stage turbo-machine configurations.

2 Nonlinear Harmonic Method

The turbomachinery fluid flows are periodic by nature due to stationary and rotating cyclic symmetric components. This property is utilized in harmonic CFD methods where the transient flow field is approximated by the superposition of periodic oscillations. Those oscillations are given by their frequency, amplitude and phase, which in turn can be expressed as complex numbers (by Euler's formula).

The methods implemented in state-of-the-art simulation software work in the frequency domain and use finite Fourier series (Nonlinear Harmonic method [6]) or discrete Fourier transforms (Time Spectral method [10] and Harmonic Balance method [2]). A good overview on the different harmonic CFD methods is given in [5].

Here, we focus on the Nonlinear Harmonic (NLH) method implemented in NUMECA's turbine simulation software FINE/Turbo [11]. The approach uses a finite Fourier decomposition to express the steady state behavior of the transient flow field as developed in [4–6].

The state variable $u^{(\varrho)}(\mathbf{r}, t)$ at position $\mathbf{r}$ and time t in blade row ϱ is split into the time-averaged variable $\overline{u}^{(\varrho)}(\mathbf{r})$ and a periodic fluctuation around this mean value. The periodic oscillation is decomposed into harmonic terms,

$$u^{(\varrho)}(\mathbf{r}, t) = \overline{u}^{(\varrho)}(\mathbf{r}) + \underbrace{\sum_{k_1=-K_1}^{K_1} \cdots \sum_{k_N=-K_N}^{K_N} \tilde{u}_{\mathbf{k}}^{(\varrho)}(\mathbf{r}) \cdot \mathrm{e}^{\mathrm{i}\,\omega_{\mathbf{k}}^{(\varrho)} t}}_{\mathbf{k} \neq \mathbf{0}}. \tag{1}$$

Here, N is the number of blade rows and $\mathbf{k} = (k_j)_{j=1}^N$.

The factor $\tilde{u}_{\mathbf{k}}^{(\varrho)}(\mathbf{r}) \in \mathbb{C}$ denotes the harmonic's complex amplitude. It can be reformulated as magnitude and phase-shift of the oscillation (using Euler's formula).

The corresponding frequency $\omega_{\mathbf{k}}^{(\varrho)}$ is built from the blade passing frequencies $\omega^{(j,\varrho)}$ generated by the influencing adjacent blade rows $j = 1, \ldots, N$, where

$$\omega_{\mathbf{k}}^{(\varrho)} = \sum_{j=1}^{N} k_j \cdot \omega^{(j,\varrho)} = \sum_{j=1}^{N} k_j \cdot n_j \cdot (\Omega_\varrho - \Omega_j), \tag{2}$$

with the blade count n_j and the rotation speed Ω_j of row j.

Since the time dependent state variable $u^{(\varrho)}$ is a real number, $\tilde{u}_{\mathbf{k}}^{(\varrho)}$ is complex conjugated to $\tilde{u}_{-\mathbf{k}}^{(\varrho)}$. So the number of free variables is $(\prod_{j=1}^{N}(2K_j+1)-1)/2+1$. They are reduced by certain blade number constellations or additional restrictions on the valid combinations in $\mathbf{k}$.

FINE/Turbo specifies the harmonic $\tilde{u}_{\mathbf{k}}^{(\varrho)}$ by notation $(\text{sign}_+(\Omega_\varrho - \Omega_j)k_j \text{ G}j)_{j=1}^{N}$ where "G" stands for "group" and sign_+ is the sign function where the value at 0 is $+1$. The employed software (version 10.1) allows a maximum of two non-zero entries of $\mathbf{k}$ ("rank-2") which represent adjacent blade rows of row ϱ and a maximum of $K_j = 10$ fundamental harmonics per row.

Moreover, interactions between upstream and downstream influencing rows can be restricted. With a given order of interaction χ, a valid $\mathbf{k}$, having at least two non-zero entries corresponding to both upstream and downstream rows, fulfills $-\chi \leq k_j \leq \chi, j = 1, \ldots, N$.

In the NLH method this decomposition is used for time-averaging the unsteady Navier-Stokes equations in order to compute an approximation of the steady state transient solution of the turbomachinery problem. This approach is analogous to Reynolds averaging, except that the periodic fluctuations are assumed to predominate over the turbulent fluctuations [11]. As in the concept of turbulence modeling, the periodic fluctuations contribute additional terms to the time-averaged Navier-Stokes equation, referred to as deterministic stresses. For the model closure, a transport equation for the unsteady perturbations is obtained by retaining the first-order terms in the basic unsteady flow equations. Casting this first order linearized equation into the frequency domain gives the remaining equations (besides the turbulence model) to close the problem [11]. The reader is referred to [6] and [11] for a detailed explanation of the NLH method.

As shown in [8] this approach is more efficient than the classical transient simulation since the computationally expensive calculation of the initial transient response is avoided. Also, computer storage is minimized since the transient information is represented by a small amount of complex flow data, which needs not to be saved for each time step.

Similar harmonic CFD methods are provided in commercial software tools e.g. STAR-CCM+ (Harmonic Balance method).

3 Mapping of Pressure Excitations

For the computation of structural vibrations harmonic Finite Element methods (FEM) are used to solve the linear equations of motion. The excitation is split into its frequency components which are considered independently during the simulation. In the steady state behavior, each response vibrates with its exciting frequency. With this ansatz the equations of motion are transferred to the frequency domain. Modal analyses can be used to reduce the number of degrees of freedom by computing the basic linear independent motion shapes of the structure.

The system loaded by the overall pressure fluctuation given by

$$p(\mathbf{r}, t) = \overline{p}(\mathbf{r}) + \sum_{\substack{k=-K \\ k\neq 0}}^{K} \tilde{p}_k(\mathbf{r}) \cdot \mathrm{e}^{\mathrm{i}\omega_k t} \tag{3}$$

responds by linear superposition (inverse Fourier decomposition) of the single responses to the time-averaged pressure and the harmonic pressure fluctuations, as given by

$$\mathbf{x}(\mathbf{r}, t) = \overline{\mathbf{x}}(\mathbf{r}) + \sum_{\substack{k=-K \\ k\neq 0}}^{K} \tilde{\mathbf{x}}_k(\mathbf{r}) \cdot \mathrm{e}^{\mathrm{i}\omega_k t} \tag{4}$$

for the deformation vector $\mathbf{x}$. This is possible due to the linearity of the equations of motion which are solved in the frequency response analysis.

To simulate the flow-induced vibrations the pressure excitations are transferred to the structural mesh using the Fraunhofer SCAI software MpCCI FSIMapper. The basic procedure is shown schematically in Fig. 1.

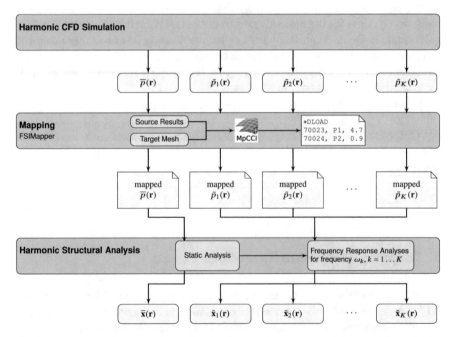

Fig. 1 Scheme of mapping procedure. The time-averaged pressure $\overline{p}$ and the complex amplitudes $\tilde{p}_k, k = 1, \ldots, K$ as result of the harmonic Computational Fluid Dynamics (CFD) simulation are transferred via the software MpCCI FSIMapper to the target structural mesh. The resulting files contain the boundary conditions of the structural model in the target code format. They are used in the harmonic structural analysis in order to compute the time-averaged displacement $\overline{\mathbf{x}}$ and the complex displacements $\tilde{\mathbf{x}}_k, k = 1, \ldots, K$ for each considered harmonic excitation [15]. Published with kind permission of ©Fraunhofer SCAI 2016. All Rights Reserved

In the first step, the excitations are provided by a harmonic turbine flow simulation in terms of the time-averaged pressure $\bar{p}$ and a certain number of harmonic pressures $\tilde{p}_k \in \mathbb{C}, k = 1, \ldots, K$ (frequency components).

The harmonic result file is read by MpCCI FSIMapper and the harmonic data is mapped to the structural target mesh. The two meshes do not necessarily need to coincide but represent approximately the same geometric shape. The MpCCI FSIMapper algorithms are able to handle different mesh densities or element formulations [1].

MpCCI FSIMapper exports for each harmonic a loading file which contains the corresponding complex excitation pressures on the target mesh. In addition, it exports a file with the time-averaged load. The data contained in these files is used in the harmonic structural simulation (stage three in Fig. 1).

The first step of the structural vibration analysis is the computation of the static deformation $\bar{\mathbf{x}}$ at the time-averaged pressure loading. This serves as the base state for the subsequent frequency response steps. In each of those steps, the mapped real and imaginary load data are included and the response (complex deformation $\tilde{\mathbf{x}}_k, k = 1, \ldots, K$) at the corresponding frequency is calculated. Here, also resulting quantities like stress, strain, etc. responses are available.

MpCCI FSIMapper is able to read the harmonic results from FINE/Turbo's CFD General Notation System (*.cgns) file and the EnSight Gold *.case format which can be exported by STAR-CCM+. On the target simulation side, MpCCI FSIMapper supports the simulation definition syntax of Abaqus, Ansys Mechanical APDL and Nastran.

3.1 Periodic Models and Nodal Diameters

In turbomachinery simulations, periodic models are often used in order to reduce computation times. Usually, the mapping algorithms need a more-or-less coarse match between the source and target meshes.

In order to offer flexible modeling of source and target mesh the procedure presented here provides the possibility of mapping between different periodic sections which in fact represent the same full model.

Figure 2 shows schematically two different cyclic symmetric meshes (black lines) for the use in a data mapping. The mapping algorithm uses the periodicity information to map the data from the source mesh to the—at first glance non-matching—target mesh by "rotating" the data to be present on the virtual full source model (gray lines).

In general, dynamic excitations of cyclic symmetric systems—as the harmonic turbo-machine pressure fields $\tilde{p}_k \in \mathbb{C}, k = 1, \ldots, K$—cannot be described by a simple passage-to-passage periodicity. Moreover, "phase-shift" periodicities occur,

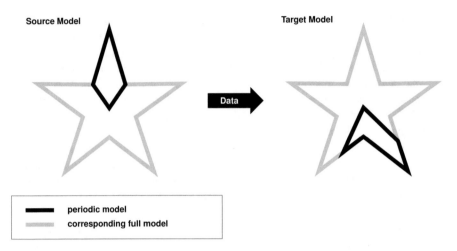

Fig. 2 Mapping of data between different periodic sections (*black lines*) which represent the same full model (*gray lines*). The data is mapped from the *left* source model to the *right* target model by "rotating" the data to be present on the virtual full source model [15]. Published with kind permission of ©Fraunhofer SCAI 2016. All Rights Reserved

which means that in adjacent passages the dynamic amplitude is the same, but there is a constant temporal phase angle difference, the inter-blade phase angle [5, 12–14]:

$$\sigma_{ND}^{\text{forward}} = \frac{2\pi \cdot ND}{n} \tag{5}$$

$$\sigma_{ND}^{\text{backward}} = \frac{-2\pi \cdot ND}{n}. \tag{6}$$

Here, n denotes the number of periodic sections and the variable $ND \in \mathbb{N}$ is called nodal diameter (also known as circumferential wave number or cyclic symmetry mode). The maximum valid nodal diameter ND_{max} is given by

$$ND_{\text{max}} = \begin{cases} \frac{n}{2} & \text{if } n \text{ even} \\ \frac{n-1}{2} & \text{if } n \text{ odd.} \end{cases} \tag{7}$$

In the following we abbreviate a shape of nodal diameter $y \in \{0, 1, \ldots, ND_{\text{max}}\}$ by NDy.

Figure 3 shows a schematic example of a six bladed disk where $ND_{\text{max}} = 3$. In the ND0 shape all blades are deflected exactly in phase (shown in Fig. 3a). For an even number of blades (as it is the case here) the shape with $ND = ND_{\text{max}}$ is characterized by an inter-blade phase angle of π, i.e. each blade is deflected in the opposite direction to its neighbors (Fig. 3d). The ND1 resp. ND2 shape is characterized by two resp. four direction changes in the blade row, see Fig. 3b, c.

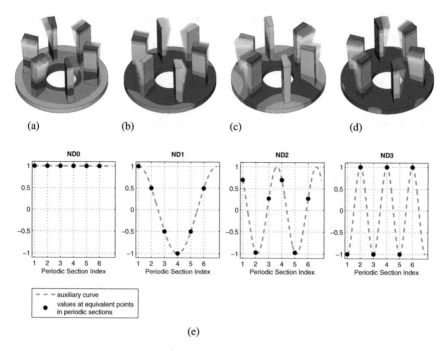

Fig. 3 First bending mode shapes of a six bladed disk for all valid nodal diameters *ND* [15]. Published with kind permission of ©Fraunhofer SCAI 2016. All Rights Reserved. (**a**) ND0, $\sigma_0 = 0$. (**b**) ND1, $\sigma_1 = 1/3 \cdot \pi$. (**c**) ND2, $\sigma_2 = 2/3 \cdot \pi$. (**d**) ND3, $\sigma_3 = \pi$. (**e**) zero crossings

The nodal diameter is equal to half of the number of circumferential zero crossings, which can be seen as radial lines with zero displacements.

To provide the complex data $d^{(1)} = \Re(d^{(1)}) + i \cdot \Im(d^{(1)})$ from a periodic source model (section $s = 1$) to its remaining sectors $s = 2, \ldots, n$, Eq. (8) needs to be applied:

$$\begin{bmatrix} \Re(d^{(s)}) \\ \Im(d^{(s)}) \end{bmatrix} = \begin{bmatrix} \cos(\sigma_{ND} \cdot (s-1)) & \sin(\sigma_{ND} \cdot (s-1)) \\ -\sin(\sigma_{ND} \cdot (s-1)) & \cos(\sigma_{ND} \cdot (s-1)) \end{bmatrix} \cdot \begin{bmatrix} \Re(d^{(1)}) \\ \Im(d^{(1)}) \end{bmatrix}. \quad (8)$$

Here, σ_{ND} is referred to as forward or backward inter-blade phase angle. See Sect. 3.2 for the derivation of an excitation's nodal diameter in turbomachinery.

Equation (8) corresponds to a rotation in the complex plane: the amplitude of the complex number stays the same and its phase is reduced by $\sigma_{ND} \cdot (s-1)$ and thus it defines the phase-shift periodicity mathematically.

If the excitation has ND0 shape, the data can be simply copied from one section to another because the matrix from Eq. (8) becomes the unit matrix. For an even number of blades and nodal diameter equals $ND_{\max}$ the data on the full model can be created by copying the data and applying a sign-change alternatingly. In both cases the matrix from Eq. (8) is diagonal, so the real and the imaginary part are decoupled and can be treated independently.

For vector quantities the derived values from Eq. (8) have to be additionally rotated by the angular pitch $2\pi(s-1)/n$ around the cyclic symmetry axis.

3.2 Deriving Excitation and Responding Shape

The specific shape of an excitation, i.e. nodal diameter and forward or backward mode, can be derived by the periodicity of the components in the turbomachinery system [12–14]. A component of periodicity n_j and rotation speed Ω_j contributes multiples of the blade passing frequency $\omega^{(j,\varrho)} = n_j \cdot (\Omega_\varrho - \Omega_j)$ to the excitation frequencies in row ϱ. In multi-stage machines the contributions are combined by linear combination in order to build the resulting harmonic frequencies [see Eqs. (1) and (2)].

The harmonic frequency $\omega_{\mathbf{k}}^{(\varrho)}$ is associated with the engine order

$$E_{\mathbf{k}} = \sum_{j=1}^{N} k_j \cdot n_j. \tag{9}$$

An engine order E excites a nodal diameter $0 \leq y \leq \lfloor n/2 \rfloor$ in forward resp. backward mode

$$\text{if} \begin{cases} E \geq 0 \quad \text{and} \quad \begin{cases} \exists\, a \geq 0 : y = E - a \cdot n & \Rightarrow \text{backward ND}y \\ \exists\, b > 0 : y = -(E - b \cdot n) & \Rightarrow \text{forward ND}y \end{cases} \\[2em] E < 0 \quad \text{and} \quad \begin{cases} \exists\, a \geq 0 : y = |E| - a \cdot n & \Rightarrow \text{forward ND}y \\ \exists\, b > 0 : y = -(|E| - b \cdot n) & \Rightarrow \text{backward ND}y. \end{cases} \end{cases} \tag{10}$$

The corresponding forward or backward inter-blade phase angle σ_{ND} is used in Eq. (8).

These conditions correspond to the ZZENF (Zig Zag shaped Excitation line in the Nodal diameter versus Frequency) diagram developed in [12] which is shown in Fig. 4 as nodal diameter versus engine order diagram. It shows which engine order $E > 0$ causes which excitation shape on a n-periodic part for two examples.

The property $\overline{\tilde{u}_{\mathbf{k}}^{(\varrho)}} = \tilde{u}_{-\mathbf{k}}^{(\varrho)}$ is directly associated with the forward/backward mode. If $\tilde{u}_{\mathbf{k}}^{(\varrho)}$ has a shape of forward (resp. backward) NDy, its complex conjugate $\tilde{u}_{-\mathbf{k}}^{(\varrho)}$ has a backward (resp. forward) NDy shape. This can be seen by the relationship $E_{-\mathbf{k}} = -E_{\mathbf{k}}$ and Eq. (10).

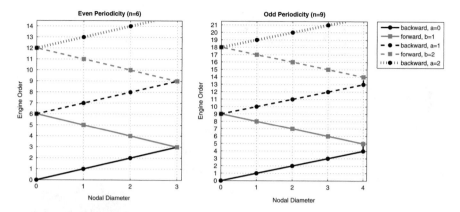

Fig. 4 Derived Zig Zag shaped Excitation line in the Nodal diameter versus Frequency (ZZENF) diagram for a six (*left*) and nine (*right*) bladed disk. Using the diagram the excitation shape (nodal diameter in backward (*black lines*) or forward (*blue lines*) mode) on a *n*-periodic component caused by a certain engine order can be determined. Published with kind permission of ©Fraunhofer SCAI 2016. All Rights Reserved

3.3 Summary

In summary, with the described tools and formulas it is possible to derive the pressure excitation shape (as forward or backward nodal diameter) from the turbomachinery configuration and Eq. (10). This information is used to provide the complex data (e.g. exciting pressures) from one periodic section to the corresponding full source model [Eq. (8)]. Thus MpCCI FSIMapper is enabled to map data between cyclic symmetric meshes which in fact represent the same full model. The resulting deformation shape (forward/backward nodal diameter) is the same as the excitation shape.

4   Application Example

The described procedure is applied to a stator-rotor-stator-rotor turbomachinery configuration, shown in Fig. 5. The axis of rotation is the positive z-axis, which also coincides with the rough flow direction. The information about rotation speeds Ω and blade numbers n is listed in Table 1.

Exemplarily, the flow-induced vibrations on row $\varrho = 2$ are computed. For this purpose, the rank-2 NLH method in FINE/Turbo calculates the time-averaged pressure and 16 complex pressure harmonics where $2 = K_1 = K_3 = K_4$ basic harmonics are used per row. Only interactions between up- and downstream rows of maximal order $\chi = 1$ are considered. As stated in Eq. (3), the conjunction of these contributions approximates the transient steady state pressure behavior. A single

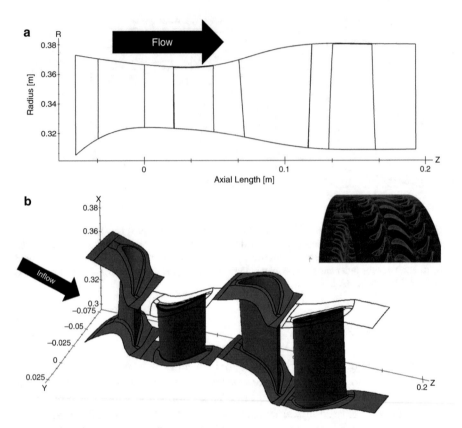

Fig. 5 Considered two-stage axial turbomachinery configuration, where only one blade passage is meshed per row. The first and third blade row (*from left*) build the stationary guide vanes. The unshrouded rows are rotating around the *z*-axis. Published with kind permission of ©Fraunhofer SCAI 2016. All Rights Reserved. (**a**) Aerodynamic flow path. (**b**) FINE/Turbo's periodic mesh and its corresponding replicated model in red

Table 1 Blade count and rotation speeds for the considered turbomachinery configuration

Row j	Ω_j [Hz]	n_j
1	000.00	46
2	138.05	76
3	000.00	48
4	138.05	70

blade passage is used in each row with the corresponding phase-shift periodic boundary condition. The flow channel is discretized by a structured mesh, where the boundary layer is resolved.

The vibration analysis is done in the structural simulation software Abaqus where a modal steady state dynamics analysis is performed. The periodic target model of row 2 is meshed by first order hexa elements. Compared to the CFD mesh, it exhibits

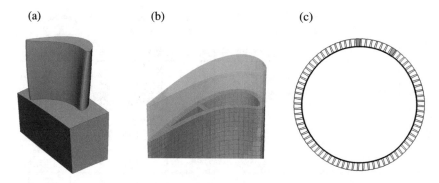

Fig. 6 Structural Abaqus hexa mesh of one passage of row $\varrho = 2$ with cavities. Compared to the CFD model, the position and section shape differs. Published with kind permission of ©Fraunhofer SCAI 2016. All Rights Reserved. (**a**) Overview. (**b**) Interior. (**c**) Position of target blade in row 2 compared to periodic CFD mesh (*gray*)

a different section shape (with planar cutting faces) and represents a different blade in the row, cf. Fig. 6.

So, the methods from Sects. 3.1 and 3.2 have to be applied in the mapping process shown in Fig. 1.

4.1 Harmonic CFD Simulation

Table 2 shows all combinations of k_1, k_3 and k_4 resulting in a positive harmonic frequency $\omega_k^{(2)}$, with the corresponding engine orders E, nodal diameters ND and inter-blade phase angles σ_{ND}.

The calculated time-averaged pressure is shown in Fig. 7. It corresponds to the pressure level around which the harmonics oscillate. The maximum value with 2.22bar is located at the leading edge on the blade's pressure side.

The pressure excitations of the harmonics 3, 4, 8 and 11 have the highest influence to the total pressure oscillation. Their magnitudes are shown in Fig. 8.

4.2 Mapping

For the flow-induced vibration analysis the mean pressure and the 16 complex pressure amplitudes are transferred to the structural mesh.

In order to permit a mapping between the different fluid and structure sections, the exciting pressure harmonics are provided on the full source model using Eq. (8) and the inter-blade phase angles from Table 2.

Table 2 Calculated harmonics on row $\varrho = 2$ in a rank-2 NLH simulation with $K_1 = K_3 = K_4 = 2$ and maximal order of interaction $\chi = 1$ between up- and downstream rows

k	k_1	k_3	k_4	$\omega_k^{(2)}$ [Hz]	E	ND	σ_{ND} [rad]
1	0	0	1	0.0	70	ND6 forward	0.49604
2	0	0	2	0.0	140	ND12 forward	0.99208
3	1	0	0	6350.3	46	ND30 forward	2.48020
4	2	0	0	12700.6	92	ND16 backward	−1.32278
5	−1	1	0	276.1	2	ND2 backward	−0.16535
6	0	1	−2	6626.4	−92	ND16 forward	1.32278
7	0	1	−1	6626.4	−22	ND22 forward	1.81882
8	0	1	0	6626.4	48	ND28 forward	2.31486
9	0	1	1	6626.4	118	ND34 forward	2.81090
10	0	1	2	6626.4	188	ND36 backward	−2.97625
11	1	1	0	12,976.7	94	ND18 backward	−1.48812
12	0	2	−2	13,252.8	−44	ND32 backward	−2.64555
13	0	2	−1	13,252.8	26	ND26 backward	−2.14951
14	0	2	0	13,252.8	96	ND20 backward	−1.65347
15	0	2	1	13,252.8	166	ND14 backward	−1.15743
16	0	2	2	13,252.8	236	ND8 backward	−0.66139

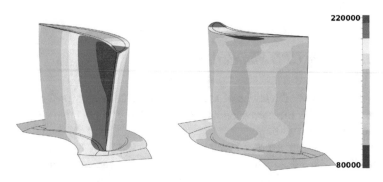

Fig. 7 Time-averaged pressure distribution on the pressure and suction side of row 2. Result of FINE/Turbo. Published with kind permission of ©Fraunhofer SCAI 2016. All Rights Reserved

Figure 9 visualizes this cyclic symmetric mapping of the third harmonic complex pressure from the source to the target mesh. The data on the periodic source model (Fig. 9 left) exhibits a forward ND30 periodicity and is "rotated" continuously over the periodic boundaries (Fig. 9 middle). These data are mapped to the periodic target model (Fig. 9 right).

The mapping process uses simultaneously the real and imaginary parts of the data, since the matrix of Eq. (8) is dense and couples them.

The complex data differs on the periodic source and target model since they correspond to different blades. This difference is simply induced by the phase-shift between the blades resulting from the forward inter-blade phase angle. The corresponding magnitude of the pressure excitation is identical.

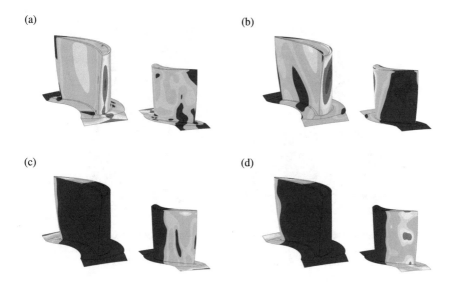

Fig. 8 Pressure excitation magnitudes on the pressure and suction side of row $\varrho = 2$ for the four harmonics with the highest magnitude. Contour plot of result of FINE/Turbo. Published with kind permission of ©Fraunhofer SCAI 2016. All Rights Reserved. (**a**) Harmonic 3 at frequency 6350.3Hz with a maximum of 5773Pa. (**b**) Harmonic 4 at frequency 12700.6Hz with a maximum of 1161Pa. (**c**) Harmonic 8 at frequency 6626.4Hz with a maximum of 934Pa. (**d**) Harmonic 11 at frequency 12976.7Hz with a maximum of 721Pa

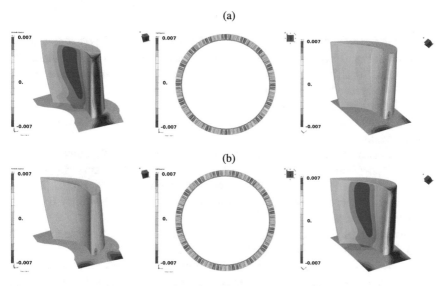

Fig. 9 Cyclic symmetry mapping procedure demonstrated for the third harmonic pressure with MpCCI FSIMapper. The data on the periodic source model (*left*) is "rotated"İ using the excitation shape information to build the full source mesh (*middle*). These data are mapped to the periodic target model (*right*). Published with kind permission of ©Fraunhofer SCAI 2016. All Rights Reserved. (**a**) Real part. (**b**) Imaginary part

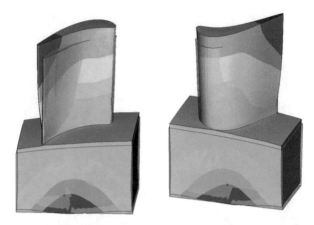

Fig. 10 Deformation resulting from centrifugal load and time-averaged pressure on row 2, scaled by a factor of 300. The blade is deflected in the direction of the suction side. Result of Abaqus. Published with kind permission of ©Fraunhofer SCAI 2016. All Rights Reserved

4.3 Harmonic Structural Analysis

In the mapping process, include files in Abaqus input format are created to define the complex loading for the frequency response analyses. For each considered harmonic a frequency response step is defined which uses the complex loading files. Abaqus uses the information about the excitation shape in order to build the periodic phase-shift boundary conditions.

The time-averaged pressure and the centrifugal load result in a ND0 deformation as shown in Fig. 10 (scaled by a factor of 300). As expected, the blade is elongated by the centrifugal load and deflected in the direction of the suction side by the pressure difference between both blade sides. The maximal deformation is located at the blade tip at trailing edge.

In the modal based frequency response simulations a modal damping of 5% has been assumed.

For the two blade tips at leading and trailing edge the displacement response frequency spectrum is shown in Fig. 11 (without the time-averaged deformation at 0Hz). The blade vibration excited by the pressure harmonic 3 at 6350.3Hz has the highest influence to the total deformation. It is created by the 46 blades of the upstream stator row 1.

The resulting deformation magnitudes for the harmonics 3, 4, 8 and 11 are shown in Fig. 12. Each color range is scaled to its maximal amplitude.

The inverse Fourier transformation of the spectra shown in Fig. 11 (plus the time-averaged deformation from Fig. 10) leads to the transient steady state deflection of the blade tips in Fig. 13. As already seen in the spectra, the frequency 6350.3Hz

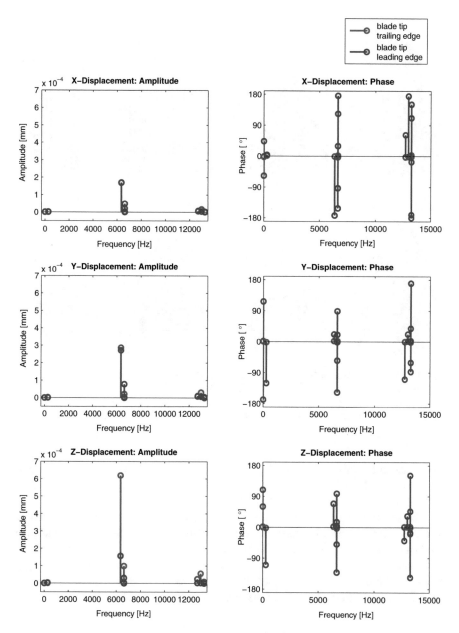

Fig. 11 Frequency spectrum of the blade tip's displacements. The third harmonic at 6350.3Hz has the highest contribution to the total vibration. Result of frequency response simulation in Abaqus. Published with kind permission of ©Fraunhofer SCAI 2016. All Rights Reserved

(a)   (b)

(c)   (d)

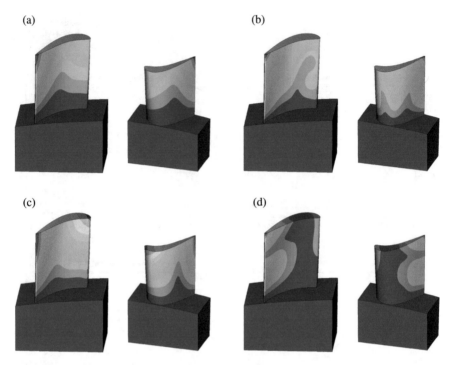

Fig. 12 Displacement magnitudes in row 2 for the four selected harmonics, view from pressure and suction side. Result of Abaqus. Published with kind permission of ©Fraunhofer SCAI 2016. All Rights Reserved. (**a**) Harmonic 3 at frequency 6350.3Hz. (**b**) Harmonic 4 at frequency 12700.6Hz. (**c**) Harmonic 8 at frequency 6626.4Hz. (**d**) Harmonic 11 at frequency 12976.7Hz

dominates the oscillation in all three degrees of freedom. A coordinate transformation leads to the transient steady state behavior in cylindrical coordinates, i.e. in radial (r), circumferential (ϕ) and in axial (z) direction.

The resulting amplitudes of stress or strain cycles can be used for a fatigue analysis [3, 9]. Figure 14 shows hot spots of high stress oscillation magnitudes at frequency 6350.3Hz (i.e. at the third harmonic) where failure-probability is highest. They can be found at the chamfer where the hub and the leading and trailing edges meet. The final durability assessment uses fatigue principles and material properties in order to estimate the lifetime of the dynamically loaded part.

5   Conclusion

In this paper a workflow was presented to simulate flow-induced vibrations of turbomachinery blades. The excitatory pressure fluctuations were calculated as complex data in the frequency domain, so no data conversion was needed in order

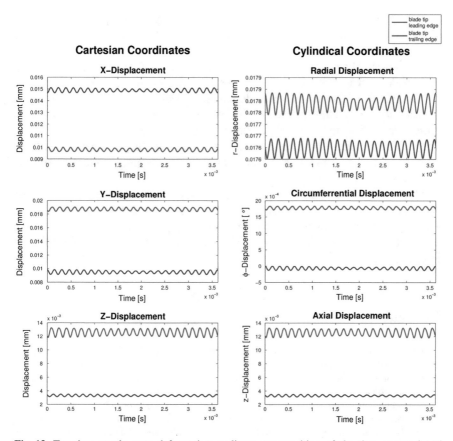

Fig. 13 Transient steady state deformation as linear superposition of the time-averaged and harmonic responses. The main frequency of the signals is 6350.3Hz. Published with kind permission of ©Fraunhofer SCAI 2016. All Rights Reserved

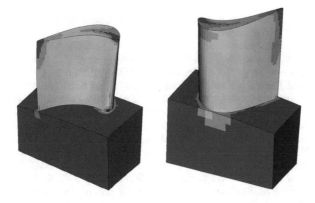

Fig. 14 Stress cycle hot spots at 6350.3 Hz. Coloring by the magnitude of the maximum von Mises Stress of all angles. Result of Abaqus. Published with kind permission of ©Fraunhofer SCAI 2016. All Rights Reserved

to define the loading in the vibration analysis. The mapping procedure allowed the data transfer between differently shaped periodic sections of a cyclic symmetric geometry.

The presented procedure can be easily transferred to different application areas such as electromagnetic induced vibrations and noise in motors or generators.

For CFD simulation codes which do not provide harmonic but transient analyses the presented procedure is also applicable. After a transient simulation on the source model, the time steps which build the steady state behavior are converted using a Fourier transformation to frequency dependent complex amplitudes. They correspond to the harmonic amplitudes used in the mapping method, where the procedure can be continued.

The results of the application example do not encounter aerodynamic damping since the influence of the structural vibration to the flow is not considered. Here, a complete coupling of the complex quantities is planned as future work using the vendor-neutral coupling interface MpCCI CouplingEnvironment developed at Fraunhofer SCAI.

References

1. Fraunhofer Institute for Algorithms and Scientific Computing SCAI, MpCCI 4.4.1 Documentation, Part X FSIMapper, Sankt Augustin (2015)
2. A.K. Gopinath, E. van der Weide, J.J. Alonso et al., Three-dimensional unsteady multi-stage turbomachinery simulations using the harmonic balance technique, in 45th AIAA Aerospace Sciences Meeting and Exhibit, Reno, Nevada, AIAA Paper 2007-0892 (2007)
3. A. Halfpenny, A frequency domain approach for fatigue life estimate from finite element analysis, in *International Conference on Damage Assessment of Structures (DAMAS 99)*, Dublin (1999)
4. L. He, Method of simulating unsteady turbomachinery flows with multiple perturbations. AIAA J. **30**, 2730–2735 (1992)
5. L. He, Fourier methods for turbomachinery applications. Prog. Aerosp. Sci. **46**, 329–341 (2010)
6. L. He, W. Ning, Efficient approach for analysis of unsteady viscous flows in turbomachines. AIAA J. **36**, 2005–2012 (1998)
7. C. Lechner, *Stationäre Gasturbinen*, 2nd edn. (Springer, Berlin, 2010)
8. M. Membera, A. Loos, A. Kührmann et al., Validation of the non-linear harmonic approach for quasi-unsteady simulations in turbomachinery, GT2009-59933, in ASME Turbo Expo 2009: Power for Land, Sea and Air, vol. 7, Orlando, June 2009, pp. 567–577
9. S. Suresh, *Fatigue of Materials*, 2nd edn. (Cambridge University Press, Cambridge, 1998)
10. E. van der Weide, A.K. Gopinath, A. Jameson, Turbomachinery applications with the time spectral method, in 17th AIAA Computational Fluid Dynamics Conference, Toronto, ON, 2005, American Institute of Aeronautics and Astronautics
11. S. Vilmin, E. Lorrain, C. Hirsch, M. Swoboda, Unsteady flow modeling across the rotor/stator interface using the nonlinear harmonic method, GT2006-90210, in ASME Turbo Expo 2009: Power for Land, Sea and Air, vol. 6, Barcelona, May 2006, pp. 1227–1237
12. J. Wildheim, Excitation of rotating circumferentially periodic structures. J. Sound Vib. **75**, 397–416 (1981)

13. J. Wildheim, Excitation of rotationally periodic structures. J. Appl. Mech. **46**, 878–882 (1981)
14. J. Wildheim, Vibrations of rotating circumferentially periodic structures. Q. J. Mech. Appl. Math. **34**, 213–229 (1981)
15. N. Wirth, A. Oeckerath, Analysis of flow-induced vibrations in turbomachinery by mapping of complex fluid pressures. Int. J. Multiphys. **9**, 195–208 (2015)

Molecular Dynamics Simulation of Membrane Free Energy Profiles Using Accurate Force Field for Ionic Liquids

Thorsten Köddermann, Martin R. Schenk, Marco Hülsmann,
Andreas Krämer, Karl N. Kirschner, and Dirk Reith

1 Introduction

Understanding the interactions and dynamics that occur within biological membranes is important in drug development, toxicological research, and metabolism research. Subsequently, there is a significant amount of experimental research focused on determining partition coefficients of molecules between lipid membranes and water, which is a way to assess the bioavailability of solutes. Partition coefficients are a measurement for the distribution of a solute between two phases that is in thermodynamic equilibrium. Experimentally, liposomes are considered good models for biological membranes [10, 11, 27]. Since phosphatidylcholine (PC) lipids are the most abundant lipids in mammalian membranes [55] and are an experimentally convenient substance [10], they are most often used as membrane models. Most methods to calculate membrane/water partition coefficients are empirical and based on octanol/water partitioning as a standard system [31]. Theoretical methods like molecular dynamics (MD) can provide an atomistic and molecular understanding of the partitioning process. MD simulations have been

T. Köddermann
Fraunhofer Institute for Algorithms and Scientific Computing SCAI, Schloss Birlinghoven,
53757 Sankt Augustin, Germany

M.R. Schenk • A. Krämer • K.N. Kirschner
Bonn-Rhein-Sieg University of Applied Sciences, Grantham-Allee 20, 53757 Sankt Augustin,
Germany

M. Hülsmann • D. Reith (✉)
Fraunhofer Institute for Algorithms and Scientific Computing SCAI, Schloss Birlinghoven,
53757 Sankt Augustin, Germany

Bonn-Rhein-Sieg University of Applied Sciences, Grantham-Allee 20, 53757 Sankt Augustin,
Germany
e-mail: marco.huelsmann@h-brs.de; dirk.reith@scai.fraunhofer.de

© Springer International Publishing AG 2017
M. Griebel et al. (eds.), *Scientific Computing and Algorithms in Industrial
Simulations*, DOI 10.1007/978-3-319-62458-7_14

used for many different molecules in different lipid bilayers [24, 42, 52]. These simulations were able to reproduce experimentally determined structural properties of membranes [45].

One advantage that MD methods offer is the ability to calculate free energy profiles. They provide insight into preferred membrane regions for specific solutes. Moreover, they can be used to calculate partition coefficients. In contrast, free energy profiles of solutes in lipid membranes are experimentally not accessible.

Some authors transferred experimental partition coefficients into a free energy of partitioning and compared them to transfer energies deduced from calculated free energy profiles [24, 44]. Only a few authors calculated partition coefficients from simulated free energy profiles and compared them to experimental data [4, 56]. Notably, [51] calculated partition coefficients from equilibrium MD simulations.

The goal of this study is to predict the ability of three different ionic liquids (ILs) to penetrate a 1-palmitoyl-2-oleoyl-sn-glycero-3-phosphocholine (POPC) membrane. We simulated the free energy profile for two ion pairs in the membrane in order to investigate the cation/anion effect and to better understand which structural features of the IL facilitate or hinder membrane penetration.

2 Computational Methods

2.1 Simulation Details

2.1.1 Technical Details

MD simulations were used to calculate the free energy profiles, partition coefficients, and positions/orientations of all cations and anions in the system for three different ILs: [C$_2$MIM][NTf$_2$], [C$_{12}$MIM][NTf$_2$] [C$_2$MIM][EtSO$_4$] in POPC. In the simulation box (6.5 nm × 6.5 nm × 10.0 nm) 128 POPC, 8864 water and 2 ion pairs were simulated at 310 K and 1 bar. The bilayer normal was used as a reaction coordinate for the profiles. All MD simulations were carried out using the GROMACS 4.5.5 package [3, 14, 46]. Unless otherwise stated, isobaric-isothermal (NpT) ensemble conditions were used. The used force field equation is a sum of pair-wise additive inter- and intraatomic Lennard-Jones and Coulombic potentials, as well as angle and torsion potentials (i.e. bond energies were not considered):

$$U = \sum_{ijk}^{\#angles} \frac{k_a}{2} \left(\theta_{ijk} - \theta_0\right) + \sum_{ijkl}^{\#dihedrals} \sum_{m=1}^{n} k_d \left(1 + \cos(m\psi_{ijkl} - \psi_0)\right) +$$

$$\sum_{i=1}^{N-1} \sum_{j>1}^{N} \left[U^{LJ}\left(r_{ij}\right) + U^{Coul}\left(r_{ij}\right)\right] \tag{1}$$

with

$$U^{LJ}\left(r_{ij}\right) = 4\varepsilon_{ij}\left[\left(\frac{\sigma_{ij}}{r_{ij}}\right)^{12} - \left(\frac{\sigma_{ij}}{r_{ij}}\right)^{6}\right]$$

and

$$U^{Coul}\left(r_{ij}\right) = \frac{q_i q_j}{4\pi\varepsilon_0 r_{ij}},$$

where $k_{a/d}$ are the angle force constants, θ is the bond angle, ψ is the dihedral angle, ε_{ij} is the LJ diameter, σ_{ij} is the LJ well depth, q_i, q_j are the charges, r_{ij} is the distance between two atom sites and ε_0 is the permittivity of free space.

The CHARMM36 force field [32] was used for the POPC membrane and the TIP4P-Ew model [16] was used for water. For the [NTf$_2$] ILs our previously published force field [34] was chosen and the force field for [C$_2$MIM][EtSO$_4$] was parameterized in this work.

For the viscosity simulations the canonical ensemble (NVT) is used since the Green-Kubo equations are only defined for constant volumes.

The pure [C$_2$MIM][EtSO$_4$] simulation box, used in the force field development, consisted of 216 ion pairs and was simulated at 1 bar. Electrostatic interactions were computed using the particle mesh Ewald summation method with a real space cutoff of 1.2 nm, a mesh spacing of approximately 0.12 nm, and fourth order interpolations [12]. For simplicity reasons the 1–4 interactions of the ionic liquids are switched off. The OPLS-combining rules are applied. Temperature control was achieved using a Nose-Hoover thermostat [15, 40] and the Rahman-Parrinello barostat [41, 43] with coupling times $\tau_T = 0.5$ ps and $\tau_p = 2.0$ ps, respectively.

The heat of vaporization ($\Delta_v H$) was obtained from the calculated internal energies of the liquid ILs and the vacuum ion pairs at 298 K. It was assumed that the vapor phase is formed by isolated, neutrally charged contact ion pairs, which were simulated in vacuum without periodic boundary conditions and with translation and rotation degrees of freedom switched off. The transport properties such as the self-diffusion coefficients (D) and the shear viscosities (η) were calculated using the Einstein relation and the Green-Kubo relation, respectively.

2.1.2 Force Field Development for [C$_2$MIM][EtSO$_4$]

The purpose of this force field development is to parameterize a united atom force field for [C$_2$MIM][EtSO$_4$] that describes experimental properties as accurately as possible. Since ionic liquids exhibit slow microscopic dynamics, equilibrium runs took 2 ns and were followed by trajectories of 10 ns long with a 2 fs time step. We adopted all cation parameters from our previously published united atom force field for the IL family [C$_n$mim][NTf$_2$] [34]. The structure of the EtSO$_4$ anion was calculated quantum mechanically at the MP2/6-31+G* level of theory with the

Gaussian03 package [13]. The electrostatic potential around the optimized structure was computed at the MP2/cc-pVTZ level afterwards. The ESP point charges [2, 50] placed at the center of mass of each atom were then calculated with Gaussian03 to reproduce the electrostatic potential (Table 1). All bonds were kept fixed to speed up the simulations (Table 2). This should have little affect on investigated properties. The 1–4 interactions are switched off completely. All angle potentials and the CA_1-Oet-S-OS dihedral angle (Fig. 1 and Tables 3, 4, 5) were taken from the force field of [38]. The dihedral angle CA_2-CA_1-Oet-S (Fig. 1 and Table 5) was parameterized to a quantum mechanically computed torsion profile done at the MP2/6-31+G* level (Fig. 2). The calculations were done using the Gaussian03 package [13]. The LJ

Table 1 LJ parameter and Coulombic charges of the force field for $[C_nMIM][EtSO_4]$

Atom	q/e	σ (Å)	ε (kJ/mol)
CN3	0.22	3.8187	0.81482
CN2	0.09	3.9915	0.38247
CT2	0.13	3.9500	0.38247
CTE	0.13	3.7500	0.81482
C_R	−0.11	1.8147	0.43407
CTS	0.0	3.9500	0.38247
CT3	0.0	3.7500	0.81482
C_W	−0.13	2.5915	0.21583
H_A	0.21	1.4520	0.18066
H_B	0.21	2.0570	0.09563
N_A	0.15	3.7899	0.72609
O_S	−0.57	3.6000	0.26319
S_O	0.96	3.5500	0.31373
Oet	−0.44	3.7888	0.26353
CA_1	0.335	3.1575	0.68247
CA_2	−0.145	2.9875	0.81482

Table 2 Bond distance parameter of the force field for $[C_nMIM][EtSO_4]$

Bonds	r (Å)
C_R-N_A	1.315
C_W-N_A	1.378
C_W-C_W	1.341
N_A-CN2	1.430
N_A-CN3	1.430
C*-C*	1.54
C_R-H_A	1.08
C_W-H_B	1.08
CA_1-CA_2	1.540
CA_1-Oet	1.424
Oet-S_O	1.715
O_S-S_O	1.483

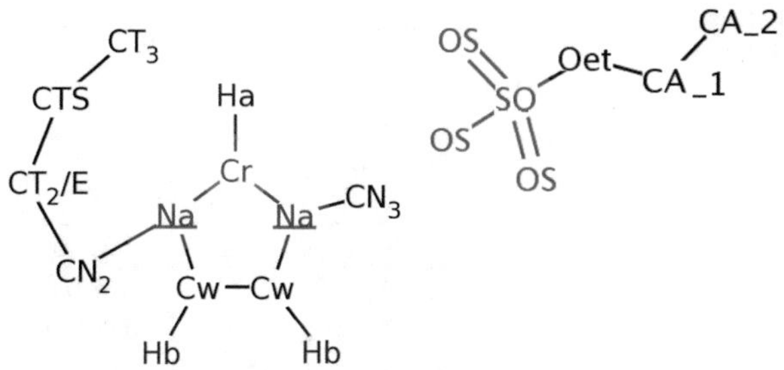

Fig. 1 Nomenclature of the interaction sites of the IL cation and anion. Published with kind permission of ©Fraunhofer SCAI 2016. All Rights Reserved

Table 3 Bond angle parameter of the force field for [C$_n$MIM][EtSO$_4$]

Angle	Φ (°)	k_a (kJ mol^{-1} rad^{-2})
C$_W$-N$_A$-C$_R$	108.0	292.6
C$_W$-N$_A$-CN3	125.6	292.6
C$_W$-N$_A$-CN2	125.6	292.6
C$_R$-N$_A$-CN3	126.4	292.6
C$_R$-N$_A$-CN2	126.4	292.6
N$_A$-C$_R$-H$_A$	125.1	146.3
N$_A$-C$_R$-N$_A$	109.8	292.6
N$_A$-C$_W$-C$_W$	107.1	292.6
N$_A$-C$_W$-H$_B$	122.0	146.3
C$_W$-C$_W$-H$_B$	130.9	146.3
N$_A$-C*-C*	109.47	419.0
C*-C*-C*	114.0	519.6
CA$_2$-CA$_1$-Oet	109.47	419.0
CA$_1$-Oet-S$_O$	116.6	300.5
O$_S$-S$_O$-O$_S$	114.0	969.0
O$_S$-S$_O$-Oet	105.1	1239.6

parameters of the alkyl chain of the anion were taken from the united atom force field of [6]. The LJ parameters of the sulfur and oxygen atom sites were taken from our force field for NTf$_2$ [33].

2.2 Umbrella Sampling

The umbrella sampling (US) method was employed to calculate the free energy profiles [49, 53, 54]. As a biasing potential is used, an unbiasing procedure must be applied to calculate the correct free energy profiles. Here, the *weighted histogram analysis method* [36] (WHAM) was used as implemented in the g_wham program [17]. In each simulation two ion pairs were present. One was kept fixed at the

Table 4 Dihedral angle parameter of the cation

Dihedrals	k_d (kJ/mol)	m	ψ (deg)
X-N_A-C_R-X	9.73	2	180
X-C_W-C_W-X	22.49	2	180
X-N_A-C_W-X	6.275	2	180
C_R-N_A-CN2-CN2/E	−1.7663690	1	0
C_R-N_A-CN2-CN2E	0.2542844	2	0
C_R-N_A-CN2-CN2E	−1.5838059	3	0
C_R-N_A-CN2-CN2E	0.0043330	1	90
C_R-N_A-CN2-CN2/E	0.0165786	2	90
C_R-N_A-CN2-CN2E	0.0712581	3	90
N_A-CN2-CT2-CT3/S	1.469	1	0
N_A-CN2-CT2-CT3/S	−0.443	2	0
N_A-CN2-CT2-CT3/S	6.402	3	0
C*-C*-C*-C*	2.786	1	0
C*-C*-C*-C*	−0.567	2	0
C*-C*-C*-C*	6.579	3	0
Improper dihedral			
X-N_A-X-X	4.185	2	180
X-$C_{W/R}$-X-X	4.1	2	180

Table 5 Dihedral angle parameter of the anion

Dihedral	k_d (kJ/mol)	m	Ψ (°)
CA_1-Oet-S-O_S	1.2408	3	0
CA_2-CA_1-Oet-S_O	−10.5178	1	0
CA_2-CA_1-Oet-S_O	2.0219	2	0
CA_2-CA_1-Oet-S_O	3.0907	3	0
CA_2-CA_1-Oet-S_O	0.1785	1	90
CA_2-CA_1-Oet-S_O	0.2612	2	90
CA_2-CA_1-Oet-S_O	0.1509	3	90

Fig. 2 Torsion energy profile of the dihedral angle CA_2-CA_1-Oet-S of the anion calculated quantum mechanically compared with the simulated energy profile. The QM level has been MP2/6-31+G*. Published with kind permission of ©Fraunhofer SCAI 2016. All Rights Reserved

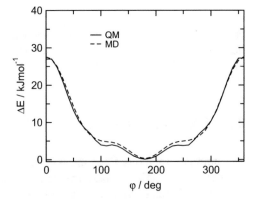

z-position by the umbrella force, the other one was able to diffuse freely. This represents the experimental situation were the formation of ion pair aggregates is possible, in order to decrease the ion charge. The reference state was chosen to be the center of the membrane with a free energy of zero. Starting structures for the lipid bilayers were obtained from the CHARMM-GUI project [25, 26]. Each US simulation was simulated for the same simulation time, such that the first nanosecond of each simulation was treated as equilibration period and was not included in the analysis.

3   Results and Discussion

3.1   Force Field Development for [C₂MIM][EtSO₄]

3.1   Force Field Development for [C$_2$MIM][EtSO$_4$]

In the parametrization procedure the LJ parameters of the anion were changed manually to reproduce experimental densities and self diffusion coefficients. Since the LJ intramolecular interactions alter the dihedral angle CA_2-CA_1-Oet-S, it had to be reparameterized as well. This parametrization was non-systematic, but the obtained results justify this procedure. The new LJ parameters are all given in Table 1. The nomenclature for the interaction sites is shown in Fig. 1.

3.1.1   Density

In Fig. 3 our simulated data is compared to two experimental data sets. In general all densities are decreasing almost linearly with increasing temperature. The error in our simulated data is approximately 0.3%. Our simulated data exactly reproduce the experimental data from [48]. The experimental densities from [57] are shifted

Fig. 3 Densities of [C$_2$MIM][EtSO$_4$] as a function of temperature (measured data [48, 57]: *filled symbols*; simulated data: *open symbols*). Published with kind permission of ©Fraunhofer SCAI 2016. All Rights Reserved

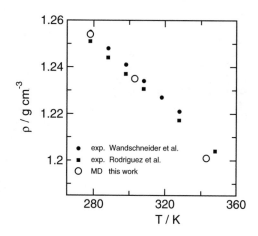

to higher values. These shifts may be caused by a different water concentration of the measured samples. Therefore, the new parameterized force field provides to a good description of the liquid density of [C$_2$MIM][EtSO$_4$].

3.1.2 Self-Diffusion Coefficients

In the context of self-diffusion it is crucial to simulate over sufficiently long times to be in the diffusive regime. As we already discussed in our previous publication [33] for the ionic liquid [C$_2$MIM][NTf$_2$] the mean square displacements of both ions at 303 K after 10 ns are already much larger than the actual size of an ion. The same is true for [C$_2$MIM][EtSO$_4$]. Thus, we can be assured to be in the diffusive regime.

In Fig. 4 the self-diffusion coefficients simulated in this work are compared with experimental pulsed field gradient nuclear magnetic resonance values [39]. The experimental data differ by a factor of about 1.3 from each other. This factor is much larger than the experimental errors and gives an estimate of the accuracy up to which self-diffusion coefficients can be determined experimentally. This estimate is important for deciding how accurate the force field parameterization has to be done, as it makes no sense trying to be more exact in simulation than the experimental variations. Our simulated data differ at most by a factor of 1.5 which is in the range of the uncertainties of the experimental data. The results show that our new force field is also able to describe the self-diffusion of [C$_2$MIM][EtSO$_4$].

3.1.3 Heat of Vaporization

The heat of vaporization $\Delta_v H$ has been computed from the difference between the molar internal energy of the gas and the liquid phase. Table 6 compares the

Fig. 4 Self-diffusion coefficients of [C$_2$MIM][EtSO$_4$] as a function of temperature (measured data [39]: *filled symbols*; simulated data: *open symbols*). Published with kind permission of ©Fraunhofer SCAI 2016. All Rights Reserved

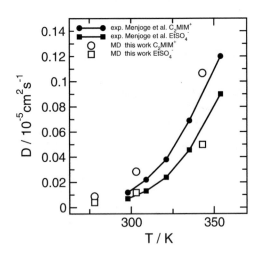

Table 6 Measured [1] as well as simulated heat of vaporization (in kJ/mol) of [C$_2$MIM][EtSO$_4$]

T (K)	$\Delta_v H_{Armstrong}$	$\Delta_v H_{sim}$
298	164.0	163.6

Fig. 5 Shear viscosities of [C$_2$MIM][EtSO$_4$] as a function of temperature (measured data [37]: *filled symbols*; simulated data: *open symbols*). Published with kind permission of ©Fraunhofer SCAI 2016. All Rights Reserved

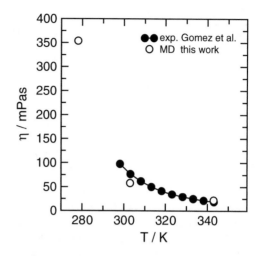

simulated value with the experimental one at 298 K. The heat of vaporizations only differ by 0.4 kJ/mol. Clearly the refined force field is capable to describe correctly the internal energy of the liquid and the interaction of ion pairs in the gas phase.

3.1.4 Shear Viscosity

Although the viscosity of highly viscous fluids is hard to simulate with classical MD, the simulated data agree excellently with the experimental ones measured from [37] (see Fig. 5).

3.2 Free Energy Profiles

Figure 6 shows the free energy profile of [C$_2$MIM][NTf$_2$] with the different ions that are forced to move through the membrane due to the umbrella sampling. The free energy is normalized to 0 kJ/mol at the center of the membrane ($r = 0$ nm). At $r = 4$ nm the center of the water phase is located. The green profile is the free energy of one isolated cation, the black profile is the free energy when one anion accompanied by a single cation and the red profile is the free energy when one ion pair accompanied by a second one is moving through the membrane.

There are significant differences between the three profiles. For the single cation, the global free energy minimum in the water phase is located at $r \approx 4$ nm. This is

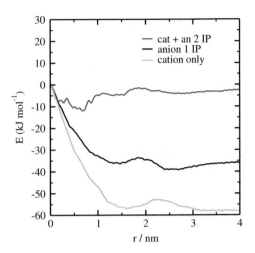

understandable because a single cation cannot shield its charge inside the membrane
and can minimize its free energy best in the water phase together with the anion.
However, there is also a second minimum inside the polar part of the membrane
($r \approx 1.7$ nm). It is very unlikely that a single cation could penetrate the nonpolar
part of the membrane because it would have to overcome a barrier of 58 kJ/mol. The
single ion pair is favorably located on the surface of the membrane at $r \approx 2.5$ nm,
but can also be situated within the polar part of the membrane ($r \approx 1.5$ nm). The
energy barrier of the ion pair to cross the membrane decreases to 39 kJ/mol. In
contrast to the first two profiles the global minimum for the two ion pairs is between
the nonpolar center and the polar part of the membrane ($r \approx 0.7$ nm). The profile is
relatively flat and exhibits only a small energy barrier of 10 kJ/mol at the center of
the membrane.

This data suggests that as more ions are able to cluster inside the membrane,
the smaller the energy barrier at the center of the membrane becomes. This is due
to the fact that the ions can shield their charge better in the nonpolar region of the
membrane. The system with two ion pairs represents the most realistic situation,
since the IL concentration in the experiment allows the formation of ion pair cluster.
Hence for further investigations we always simulated systems with two ion pairs.

In Fig. 7 the free energy profile of [C$_2$MIM][NTf$_2$] is compared to the ones
of [C$_{12}$MIM][NTf$_2$] and [C$_2$MIM][EtSO$_4$]. The profiles for [C$_2$MIM][NTf$_2$] and
[C$_{12}$MIM][NTf$_2$] are similar in that both IL profiles have a minimum located inside
the membrane at $r \approx 1.8$ nm. Their only difference is seen in the free energy in the
water phase. Due to the nonpolar chain of [C$_{12}$MIM][NTf$_2$] the free energy in the
water phase is much lower and thus it is less soluble in water than [C$_2$MIM][NTf$_2$].
The profile of [C$_2$MIM][EtSO$_4$] looks quite different. The free energy in the water
phase is significantly higher compared to the other ILs, which means it is more
water soluble. Consequently the energy barrier at the center of the membrane rises
to 20 kJ/mol. This shows that the high polarity of the [EtSO$_4$] anion compared to

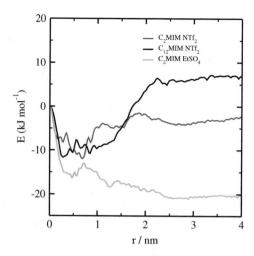

Fig. 7 Profile of the free energy of the ionic liquids [C$_2$MIM][NTf$_2$], [C$_{12}$MIM][NTf$_2$] [C$_2$MIM][EtSO$_4$] as a function of the distance from the center of the membrane. Published with kind permission of ©Fraunhofer SCAI 2016. All Rights Reserved

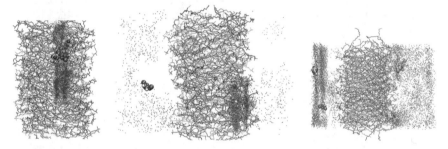

Fig. 8 Position of the cation ring and the center of mass of the anion cumulative at every time step for [C$_2$MIM][NTf$_2$]. Published with kind permission of ©Fraunhofer SCAI 2016. All Rights Reserved

the [NTf$_2$] anion leads to a major shift in the properties of the IL. Therefor, the [NTf$_2$] ILs accumulate inside the membrane, while the [EtSO$_4$] IL is more likely to be found in the water phase.

The free energy profile gives important information about the thermodynamic properties of the system but lacks the atomistic interaction details. One major advantage of molecular dynamics is the possibility to analyze the structure of the system as well. Figures 8 , 9 and 10 show the simulation boxes for the ion pair fixed at $r = 0$, 1 and 4 nm. For clarity, the water is not shown. The position of the cation C$_r$ atom site and the anion's N/SO atom site are shown as a point at every time step. This gives the information about the location of each ion for the total simulation time. For [C$_2$MIM][NTf$_2$] and [C$_{12}$MIM][NTf$_2$] all four ions form a cluster at $r \approx 0$ nm. This is according to the free energy profile, since the ions

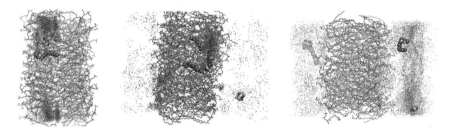

Fig. 9 Position of the cation ring and the center of mass of the anion cumulative at every time step for [C$_{12}$MIM][NTf$_2$]. Published with kind permission of ©Fraunhofer SCAI 2016. All Rights Reserved

Fig. 10 Position of the cation ring and the center of mass of the anion cumulative at every time step for [C$_2$MIM][EtSO$_4$]. Published with kind permission of ©Fraunhofer SCAI 2016. All Rights Reserved

have to shield themselves the most while in the nonpolar part of the membrane. For [C$_2$MIM][EtSO$_4$] the position at the center of the membrane is very unfavorable, and leads to one anion leaving the membrane. In all systems the ions arrange in a sandwich-like structure (i.e. cation, anion, cation). If the fixed ion pair moves to $r = 1$ nm, then more ions can leave the membrane: One cation in the case of [C$_2$MIM][NTf$_2$], one anion in the case of [C$_{12}$MIM][NTf$_2$] and one ion pair in the case of [C$_2$MIM][EtSO$_4$]. Here, again more [C$_2$MIM][EtSO$_4$] ions leave the membrane because the global minimum is located in the water phase. At the center of the water phase the situation for all IL is the same. The second ion pair is moving freely through the water phase, sometimes touching the membrane surface.

4 Outlook and Conclusion

4.1 Outlook: Towards Fully Automated Force Field Development

Force field parameterization is often a very tedious and time-consuming task, especially in the case of intermolecular interactions. In the present work, the LJ parameters σ and ε were manually adjusted to reproduce experimental data, as it could be done quickly by experience and chemical intuition. However, in the future we intend to parameterize force fields in a more systematic and user-friendly way. In recent years, two optimization toolkits were developed at SCAI, which can be efficiently combined: The global optimization toolkit *CoSMoS* [35] and the local optimization techniques implemented in the software package *GROW* [19–21], see Fig. 11.

The utilization of a global optimization followed by a local optimization is expected to be the best way to parameterize intermolecular force fields. Due to the global pre-optimization, no initial force field parameters (e.g. from literature) have to be supplied. Thus, this combination workflow can be used in a generic way independent of the optimization problem and the targeted physical properties. Moreover, computing time can be reduced due to efficient gradient and Hessian calculations within GROW [22] or by replacing GROW with an alternative derivative-free method based on sparse grids, called *SpaGrOW* [18]. The proposed methodology turned out to be reliable for many chemical systems, especially for pure systems like phosgene, methanol, ethylene oxide, and $[C_nMIM][NTf_2]$ [34].

The next step will be to apply the combined toolkits to additional ionic liquids and larger molecular systems like fluorinated alcohols. Thereby, many physical properties like densities, VLE data, and transport properties will be simultaneously fitted to experimental data over a wide range of temperatures and pressures. Recently, the packages have been re-implemented into a user-friendly, modular, and easily extendible toolkit [23] so that they are ready to be applied to a large number of force field parameterization tasks.

While CoSMoS and GROW can efficiently optimize intermolecular parameters, $Wolf_2Pack$ (Fig. 12) was developed to optimize intramolecular parameters and to further validate already existing force fields [30, 47]. $Wolf_2Pack$ uses quantum mechanical target observables for optimizing bond, angle and torsion parameters. Consequently, a database has been constructed that contains a variety of molecules, their relative potential energies and geometries as computed using different theory levels (e.g. HF/6-31G(d), MP2/aug-cc-pVDZ), partial atomic charges as computed using different approaches (e.g. AMBER [7], OPLS [28], Glycam06 [29]), and force fields (e.g. Gaff [58], Parm14SB [5], Glycam06 [29], ExTrM [30]). The size of this database increases over time as new molecules are investigated, often due to new atom type combinations for modeling internal coordinates, and as additional theory levels are employed.

A complete force field includes both the intra- and intermolecular parameters, which are closely coupled to one another. Therefore, to fully optimize a force field one must utilize a workflow where both $Wolf_2Pack$ and CoSMoS/GROW are integrated together in a cyclic nature (Fig. 13). Currently an implemented realization of this workflow is being sought, as well as a guideline for how often the intramolecular-intermolecular cycle should be performed for obtaining overall converged parameters.

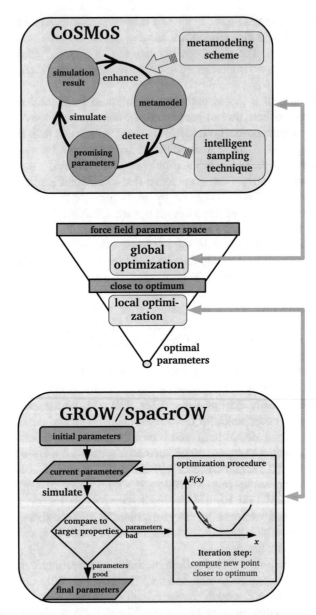

Fig. 11 Combination of the global optimization toolkit CoSMoS with the local optimization package GROW for the systematic parameterization of force fields. Published with kind permission of ©Fraunhofer SCAI 2016. All Rights Reserved

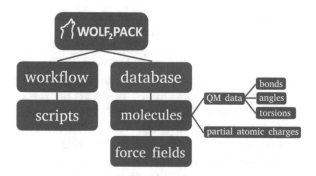

Fig. 12 Component overview of Wolf$_2$Pack. Published with kind permission of ©Fraunhofer SCAI 2016. All Rights Reserved

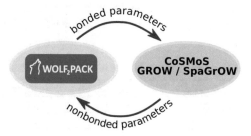

Fig. 13 Complete workflow for optimizing intra- and intermolecular parameters. Published with kind permission of ©Fraunhofer SCAI 2016. All Rights Reserved

4.1.1 Case Study: Automated Parameterization of Ethylene Oxide

CoSMoS and GROW have been tested for several small molecules [21, 35]. In this outlook, we summarize the parameterization of ethylene oxide as an example. Having been a challenge molecule in IFPSC IV, ethylene oxide is an intricate but viable example for force field parameterization. A good force field proposed in [9] is already available that can be used as a benchmark.

The simulation program ms2 [8] was used to compute densities, enthalpies of vaporization and vapor pressures at temperatures ranging from 260 to 430 K in vapor-liquid equilibrium. The results were obtained by Grand Equilibrium Monte Carlo simulations with Widom's test molecule method. The simulation box contained 500 molecules that were modeled as planar triangles with fixed angles and bond lengths. The methylene groups were coarse-grained, with 6-12-Lennard-Jones centers and point charges at each united atom. The LJ parameters were optimized, with all other model parameters set to those in the benchmark force field. The simulation of the liquid and gas phase ran for 200,000 and 100,000 time steps, respectively.

For optimizing the parameters, a global run was performed using CoSMoS. Incorporating the observables for three different temperatures (260, 330 and 430 K) the difference between simulated and experimental data was minimized. The considered parameter ranges were $\varepsilon_{CH_2}, \varepsilon_O \in [0.0\,\text{kJ/mol}, 1.2\,\text{kJ/mol}]$, $\sigma_{CH_2}, \sigma_O \in [0.0\,\text{nm}, 0.4\,\text{nm}]$. Starting from the final parameter set by CoSMoS, a local conjugate gradient optimization was done using GROW. The local optimization incorporated six temperatures (260, 300, 330, 375, 400 and 430 K) for a more reliable fine tuning.

The final force field is given in Table 7. It differs considerably from the benchmark force field in [9], yet has comparable quality (Fig. 14). While the density is less accurate at high temperatures, all three observables are exceptionally accurate from 260 to 360 K. After succeeding in the automated parameterization of ethylene oxide and other benchmark molecules, our optimization workflow is ready for force field development of larger molecules like the one presented earlier in this paper. Thus, the tedious hand parameterization that can take months or even years will be automated to generate high-quality force fields within a few weeks.

Table 7 Final Lennard-Jones parameters compared to the benchmark force field

	ε_{CH_2} (kJ/mol)	ε_O (kJ/mol)	σ_{CH_2} (nm)	σ_O (nm)
Eckl et al. [9]	0.705	0.517	0.353	0.309
CoSMoS + GROW	0.62833	0.80463	0.35972	0.29227

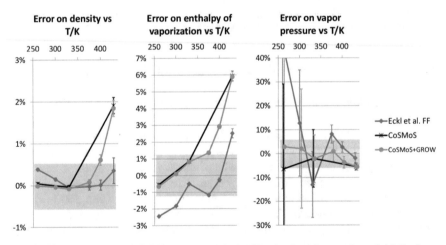

Fig. 14 Percental errors on all three simulated observables for the reference force field, the force field after the global optimization (CoSMoS) and the combined workflow (CoSMoS+GROW). The estimated accuracy of the experiment is displayed in *green*. Published with kind permission of ©Fraunhofer SCAI 2016. All Rights Reserved

4.2 Conclusion

We presented the molecular dynamics simulation of the free energy profile of the ionic liquids $[C_2MIM][NTf_2]$, $[C_{12}MIM][NTf_2]$ and $[C_2MIM][EtSO_4]$ diffusing through a POPC membrane using the umbrella-sampling technique. To be as accurate as possible we parameterized in this work a new united atom force field for the ionic liquid of type 1-alkyl-3-methylimidazoliumethylsulfate $[C_nMIM][EtSO_4]$ with $n = 1, 2, 4, 6, 8$. The proposed force field was derived to reproduce available experimental properties such as density and self-diffusion coefficients for cation and anion. We have achieved very good agreement between simulated results and experimental findings for the heats of vaporization and shear viscosities. All properties are crucial for understanding the nature and interaction of ionic liquids. Starting from $[C_2MIM][NTf_2]$ the variation of cation and anion gives insight into the cation and anion effect. Due to conformational alignment, the profile of $[C_2MIM][NTf_2]$ does not change very much and shows a significant increase in free energy with increasing chain length only in the water phase. However, there is a large decrease in free energy of the whole profile with the exchange of the anion from $[NTf_2]$ to $[EtSO_4]$. In the future we will parameterize force fields in a more systematic and user-friendly way. The global optimization toolkit *CoSMoS* [35] and the local optimization techniques implemented in the software package *GROW* [19–21] will help us to do force filed parameterization much more efficient.

Acknowledgements We thank Stefan Grundei and the Klüber Lubrication KG for continued interest and the German Federal Ministry for Education and Research (BMBF) for financial support through the *SchmiRMal* grant No. 03X4009G.

References

1. J.P. Armstrong, C. Hurst, R.G. Jones, et al., Vaporization of ionic liquids. Phys. Chem. Chem. Phys. **9**, 982–990 (2007)
2. B.H. Besler, K.M. Merz Jr., P.A. Kollman, Atomic charges derived from semiempirical methods. J. Comput. Chem. **11**, 431–439 (1990)
3. P. Bjelkmar, P. Larsson, M. A. Cuendet, B. Hess, E. Lindahl, Implementation of the CHARMM force field in GROMACS: analysis of protein stability effects from correction maps, virtual interaction sites, and water models. J. Chem. Theory Comput. **6**, 459–466 (2010)
4. M.B. Boggara, R. Krishnamoorti, Partitioning of nonsteroidal antiinflammatory drugs in lipid membranes: a molecular dynamics simulation study. Biophys. J. **98**, 586–595 (2010)
5. D.A. Case, V. Babin, J.T. Berryman, et al., *AMBER 14* (University of California, San Francisco, 2014)
6. B. Chen, J.J. Potoff, J.I. Siepmann, Monte Carlo calculations for alcohols and their mixtures with alkanes. Transferable potentials for phase equilibria. 5. United-atom description of primary, secondary, and tertiary alcohols. J. Phys. Chem. B **105**, 3093–3104 (2001)
7. W.D. Cornell, P. Cieplak, C.I. Bayly, et al., A second generation force field for the simulation of proteins, nucleic acids, and organic molecules. J. Am. Chem. Soc. **117**, 5179–5197 (1995)
8. S. Deublein, B. Eckl, J. Stoll, et al., ms2: a molecular simulation tool for thermodynamic properties. Comput. Phys. Commun. **182**, 2350–2367 (2011)

9. B. Eckl, J. Vrabec, H. Hasse, On the application of force fields for predicting a wide variety of properties: Ethylene oxide as an example. Fluid Phase Equilib. **274**, 16–26 (2008)

10. S. Endo, B.I. Escher, K.-U. Goss, Capacities of membrane lipids to accumulate neutral organic chemicals. Environ. Sci. Technol. **45**, 5912–5921 (2011)

11. B.I. Escher, R.P. Schwarzenbach, Partitioning of substituted phenols in liposome-water, biomembrane-water, and octanol-water systems. Environ. Sci. Technol. **30**, 260–270 (1996)

12. U. Essmann, L. Perera, M.L. Berkowitz, et al., A smooth particle mesh Ewald method. J. Chem. Phys. **103**, 8577–8593 (1995)

13. M.J. Frisch, G.W. Trucks, H.B. Schlegel, et al., *Gaussian 03 Revision C.02* (Gaussian, Inc. Wallingford, CT, 2004)

14. B. Hess, C. Kutzner, D. van der Spoel, E. Lindahl, GROMACS 4: algorithms for highly efficient, load-balanced, and scalable molecular simulation. J. Chem. Theory Comput. **4**, 435–447 (2008)

15. W.G. Hoover, Canonical dynamics: equilibrium phase space distributions. Phys. Rev. A **31**, 1695–1697 (1985)

16. H.W. Horn, W.C. Swope, J.W. Pitera, et al., Development of an improved four-site water model for biomolecular simulations: TIP4P-Ew. J. Chem. Phys. **120**, 9665–9678 (2004)

17. J.S. Hub, B.L. de Groot, D. van der Spoel, g_wham – a free weighted histogram analysis implementation including robust error and autocorrelation estimates. J. Chem. Theory Comput. **6**, 3713–3720 (2010)

18. M. Hülsmann, D. Reith, SpaGrOW – a derivative-free optimization scheme for intermolecular force field parameters based on sparse grid methods. Entropy **15**, 3640–3687 (2013)

19. M. Hülsmann, J. Vrabec, A. Maaß, D. Reith, Assessment of numerical optimization algorithms for the development of new molecular models. Comput. Phys. Commun. **18**, 887–905 (2010)

20. M. Hülsmann, T. Köddermann, J. Vrabec, D. Reith, Grow: a gradient-based optimisation workflow for the automated development of molecular models. Comput. Phys. Commun. **181**, 499–513 (2010)

21. M. Hülsmann, T.J. Müller, T. Köddermann, D. Reith, Automated force field optimisation of small molecules using a gradient-based workflow package. Mol. Simul. **36**, 1182–1196 (2011)

22. M. Hülsmann, S. Kopp, M. Huber, D. Reith, Utilization of efficient gradient and hessian computations in the force field optimization process of molecular simulations. Comput. Sci. Discovery **6**, 15005 (2013)

23. M. Hülsmann, K.N. Kirschner, A. Krämer, et al., Optimizing molecular models through force-field parameterization via the efficient combination of modular program packages, in *Foundations of Molecular Modeling and Simulation: Select Papers from FOMMS 2015*, ed. by R.Q. Snurr, C.S. Adjiman, D.A. Kofke (Springer, Singapore, 2016), pp. 53–77

24. J.P.M. Jämbeck, A.P. Lyubartsev, Exploring the free energy landscape of solutes embedded in lipid bilayers. J. Phys. Chem. Lett. **4**, 1781–1787 (2013)

25. S. Jo, T. Kim, W. Im, Automated builder and database of protein/membrane complexes for molecular dynamics simulations. PLoS One **2**, 880 (2007)

26. S. Jo, J.B. Lim, J.B. Klauda, W. Im, CHARMM-GUI membrane builder for mixed bilayers and its application to yeast membranes. Biophys. J. **97**, 50–58 (2009)

27. M.T.O. Jonker, S.A. van der Heijden, Bioconcentration factor hydrophobicity cutoff: an artificial phenomenon reconstructed. Environ. Sci. Technol. **41**, 7363–7369 (2007)

28. W.L. Jorgensen, D.S. Maxwell, J. Tirado-Rives, Development and testing of the OPLS all-atom force field on conformational energetics and properties of organic liquids. J. Am. Chem. Soc. **118**, 11225–11236 (1996)

29. K.N. Kirschner, A.B. Yongye, S.M. Tschampel, et al., GLYCAM06: a generalizable biomolecular force field. Carbohydrates. J. Comput. Chem. **29**, 622–655 (2008)

30. K.N. Kirschner, D. Reith, O. Jato, A. Hinkenjann, Visualizing potential energy curves and conformations on ultra high-resolution display walls. J. Mol. Graph. Model. **62**, 174–180 (2015)

31. A. Klamt, U. Huniar, S. Spycher, J. Keldenich, COSMOmic: a mechanistic approach to the calculation of membrane-water partition coefficients and internal distributions within membranes and micelles. J. Phys. Chem. B **112**, 12148–12157 (2008)

32. J.B. Klauda, R.M. Venable, J.A. Freites, et al., Update of the CHARMM all-atom additive force field for lipids: validation on six lipid types. J. Phys. Chem. B **114**, 7830–7843 (2010)

33. T. Köddermann, D. Paschek, R. Ludwig, Molecular dynamics simulations of ionic liquids: a reliable description of structure, thermodynamics and dynamics. ChemPhysChem **8**, 2464–2470 (2007)

34. T. Köddermann, D. Reith, R. Ludwig, Force field comparison on various model approaches – how to design the best model for the ionic liquid family [cnmim][ntf2]. ChemPhysChem **14**, 3368–3374 (2013)

35. A. Krämer, M. Hülsmann, T. Köddermann, D. Reith, Automated parameterization of intermolecular pair potentials using global optimization techniques. Comput. Phys. Commun. **185**, 3228–3239 (2014)

36. S. Kumar, J.M. Rosenberg, D. Bouzida, R.H. Swendsen, P.A. Kollman, The weighted histogram analysis method for free-energy calculations on biomolecules. I. The method. J. Comput. Chem. **13**, 1011–1021 (1992)

37. J.N.C. Lopes, M.F.C. Gomes, A.A.H. Padua, Nonpolar, polar, and associating solutes in ionic liquids. J. Phys. Chem. B **110**, 16816–16818 (2006)

38. J.N.C. Lopes, A.A.H. Padua, K. Shimizu, Molecular force field for ionic liquids IV: trialkylimidazolium and alkoxycarbonyl-imidazolium cations; alkylsulfonate and alkylsulfate anions. J. Phys. Chem. B **112**, 5039–5046 (2008)

39. A. Menjoge, J. Dixon, J.F. Brennecke, E.J. Maginn, S. Vasenkov, Influence of water on diffusion in imidazolium-based ionic liquids: a pulsed field gradient NMR study. J. Phys. Chem. B **113**, 6353–6359 (2009)

40. S. Nosé, A molecular dynamics method for simulating in the canonical ensemble. Mol. Phys. **52**, 255–268 (1984)

41. S. Nosé, M.L. Klein, Constant pressure molecular dynamics for molecular systems. Mol. Phys. **50**, 1055–1076 (1983)

42. M. Paloncyova, R. DeVane, B. Murcha, K. Berka, M. Otyepka, Amphiphilic drug-like molecules accumulate in a membrane below the head group region. J. Phys. Chem. B **118**, 1030–1039 (2014)

43. M. Parrinello, A. Rahman, Polymorphic transitions in single crystals: a new molecular dynamics method. J. Appl. Phys. **52**, 7182–7180 (1981)

44. G.H. Peters, C. Wang, N. Cruys-Bagger, et al., Binding of serotonin to lipid membranes. J. Am. Chem. Soc. **135**, 2164–2171 (2013)

45. T.J. Piggot, A. Pineiro, S. Khalid, Molecular dynamics simulations of phosphatidylcholine membranes: a comparative force field study. J. Chem. Theory Comput. **8**, 4593–4609 (2012)

46. S. Pronk, S. Pall, R. Schulz, et al., Gromacs 4.5: A high-throughput and highly parallel open source molecular simulation toolkit. Bioinformatics **29**, 845–854 (2013)

47. D. Reith, K.N. Kirschner, A modern workflow for force-field development – bridging quantum mechanics and atomistic computational models. Comput. Phys. Commun. **182**, 2184–2191 (2011)

48. H. Rodriguez, J.F. Brennecke, Temperature and composition dependence of the density and viscosity of binary mixtures of water + ionic liquid. J. Chem. Eng. Data **51**, 2145–2155 (2006)

49. B. Roux, The calculation of the potential of mean force using computer simulations. Comput. Phys. Commun. **91**, 275–282 (1995)

50. U.C. Singh, P.A. Kollman, An approach to computing electrostatic charges for molecules. J. Comput. Chem. **5**, 129–145 (1984)

51. E. Terama, O.H.S. Ollila, E. Salonen, et al., Influence of ethanol on lipid membranes: from lateral pressure profiles to dynamics and partitioning. J. Phys. Chem. B **112**, 4131–4139 (2008)

52. J. Tian, A. Sethi, B. Swanson, B. Goldstein, S. Gnanakaran, Taste of sugar at the membrane: thermodynamics and kinetics of the interaction of a disaccharide with lipid bilayers. J. Biophys. **104**, 622–632 (2013)

53. G.M. Torrie, J.P. Valleau, Monte Carlo free energy estimates using non-Boltzmann sampling: application to the sub-critical Lennard-Jones fluid. Chem. Phys. Lett. **28**, 578–581 (1974)
54. G. Torrie, J. Valleau, Nonphysical sampling distributions in Monte Carlo free-energy estimation: umbrella sampling. J. Comput. Phys. **23**, 187–199 (1977)
55. G. van Meer, D.R. Voelker, G.W. Feigenson, Membrane lipids: where they are and how they behave. Nat. Rev. Mol. Cell Biol. **9**, 112–124 (2008)
56. I. Vorobyov, W.D. Bennett, D.P. Tieleman, T.W. Allen, S. Noskov, The role of atomic polarization in the thermodynamics of chloroform partitioning to lipid bilayers. J. Chem. Theory Comput. **8**, 618–628 (2012)
57. A. Wandschneider, J.K. Lehmann, A. Heintz, Surface tension and density of pure ionic liquids and some binary mixtures with 1-propanol and 1-butanol, J. Chem. Eng. Data **53**, 596–599 (2008)
58. J. Wang, R.M. Wolf, J.W. Caldwell, P.A. Kollman, D.A. Case, Development and testing of a general amber force field. J. Comput. Chem. **25**, 1157–1174 (2004)

The cloud4health Project: Secondary Use of Clinical Data with Secure Cloud-Based Text Mining Services

Juliane Fluck, Philipp Senger, Wolfgang Ziegler, Steffen Claus, and Horst Schwichtenberg

1 Introduction

With the advance of imaging technologies in medical diagnosis and the increase of patient based molecular data for precision medicine, the sheer amount of clinical data is exponentially growing. In addition to the huge storage demand, large compute resources for patient-centric data analysis are becoming increasingly important for hospitals. An approach tackling both issues that becomes more and more popular is the usage of cloud-based resources for storage and large scale processing. The latest Health Insurance Portability and Accountability ACT (HIPAA) [20] includes cloud service providers as business associates and defines standards and penalties concerning the security and privacy of patient data. It makes healthcare industry and university clinical infrastructure cautious about adopting new technologies like public clouds. Nevertheless, in the US, based on HIPAA omnibus rules and the American Recovery and Reinvestment Act [1] requirements, the healthcare industry has begun to migrate patient records and other data into cloud environments.

In Germany, personally identifiable information underlies strong rules based on the federal data protection act and, in addition, the data protection regulations of the 16 federal states [42]. Currently, there is no generally accepted technical realization available for conforming cloud computing. New standards are necessary to enable cloud-based solutions for data storage and processing in Germany.

Another main challenge tackled in cloud4health is the information extraction from electronic health records. The availability of structured patient data is important for clinical research and for the development of new therapies and personalized

J. Fluck (✉) • P. Senger • W. Ziegler • S. Claus • H. Schwichtenberg
Fraunhofer Institute for Algorithms and Scientific Computing SCAI, Schloss Birlinghoven, 53757 Sankt Augustin, Germany
e-mail: juliane.fluck@scai.fraunhofer.de

© Springer International Publishing AG 2017
M. Griebel et al. (eds.), *Scientific Computing and Algorithms in Industrial Simulations*, DOI 10.1007/978-3-319-62458-7_15

medicine. Although electronic health record systems became a standard in hospitals, most patient information is still not available in a structured way for further electronic data analysis. In this area, information extraction methods can help to pill out relevant information from the pool of patient text documents assembled in the electronic health record. Those demands,

- a secure cloud environment and item a secure cloud environment and
- a cloud-based processing environment for information extraction of electronic patient records

have been tackled by Fraunhofer SCAI in the cloud4health project [9]. Cloud4health was funded by the German Federal Ministry of Economics and Technology as one of several TrustedCloud projects [52]. In a consortium of five different partners a cloud-based solution for secondary use of clinical data was developed.

In short, a generic architecture was designed to extract structured and unstructured patient data from electronic health record systems (see Fig. 1). For that, an

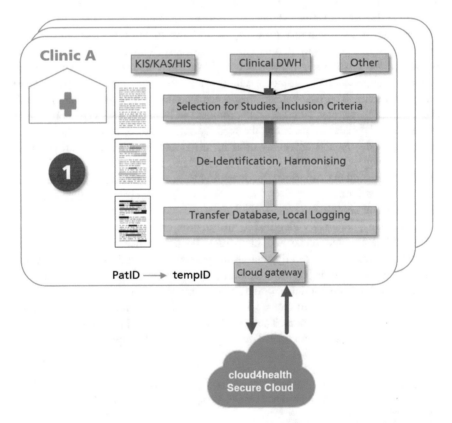

Fig. 1 cloud4health clinical bus. Published with kind permission of ©Fraunhofer SCAI 2016. All Rights Reserved

Extract, Transform and Load (ETL) process was set up by the clinical partners. Furthermore, the use case specific text mining services were set up within the cloud environment. Unstructured text documents can be sent to the cloud and structured information will be returned by the cloud. For a more detailed clinical workflow we refer to [6].

In the next section, we focus on the secure cloud implementation. After a short overview of existing solutions, we introduce the general data protection requirements for an private and highly secure cloud environment. Next, we describe our concrete implementation of this concept.

Finally, in Sect. 3 the design of a multi-purpose text mining environment for clinical information extraction is presented and explored in detail based on two application scenarios. Again, this section presides with an overview about state of the art clinical text mining solutions.

2 Developing a Secure Cloud-Solution for Medicine

2.1 Existing Cloud Solutions for Medicine

Until now, the use of a cloud infrastructure for processing medical data has been limited due to restrictions imposed by data protection laws, especially in European countries such as Germany. For example, the Cloud Standards Customer Council points out "that despite the significant advantages for the utilization of cloud computing as part of Healthcare IT (HIT), security and privacy, reliability, integration and data portability are some of the significant challenges and barriers to implementation that are responsible for its slow adoption" [8].

Public audit of cloud infrastructure is considered as one possible solution for the problem. For example [58, 59] propose to delegate risk assessment of outsourced data to an external audit party because such a trusted entity has expertise and capabilities data owners sometimes do not possess. They conclude that such an auditing service provides a transparent and cost-effective method for data owners to gain trust in cloud technologies. Unfortunately, in such a way, the main effort is shifted to establishing trust into the audit organization. Hence, the rules, principles, scope and frequency of these audits has to be defined and agreed on.

Some institutions, such as the Stanford School of Medicine, provide their users with recommendations and practices regarding the use of medical data in clouds [30]. Although these recommendations are in compliance with US data protection requirements, there are areas of security analysis and derived processes that can be applied under German data protection laws, e.g., classification of risks, data classification and data categorization.

An overview of cloud computing research solutions in healthcare is presented in [23]. In this article, several cloud computing approaches for enabling secondary use of clinical data for data analysis, text mining, or clinical research are reported.

However, only one article addresses data security by setting up and evaluating a cloud solution similar to the community cloud approach of cloud4health. In this approach, presented in [40], the authors used Amazon's Virtual Private cloud, which has SAS 70 Type II audit and ISO 27001 certification, and PCI level DSS validation as a level 1 service provider. In such a way, HIPAA compliance was achieved using an infrastructure hosted in a safeguarded cloud environment. However, this solution does not satisfy the requirements of the German Data Protection Act (BDSG Bundesdatenschutzgesetz) [19] or of clinical data protection officers. In [40], only the authors themselves audited the compliance. In contrast, in cloud4health, the hospital data protection officers evaluated the architecture and where enabled to audit at any time during processing within the cloud.

In order to avoid cloud-based data protection issues, [3] proposes to install a virtualised solution within the clinical infrastructure as strategy for cloud-based NLP services. In cloud4health, both possibilities, virtualised solutions on the Fraunhofer SCAI testbed as well as implementations within the clinical infrastructure were provided. Other publications do either not address clouds specifically [45] or do not address data protection and data security but rather omit this issue [5, 38].

Areas with related issues are the management of the collaborative usage of electronic health records across clinics [16] and the underlying security aspects of these processes [26]. A clinical data management system with a strong emphasis on data protection and security measures is sketched in [13]. Rahmouni [37] discusses challenges introduced by legal and ethical aspects—especially in multinational clinical data management.

Most of these aforementioned works and projects do not consider the German data protection framework, including its diverse regional regulations and hospital laws. Secondary use of clinical data within the German regulatory framework is investigated by Berliner Forschungsplattform Gesundheit [2]; Ganslandt et al. [22] provides a special focus on privacy enhancing tools. Projects such as Cloudi/o [46] and TRESOR [51] focus on secure ecosystems for use cases within the health sector. Other undertakings concentrate on fundamental principles of securing clinical data in cloud usage scenarios [29, 31]. To secure clinical patient data before delivering them to secondary analysis workflows in external infrastructures, [36] proposes to pre-process the patient data and to remove patient identifying information. Dankar et al. [10] presents research on potential re-identification risk of such pre-processed data.

However, those initiatives that focus on German regulations do not consider challenges that arise through the use of cloud computing services. Security measures are often only applied for a specific use case or only fit to a certain part of the whole ecosystem, including clinic infrastructures, public transport networks and clinic-external cloud infrastructures.

In cloud4health a new approach was considered. For the connection between the clinic and the cloud, a dedicated secure Cloud Gateway with multiple functions was implemented. In particular, 'on the fly' encryption and decryption of anonymized patient data or the dynamic, data-driven start and stop of cloud computing environment and analysis services are important new features. Furthermore, multiple security mechanisms are integrated in the cloud4health cloud infrastructure. In the following, the security measurements and the implemented solutions are presented.

2.2 Requirements for Cloud Infrastructures Arising from Patient Data Processing

In this section, we illustrate the requirements and boundary conditions to be considered when developing a trusted cloud infrastructure and the trustworthy processes to analyse the patient data using this cloud infrastructure. While our approach is specific for the situation in Germany it will be very similar for other European countries, especially when the new European general data protection regulation will be ratified and implemented by the member states [39].

A dedicated cloud infrastructure should be set up for handling the compute and memory-intensive extraction of patient documents. The infrastructure should be able to flexibly provide required CPU and memory to handle varying input data requirements of different clinical users and applications. Public cloud infrastructures could not be used due to German law that protects personal data: most prominent the Federal Data Protection Act (Bundesdatenschutzgesetz) [19]. Thus, a community cloud was set up at the Fraunhofer Institute SCAI.

Given the sensitive nature of personal patient data (even after data anonymization), high security requirements regarding their confidentiality and integrity have to be met. Within the cloud4health project, clinics are the main data providers for the documents to be processed in the cloud.

A first fundamental requirement is that all data transfer and processing has to be initiated and controlled by the responsible person in the clinic. Also, communication with the cloud and the text mining services has to be strictly unidirectional from the clinic to the cloud through a single interface.

Second, the concrete purpose of data processing has to be explicitly defined and followed appropriately. Thus, security measures have to guarantee that the anonymized personal patient data cannot be used for any other purpose than the predetermined one. As a consequence, data must not be accessible by other processes or other users of the cloud; multi-tenancy and separation of data processing has to be ensured on all levels during the complete lifecycle of data handling (from transfer to processing, storage and deletion).

Third, data has to be deleted after a reasonable period of time. If possible, personal data should not be stored in the cloud infrastructure at all.

Fourth, an incident response process has to be implemented, preferably integrated in an overarching security management concept.

Finally, the user of the cloud infrastructure—in case of the cloud4health project this role is taken by the clinical data provider—has to be able to check and verify the technical and organizational security measures, e.g. through a certification executed by trusted third party experts. Even though there are standards for self-assessment and self-certification, an independent verification of security measures is always preferred.

Summing up, strictly following best-practice guidelines and common personal data processing regulations has not proved to be sufficient within the cloud4health project. In most cases, cloud-specifics haven't been covered in these guidelines yet. As well, federal, national and clinic-specific requirements on personal data processing have to be harmonised and addressed appropriately. In the following section, an overview of the realized security mechanisms is given. Moreover, important parts of the implemented cloud infrastructure are explained in detail in the subsequent sections.

2.3 Security Mechanisms

Primary objective of the security measures is to be compliant with classical protective goals i.e. confidentiality, integrity, availability, legal requirements, audit capability and transparency. All protective goals are directly derived from the German Data Protection Act (BDSG Bundesdatenschutzgesetz). The security concept also includes the technical environment as mandated by the German BSI-Standard 100-2 "IT Grundschutz" [27]. Moreover, in addition to the compliance with the aforementioned BDSG the environment complies with further data protection regulations: the different Federal Data Protection Acts (Landesdatenschutzgesetz) or distinct clinical rules.

In cloud4health, only anonymized data may leave the clinics for processing. This is done as part of the workflow along the clinical bus as depicted in Fig. 1. The technology for anonymization is not in the scope of this article but we refer to [48] for detailed information of the technology used in cloud4health. Data security approaches from multiple angles were implemented in the cloud4health cloud services. Suitable data pre-processing and anonymization is combined with technical and organizational measures in the cloud environment. The following list highlights these measures (depicted in Fig. 2).

1. cloud4health implements mechanisms to prevent attacks on the anonymization of the patient data in the cloud through the secure clinical gateway to the cloud as described in Sect. 2.4.1.
2. The gateway also provides end-to-end encryption of the anonymized patient data described in Sect. 2.4.3.
3. The project provides independent virtual private networks (VPN) for every clinic, powered by the Open Source software OpenVPN. Regarding ciphers and key lengths we strictly follow the technical guidelines of the German Federal Office for Information Security (BSI). Thus, the project is able to maintain data security and confidentiality while transferring sensitive data from the clinic to the cloud over a third party, untrusted network.

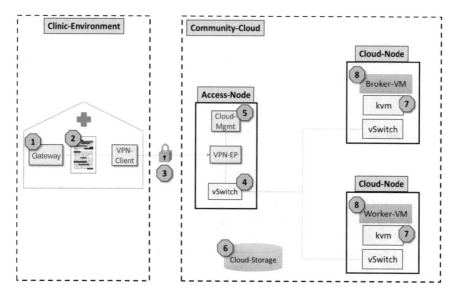

Fig. 2 Actions assuring data security and confidentiality in cloud4health. Published with kind permission of ©Fraunhofer SCAI 2016. All Rights Reserved

4. The VPN endpoint on the cloud side is via VLAN directly connected to the cloud nodes initiated by the clinic. The project uses the virtual switching Open Source software Open vSwitch [35]. Using OpenVPN and Open vSwitch, a separate virtual private network is provided for each clinic, and, in addition, directly couples the infrastructure of the clinic with the cloud resources.

5. Furthermore, the VPN endpoint is connected to the central cloud management. This OpenNebula-component manages all cloud users and handles most of the project's authentication, authorization and accounting requirements. Thus, it serves as the main cloud entry point for users by providing a graphical as well as a programming interface (GUI + API). Regarding authentication and authorization, cloud4health monitors the developments of the SkIDentity project [47] and plans to integrate suitable project outcomes in future.

6. The project adapted cloud-internal virtual machine (VM) deployment procedures. VM images are transferred from the cloud storage to cloud nodes through an encrypted channel. Encryption techniques and key lengths can be adjusted based on customers' preferences.

7. The project uses self-compiled and optimized Kernel-based Virtual Machine (KVM) kernel modules for performance improvements, especially focused on disk I/O.

8. The default shut-down procedures of OpenNebula have been extended. Once the clinic stops the virtual machines, images on cloud nodes are securely

deleted by overwriting the used disk space with random patterns multiple times.

In the following sections the different measurements and implementations are described in more detail.

2.4 Secure Cloud Infrastructure

Given the latest developments in the area of intelligence services and the ever-increasing market for software exploits, being able to find software manipulations and potential backdoors is of major importance. Thus, the usage of Open Source software (OSS) can be advantageous [28]. Of course, OSS also may still exhibit security vulnerabilities.[1] The cloud4health project uses OSS as follows: OpenNebula serves as the underlying cloud framework, the qemu/KVM software is used as the main hypervisor, OpenVPN secures the transport layer, and Open vSwitch constitutes the software solution for cloud-internal VLANs.

To successfully meet varying user demands, cloud4health set up a cloud infrastructure based on the Open Source OpenNebula framework [33]. On the development side, virtual machines with test environments can easily be set up within OpenNebula and can be replaced with updated versions if needed. Equally important, within the service provisioning cycle, virtual machines can be started and stopped immediately. Thus, the infrastructure serves as an ideal test bed for developing, stress testing and finally providing the text mining services and sufficient compute and memory to interested clients.

As services within the cloud infrastructure, the different text analysis engines are integrated within the UIMA-AS framework. Its components communicate over a message bus and each component can be started and stopped without influencing other instances. All involved components are located and executed within virtual machines that are qemu/KVM compatible. A more detailed description of the general UIMA based text mining services can be found in Sect. 3.2.

For service initialization and management, a Cloud Gateway for secure communication and data transfer between clinic and cloud has been developed. Based on user provided input data properties—such as number, size and type of text documents or the use case/scenario to be analysed—the initialization component selects suitable virtual machines. Where suitability refers to the configuration of text analysis services as well as the number of concurrently started virtual machines. The

---

[1]A recent example is the "Heartbleed bug" of the OpenSSL cryptography libraries. The vulnerability causing code was found over 2 years after its initial integration into the libraries' code base. The flawed code rendered approximately 20% of all Internet servers vulnerable to a potential theft of private data (such as private keys).

mapping from user input parameters to the initialized cloud service environment can either be based on local configuration files or a central cloud service. The initialization component directly interacts with the cloud environment's API. Thus, the whole UIMA-AS service environment can be started, configured and provided without further user interaction. The Cloud Gateway is described in more detail in the next section.

2.4.1 Secure Clinical Gateway to the Cloud

For the communication between the hospitals and the cloud containing the text mining infrastructure, a secure gateway is necessary. Figure 1 shows the Cloud Gateway positioned at the end of the clinical bus. The gateway is implemented as a web service in Java to allow easy usage in different clinics with different IT infrastructures. It has three main responsibilities:

1. It disburdens the clinician from configuring and managing the text mining and cloud infrastructure.
2. The patient records carry internal identifiers (IDs) that have to be temporarily removed for processing in the cloud. To avoid possible de-identification attacks by combining multiple documents of one patient, a new temporal random ID is selected for every single document. After ID-replacement, the document is sent to the text mining service within the cloud environment. The Cloud Gateway restores the original ID, when the results of the text mining service have been received by the gateway.
3. Encrypting all patient documents before they are sent to the cloud and decrypting all Operational Data Model (ODM) documents when they are returned from the cloud.

The already established mechanisms for user authentication in the hospitals are used, currently most often based on username and password-like credentials. Based on the authentication, the authorization for accessing services in the cloud is decided. However, the authentication interface is generic and allows use of certificates issued by a trusted Certificate Authority (CA) as well as attribute-based or role-based authorization.

Finally, the Cloud Gateway is in charge of starting the cloud environment (through an appropriate request to the cloud management system) and starting the UIMA framework with pipelines as needed by the specific study. Similarly, when all data records have been processed the gateway needs to signal the cloud management system that the resources are no longer needed, the used nodes have to be cleaned up and the cloud resources used for the study have to be shut-down. In the following the data processing flow is detailed.

2.4.2 Data Processing Flow

Initiation Phase When all clinical data has been prepared and stored within the transfer database, the Cloud Gateway can be invoked from the transfer database. First, the authentication process is conducted and the task configuration is processed in the initial phase. The configuration contains the information about the text mining application and the number of documents to be processed. With this configuration information, the Cloud Gateway is then able to trigger the Initiator component of the cloud hosting infrastructure. The Initiator component in turn starts the cloud environment by connecting to the cloud management system to start-up the Virtual Machines (VMs). Depending on the configuration, the required UIMA components for the specific text mining pipelines are started and parallel workflows are initiated. When the complete environment is up and running, the Cloud Gateway registers the UIMA pipelines at the Broker API. The Broker is responsible to pass the documents to the text mining pipelines. As result of the registration, the end point references of the pipelines are returned to the Cloud Gateway.

Parallel Processing Once the initialization phase is complete, the transfer database starts sending documents to the Cloud Gateway. In order to benefit from the computing power of the cloud resources and the inherent parallel processing capability of the UIMA framework, the Cloud Gateway would send documents asynchronously to the Broker and receives the individual results asynchronously as soon as they are ready. However, because of the restriction regarding the connections from the cloud to the clinic mentioned before, we implemented a different approach to ensure the maximum of parallel processing. Instead of sending asynchronous SOAP messages, synchronous SOAP messages are sent using an individual thread for each message. The synchronous communication allows using the same connection from the clinic to the cloud for sending one patient document and receiving the resulting ODM document, while threading allows exploiting parallel processing in the cloud.

Allocation of Temporal IDs In each document, the Cloud Gateway replaces the ID provided by the transfer database by a random temporal ID and sends the modified document to the Broker for further distribution. Once the text mining result for a document is available, this document is returned to the Cloud Gateway in form of an ODM document. The gateway replaces the temporal ID in the ODM document with the original ID and returns the data to the transfer database. By using a random temporal ID in the cloud, it is not possible to recombine multiple anonymized documents that belong to the same patient. However, when that ID is restored in the transfer database all ODM documents that belong to the same patient can be merged and analysed together.[2]

---

[2]It should be noted that the ID does not allow identifying the patient because this ID already is a result of the anonymization process. If an agreement with the responsible data protection officers can be reached, it would be desirable to only pseudonymise the documents instead, as patients could benefit from results of the text mining if these results could be mapped to a patient.

Termination Phase When all documents are processed, the transfer database indicates the end of the study. Thereupon, the Cloud Gateway requests the shutdown of the cloud infrastructure for this specific use case. All involved text analysis engines and their respective virtual machines are shut down.

In the following the end-to-end encryption and further cloud security measures are described in more detail.

2.4.3 End-to-End Encryption

The discussed cloud4health workflow applies strong encryption throughout the whole data processing (depicted in Fig. 3). By complementary measures on multiple layers, end-to-end encryption is achieved. The transport layer within the OpenVPN software [34] is secured by Transport Layer Security (TLS). Its cipher and key lengths configuration follows the regularly updated recommendations of the German BSI. Beyond this transport layer encryption, the communication of the text analysis services is secured by standard HTTPS.

Accordingly, the anonymized patient data it is encrypted on IP packet level during the transfer between the hospital and cloud through the VPN tunnel (OpenVPN).

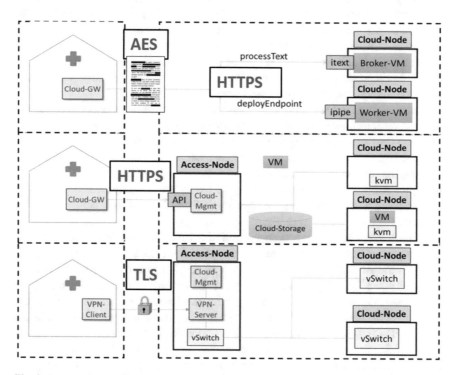

Fig. 3 Layers of encryption in cloud4health. Published with kind permission of ©Fraunhofer SCAI 2016. All Rights Reserved

However, the packets are only encrypted from the clinic tunnel endpoint to the cloud tunnel endpoint. Once the packets reach the end-point of the tunnel they are decrypted and the document would be readable within the cloud network.

In order to avoid having readable anonymized patient data in the cloud network, an additional encryption by the cloud Gateway is established. This end-to-end encryption ensures that there are no readable data in the cloud outside the UIMA text mining services. The encrypted documents can only be decrypted by the Broker which is responsible for the distribution of the documents to the UIMA pipelines. At this stage, the decrypted documents are only accessible in the memory of the compute nodes of the UIMA pipeline where they are distributed upon. It should be noted that no patient data is persisted anywhere in the cloud.

For the transport of the results back the same OpenVPN tunnel is used. Now, the Broker encrypts the results delivered by the UIMA pipelines. The encrypted results are sent back to the clinical gateway. The cloud Gateway in turn decrypts the data before sending them back to the transfer database. In the transfer database all results are gathered and merged.

Given the recent developments regarding shortcomings of encryption techniques, constant monitoring of such developments is essential. Currently, we use a state-of-the-art hybrid approach based on AES shared keys and X.509 asymmetric keys for encryption and decryption of the AES keys. X.509 certificates are created by the clinic beforehand for both the Cloud Gateway and the Broker component. To increase security, a new AES key is generated for each text mining service initiated. In case that an intruder may get access to an unencrypted AES key, the amount of compromised data is limited to the data of one text analysis (study) started from the clinic. Again, it should be noted that in the cloud4health context, the patient data always is anonymized when processed outside the clinic.

Using homomorphic encryption, where data always remains encrypted, has not been an option in cloud4health, mainly for two reasons. First, currently, the type of operations that can be applied to homomorphic encrypted data is still very limited. It does not allow complex text analysis and automated processing with load distribution within the UIMA framework. Second, the expenditure of time for homomorphic encryption is several magnitudes higher than required for the hybrid approach and would go beyond the capacity of the clinical infrastructure hosting the gateway.

In principle, encryption and decryption of the data is done using a shared AES key. To protect this key (which must be shared with the Broker), the Cloud Gateway encrypts the AES key with the public key of the X.509 certificate of the Broker. The encrypted AES key is included within the message from Cloud Gateway to the Broker. Once a message is received by the Broker, the Broker verifies the integrity of the message. When the message is valid, the Broker extracts the different components and uses its private key to decrypt the AES key. In a next step, the AES key is used to decrypt the patient document that is also included in the message. The X.509 certificate of the Broker and the X.509 certificate of the Cloud Gateway are stored in the clinic. For the Broker, the private key of certificate needs to be available

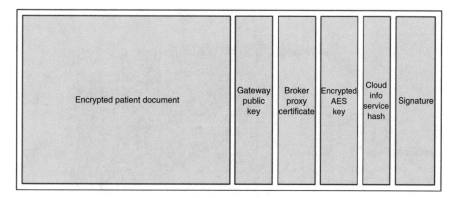

Fig. 4 Cloud Gateway XML messages to Broker. Published with kind permission of ©Fraunhofer SCAI 2016. All Rights Reserved

in the cloud and is stored within the OpenNebula Keystore. Regarding the certificate of the Cloud Gateway, only the public key needs to be available in the cloud.

As a consequence, a potential attacker could gain access to the private key of the Broker and subsequently decrypt the encrypted AES key sent to the Broker. To limit the potential damage to one compromised AES key, we do not use the original X.509 certificate created for the Broker but short-lived proxy-certificates the Cloud Gateway creates from the original X.509 certificate for each study. Thus, each AES key created for a study is encrypted using a unique proxy-certificate. In consequence, an attacker could decrypt and compromise exactly one AES key and would have at maximum access to the anonymized patient data and the corresponding text mining results of one study.

Finally, the resulting XML-document is signed using the private key of the Cloud Gateway to allow the Broker verifying sender and integrity of the messages. As an additional security measure, a hash of core information regarding the cloud infrastructure in use is included. After the communication between Cloud Gateway and Broker has been initialized, this XML-document is sent for each document with patient data including along with the encrypted patient data, the keys required for decryption, for encryption of results, for verification of the document authenticity and the target cloud infrastructure (see Fig. 4).

2.4.4 Multi-Tenancy and No Data Persistence

As different clinical users—especially when working in different clinics that use the cloud4health infrastructure—have different intentions and access rights, security measures have to guarantee data separation across all involved components and layers (hypervisor, virtual machine, storage, network, etc.). This data separation is achieved by a combination of several complementary steps. First, each customer gets access to its own exclusive text analysis environment, which consists of the text analysis services adapted to its requirements, the virtual machines and the

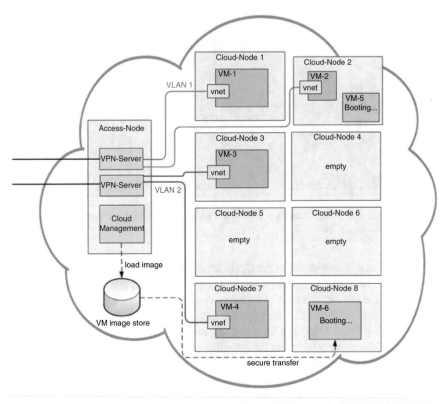

Fig. 5 Cloud architecture. Published with kind permission of ©Fraunhofer SCAI 2016. All Rights Reserved

cloud-internal network. Virtual machines are exclusively assigned to only one client, no shared services are provided. The network between the Virtual Machines (VM), the cloud-internal network is separated from other clients' communication by appropriately applying virtual networks (VLANs). A Virtual Private Network (VPN) as well as the cloud-internal network separation is sketched in Fig. 5.

In case of exceptionally high security demands (e.g. in scenarios where clinical patient data could not be sufficiently anonymized), customers can be granted exclusive execution rights on cloud nodes. This guarantees that only virtual machines of one client are executed on the respective node.

No Data Persistence Plain text, unencrypted anonymized patient data only resides in memory on the cloud nodes and no data is persisted anywhere in the cloud environment. As soon as the text analysis has been successfully completed, all involved text analysis engines and their respective virtual machines are shut down. Their remnants on the cloud hypervisor nodes (i.e. the OS disk images) are securely deleted by multiply overwriting them with random patterns. Thus, also unintentional data leftovers such as log files or memory swaps are permanently and irrecoverably removed from the cloud infrastructure.

3 Clinical Text Mining Solutions

3.1 Short Literature Overview

Natural Language Processing (NLP) in Biomedicine (BioNLP) is an active research field where a number of information retrieval and information extraction methods has been developed and adapted especially to the life science domain. In the area of clinical text mining, method development focuses on extracting relevant information from clinical text documents. These documents are a rich resource for phenotype information but represent also one of the most diverse document class. Early clinical text mining approaches aimed at the extraction of specific terminology such as ICD coding [15], the annotation of Unified Medical Language System (UMLS) (e.g. [12]) or SNOMED-CT [17]. Other approaches focus on the extraction of medication information. MedEx [60] for example, recognizes medication information with a recall and precision ≥ 0.90 for discharge summaries and clinic notes on Vanderbilt clinical documents. For a detailed overview about mining in electronic health records we refer to [11].

So far, most work has been done on patient data in English language. There are several reasons for this language focus. First, in the USA, electronic health records have been introduced much earlier compared to other countries. Second, more resources, in particular terminology resources as well as text mining workflows, are available in English. Finally, data protection laws are different in the USA and UK compared to Germany. That makes it easier to mine clinical records in these countries.

Nevertheless, even in English, the public availability of clinical records is still an issue [4]. With the purpose to support method development in clinical text mining, the Informatics for Integrating Biology and the Bedside (i2b2) NLP challenges provide annotated medical text data and tasks [49]. The 2009 i2b2 NLP challenge focused on medication extraction using de-identified discharge summaries from healthcare partners [54]. Twenty teams competed to identify medications and their signatures. The best systems achieved F-scores ≥ 0.80.

In the 2010 i2b2 shared tasks, the objective was to evaluate state of the art techniques for information extraction from patient health records [55]. The task organizers presented data for the extraction of concepts, assertions, and relations in discharge summaries. The goal of the concept task was to recognize medical problems, treatments and tests. In the assertion classification, a given medical problem needs to be classified as present, absent, hypothetically or conditionally present in the patient. Especially the absence of a clinical problem is important within patient anamneses and occurs frequent in clinical diagnosis. The relation extraction task includes medical problem–treatment relations, medical problem–test relations and medical problem–medical problem relations. Using this reference standard, 22 systems were developed for concept extraction, 21 for assertion classification, and 16 for relation classification. For assertions and relations, best systems achieved F-scores of 0,9 and 0,73 respectively. Best F-scores for concept recognition reached 0,85 for exact and 0,93 respectively for inexact match. The participating Fraunhofer system was under the top five with F-score values of 0,82 and 0,91 respectively [24].

Methodically, most participating systems were based on machine learning. Moreover, many teams used ensemble techniques of different classifiers to solve the extraction task and to increase performance. As a result, bigger compute resources are necessary to run these methods on large corpora.

Co-reference resolution, temporal relations and de-identification are other important aspects in clinical text mining. These tasks were tackled in the I2B2 2011, 2012 and 2014 challenges. Another challenge focusing on information retrieval of clinical texts is TREC Medical TREC [50]. This information retrieval assessment started in 2011 and provided training data and assessment environments for finding cohorts for comparative effectiveness research [56]. Methodically, almost all tools used query expansion based on medical terminology such as MESH and UMLS. Assertion recognition, especially for negations, had a high impact of the retrieval performance. Often, medical doctors document the absence of a symptom or a disease in medical records in form of negated statements.

Focusing on general purpose workflows in the medical area, MEDLEE is one of the first workflows developed for information extraction within clinical notes summaries [21]. One MEDLEE application example is the discovery of adverse drug events in discharge summaries [57]. Whereas not all adverse effects could be found (recall value of 0,28), the precision of the extraction system is very high (precision value of 0,985).

Based on the Strategic Health IT Advanced Research Projects (SHARP) [43] a consortium of 16 partners founded the SHARPn project [44]. In SHARPn tools and resources are developed that influence and extend secondary uses of clinical data [7]. One of the information extraction tools provided by SHARPn is cTAKES [41]. It is developed by the Mayo clinics and builds its information extraction environment on existing open-source technologies—the Unstructured Information Management Architecture (UIMA) [53] framework and OpenNLP natural language processing toolkit. In particular, the modularity of UIMA allows an easy adaptation of the workflow to different clinical applications. In SHARPn, numerous use case specific extraction solutions—named Clinical Element Models (CEMs)—were developed.

In cloud4health, a similar approach was taken for a German medical information extraction environment. A generic, UIMA based text mining framework was set up that could be adapted to different use cases or CEMs. An overview of the generic text mining service and description of two use case scenarios is given in the next section.

3.2 General Architecture of the Text Mining Services

This section gives an overview of the general architecture of the text mining services and their implementation in the cloud environment. The overall workflow of the textual resources within the cloud4health solution is the following:

- The text documents are anonymized within the ETL process and all anonymized documents are stored in a transfer database.
- When all documents are in the transfer database, the hospital can initiate the text extraction process via the Cloud Gateway.
- The text documents are sent to the cloud environment via a secure connection.
- Based on the selected text mining service, the documents are processed and the desired information is extracted.
- The structured information is translated to CDISC Operational Data Model (ODM) [32] format. OMD is a vendor neutral, platform-independent format for interchange and archive of clinical study data.
- The ODM document is sent back to the hospital via a secure connection.
- All information is merged in the transfer database and can be analysed based on the application scenario.
- After processing the cloud-based text mining service is shut down.

For the different use cases, each text mining service is a customized UIMA workflow containing several subsequent processing steps applied to the input documents. Figure 6 shows the general architecture of a text mining service. Each text mining service starts with a reader converting the input text document into a UIMA internal format using a specific type-system. The subsequent steps are a customized sequence of different text processing units. This sequence is highly use case specific. Usually, it starts with several natural language pre-processing units (e.g. tokenization, sentence detection). The recognition of various terminologies with the help of named entity recognition is the second building block often integrated into the pipeline. Relation extraction with different methods spanning from regular expressions, rules or machine learning based methods

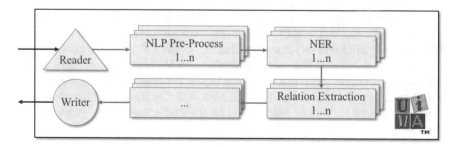

Fig. 6 General architecture of a text mining service. Published with kind permission of ©Fraunhofer SCAI 2016. All Rights Reserved

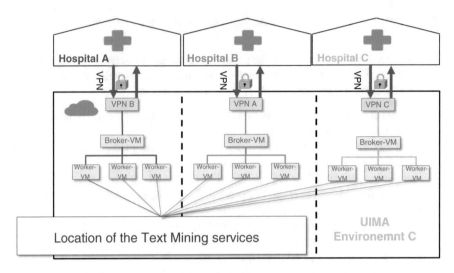

Fig. 7 Location of the use case and hospital specific text mining services within the cloud architecture. Published with kind permission of ©Fraunhofer SCAI 2016. All Rights Reserved

is a third possible block. Depending on the use case, steps can be omitted, different methods can be used in parallel or additional steps can be integrated. The last step of every text mining service is a writer encoding the processed document and the results of the respective analysis into a specific XML format, the ODM format. It is a standard format for the exchange of medical information.

In the cloud4health project, different text mining services have been developed and deployed within the cloud environment. They can be executed in parallel in order to scale the services for different sizes of document corpora. Figure 7 shows the operation area of the developed text mining services within the cloud4health architecture. Each hospital initiates its separate environment within the cloud and for each use case a separate text mining service. The environment consists of one broker node managing the in- and output data and the load balancing of each worker. The worker nodes are implemented as an independent virtual machine (VM) and contain the pipeline for the text mining services. This mechanism implements already the first level of parallelization since a different number of VMs can be started in the virtual environment. The concrete number of VMs is configured based on the size of the incoming text corpora. Depending on the complexity of the service, a defined number of CPUs are assigned to each VM. On the second parallelization level, the text mining service itself can be parallelized across these CPUs via multi threading.

3.3   Overview Use Cases

The flexible and scalable solution presented here enables a lot of different large-scale use cases not necessarily limited to the field of healthcare. Even in the medical area, various application scenarios are reasonable to support research and quality assurance of medical intervention and care. In order to cover a broad spectrum of possible applications, cloud4health selected five different application scenarios. The implemented extraction services show exemplary use cases from the area of pharmacovigilance, clinical guideline verification/plausibility checks, retrospective filling of registers and structuring of huge data resources. The following paragraphs briefly describe the requirements, the constraints, and the data of each use case. For more details we refer to the cloud4health home page [9].

Pharmacovigilance Pharmacovigilance describes the process of monitoring and the risk management of side effects of a certain drug. Retrospective analysis of unstructured patient data can help to detect and understand possible side effects of a prescribed medication. The cloud4health project enables the analysis of a huge amount of patient data and the detection of diagnosis, side effects, medication, and laboratory values. Due to the scalability of the infrastructure, it is possible to process a vast amount of different cases in order to detect unknown side effects. This system can also act as a early detection system for undetected serious side effects.

Prescription Plausibility Check With the help of a text mining service, the validity of prescriptions can be automatically evaluated. This is an important quality control to avoid false medication due to contraindications or undesirable drug—drug interactions. This use case was implemented by the cloud4health project partners. Due to the very high privacy standard for psychiatry data, a private cloud behind the hospital firewall was installed for that use case.

Retrospective Studies: Endoprosthetic Register Retrospective analysis of patient data is an important way to asses the quality of medical interventions, care and medical products. In case of the endoprosthetic register, a German wide register was implemented in 2014 [18]. Cloud4health used the register design as template to extract information about the surgery quality or the quality of the used materials, medication, or coatings. The text mining service was used to extract various information about endoprosthetic surgeries and the prostheses itself from unstructured patient records. More details of this use case are described in Sect. 3.4.

Structuring of Biobank/Pathology Information The analysis of biobank material is of high scientific interest and can be used for quality assurance, as teaching material, or for private or public research projects. The biobank or biorepository is composed of a storage unit for the biological samples (usually human or from animal models) and a database unit for its management. In the past, relevant information of the biological samples are only stored in text documents within the database. To make this information searchable, the sample information has to be extracted from text and filed in a structured manner. In the cloud4health application, TNM (Tumor Nodes Metastasis) classification from a huge corpus of unstructured pathology reports (>= 500,000) were extracted and transferred to the corresponding pathology database. More details of this use case are described in Sect. 3.5.

3.3.1 General Use Case Process Model

The development and implementation of each use case was organized in a simple process model containing the specification of the use case, the development of the software, and the testing of the solution. The four basic steps of this process model are:

1. Specification of the application scenario

   a. Task definition (together with the customer)
   b. Definition of a prototypic data set to be extracted
   c. Data release request for clinic/data protection officials

2. Data and data release

   a. Data release approval from clinic/data protection officials
   b. Release of training data (text)
   c. Extraction workflow specification
   d. Optional re-definition of the data to be extracted

3. Implementation, software development

   a. Selection of the methodical approach
   b. Implementation of the first prototype
   c. Iteration / refinement with customer
   d. Final version and export into cloud environment

4. Data extraction, results and conclusion

   a. Final analysis and documentation of the results
   b. Performance measures
   c. Reporting of the results
   d. Finale release of the use case specific text mining service

In addition to the technical integration and the extraction complexity of the task, the close collaboration with the hospital, in particular the data protection officials and the involved physicians, is one of the main factors for a successful implementation of a use case. Early involvement of data protection officials ensures a timely data release and the detailed knowledge and support of the medical specialists ensures high quality of the resulting service.

In the following, we focus on two use cases. For these cases, we describe the extraction process in detail, give insight about the unstructured input data, the used text mining approaches, and the data points extracted. The training document corpus for both use cases were provided by the university hospital of the University of Erlangen-Nürnberg[3] and hospital units of the RHÖN-KLINIKUM AG.[4]

3.4 Mining Endoprosthetic Surgery Reports

This use case was inspired by the initiative of the "Deutsches Endoprothesenregister" collecting relevant data about all kinds of endoprosthesis and their respective surgeries across Germany. The goal of the register is to monitor and improve the quality of the medical products and the way of their implantation. According to that, the use case followed the guidelines developed by the German register for enabling quality assessment of hip and knee endoprosthesis. The cloud4health consortium decided on a subgroup of data to be automatically extracted from unstructured surgery reports. The data that needs to be extracted from the unstructured free text is:

- Joint (knee/hip)
- Hemibody (right/left)
- Kind of surgery (first, revision)
- Change reasons of endoprosthesis
- Surgery history (on the respective joint)
- Secondary diagnosis
- Restriction of mobility
- Complications
- Information about pain

First, it is important to differentiate between hip and knee and the location of surgery itself (e.g. left knee). Second, another important data object is the kind of surgery done, especially if it was a first endoprosthesis implantation, a revision or another surgery. In case of a revision, it is necessary to know the change reason such as an infection or a luxation that led to the revision. Third, further important data are the information about prior surgeries on the same joint, specific secondary diagnoses and surgery complications. They all have an impact on the lifetime of the implant. Finally, as quality of life observables, we wanted to extract movement restrictions and pain before and after surgery.

The training document corpus, provided by two different hospitals, consists of 256 anonymized reports with diverse layout, length, and content. As gold standard, 70 documents were manually annotated with all specified classes for later evaluation. The chosen subset of documents in the gold standard covers most of

---

[3]in cooperation with the Chair of Medical Informatics, Friedrich-Alexander-Universität Erlangen-Nürnberg, Erlangen, Germany.

[4]RHÖN-KLINIKUM AG, Bad Neustadt/Saale, Germany.

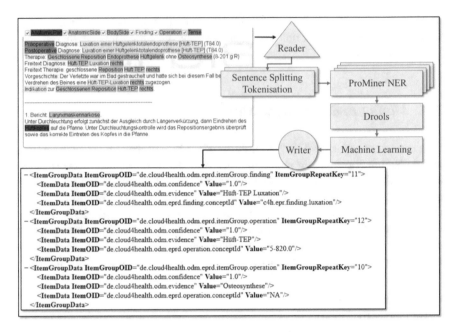

Fig. 8 Approach for extracting structured knowledge (in ODM (Operational Data Model) format [32]) from unstructured surgery reports by named entity recognition, rule-based approaches and machine learning. Published with kind permission of ©Fraunhofer SCAI 2016. All Rights Reserved

the possible variants of the data objects (e.g. joint: knee (21), hip (36), unknown (13); hemibody: right (42), left (27), unknown (3)). The comparably low number of documents annotated in this gold standard is due to the complexity of annotation.

Figure 8 shows the final text mining service for extracting the defined facts (data objects) like the joint or the hemibody from unstructured text. The text reader accepts surgery and discharge reports as input.

First, the NLP pre-processes recognise sentences and tokens. The different terminologies are recognized by the Fraunhofer SCAI system ProMiner [25]. It allows highly parallelized and efficient detection of named entities. For terminology recognition, a couple of terminologies were developed due to the fact that there are no existing ones covering all different aspects within the domain of endoprosthesis. Examples of the self-developed terminologies are the manufactures of the endoprosthesis, the list of complications, and statements about pain of the patient. If possible, we adapted already existing terminologies like OPS, ICD10, and Radlex for anatomy, surgeries, and other related topics and terminologies.

Second, based on the recognised terminology, two different relation extraction systems were applied. Rules were encoded in Domain Specific Language (DSL) within the business logic system Drools [14]. On top of the detected terms, we modeled rule-based approaches using different features of the found entities for the extraction of clearly defined facts from text. For example, the simple presence or

Table 1 Text mining results using the manually annotated data from two different hospitals

Data object to be extracted	Precision	Recall	F-Score
Joint	1.0	1.0	1.0
Hemibody	1.0	1.0	1.0
Reason for revision	0.85	0.89	0.89
Complications	0.85	0.9	0.88
Restriction of mobility	0.83	0.84	0.83
Pain	0.83	0.81	0.82

absence of terms, a special combination of terms, the frequency of terms, or the order of entities found in a sentence can encode a certain fact we want to extract from the text. This led to a set of 64 different rules to assert the relevant information.

Third, we trained a set of classifiers (Support Vector Machines) on top of the NER results in order to extract facts or data objects that cannot easily be detected by rules. The employed features correspond to those we used for the rule-based approach. The Machine Learning approach helps to decide wherever a certain surgery report talks about a surgery on the knee or the hip (even if the respective alternative joint is discussed in the same report).

Finally, the writer converts the resulting information into structured XML information in ODM format. Table 1 shows evaluation results of the endoprosthetic use case based on the gold standard. Due to the small size of the gold standard and the limited number of possible variants, the joint and the hemibody was detected correctly in all reports. All other information is found with an F-score between 0.82 and 0.89. The gold standard includes reports from two different hospitals and shows that the system is able to cope with variations coming from distinct sources with diverse layout and content.

Unfortunately, in this use case, not all desired information could be extracted from those resources. In the first version of the service, we tried to identify the endoprosthetic products implanted. Based on a terminology generated from the training corpus and from public available resources, we were able to recognise the mentioned endoprosthetic product terms in text. Nevertheless, we were hardly able to reference them to the manufacture. That was due to several reasons:

- Endoprosthesis consist of a modular system with parts in different sizes and versions that can be combined by the surgeon. This makes the product identification process complex.
- There was no exhaustive list of the endoprosthetic products publicly available. This makes it difficult to map the mentioned parts properly to the manufacturers.
- The usage of the product was not documented properly in the text and could not be accessed elsewhere in the electronic health records. This made an evaluation difficult.

Therefore, this information was omitted in the final version of the endoprosthesis text mining service. In addition, hospitals registered at the German Endoprothe-senregister currently install automatic readers for the product information. This makes the text extraction of this information obsolete. Nevertheless, it shows the restrictions occurring during set up and implementation of such a service.

3.5 Mining Pathology Reports

In order to facilitate science, clinical research, and translational studies, human tissues with high quality annotations are a central source of information especially in the field of cancer. The international standardized classification of malignant tumours (TNM), developed and maintained by the Union for International Cancer Control (UICC[5]), is a cancer staging system used for the annotation of all solid tumours. These tumour classifications are important annotations for finding adequate tumour samples in pathology departments and tumour banks. Modern routine documentation systems allow for the structured documentation of pathology samples including tumour classifications like TNM. Unfortunately, in a high number of current pathology documentation systems, the tumour classification and other important clinical (meta-) information are only available as free text information in pathology reports. In the University Hospital Erlangen for example, a new pathology documentation system is already integrated into the routine documentation system. But over 500,000 sample documentations are still only available as free text pathology reports. This makes it necessary to employ a high performance text mining approach to extract the information about the tumour classification and to convert it into a structured machine-readable representation. Therefore, the goal of this use case is to extract the complete TNM expression with pre- and suffixes. Text examples of TNM descriptions with specific tumour grading are given in Fig. 9.

TNM Description The small letters (y,p) shown in Fig. 9 are prefix modifiers of the T-component describing the diagnosis methods (p = pathologic examination) and the state of the tumour (y = status after chemotherapy). The main information is encoded by the upper case TNM letters. The T component itself describes the size of the (primary) tumour and whether it has invaded nearby tissue, the N component describes (regional) lymph nodes that are involved, and the M component describes distant metastasis. The upper case TNM letters are often followed by the class and additional information in parenthesis. The correct interpretation of the given example is $T = 2$ (tumour size $\geqslant 2$ cm), $N = 1$ (2/4) (tumour spreads to closest or small number of regional lymph nodes, 2 out of 4) and $M = 1(HEP)$ (distant metastases, liver). In addition to the main classes other information can be optionally present. In the given example, information on invasion into lymphatic vessels (L), into veins (V), and perineural invasion (P) is given. Besides this, other information like tumour grading or tumour scores (e.g. Gleason score for prostate cancer) might be given.

---

[5]http://www.uicc.org/.

Example 1:

Stadieneinteilung: ypT2 ypN1a (2/4) pM1(HEP) L1 V0 Pn0,
Tumorgraduierung: G2 (lokal, 2 + 2 + 2 nach Elston & Ellis) R-
Klassifikation: R0 (medial: 0,3 cm), RDCIS0 (lateral: 0,3 cm),
Regressionsgrading nach Sinn: Grad 1

Example 2:

Stadieneinteilung: ypT2 ypN1a (2/4) pM1(HEP) L1 V0 Pn0,
Tumorgraduierung: G1 (2 + 2 + 2 nach Elston & Ellis, klinisch) R-
Klassifikation: R2 (medial: 0,3 cm), RDCIS0 (lateral: 0,3 cm),
Regressionsgrading nach Sinn: Grad 1

Fig. 9 Two examples of TNM codifications and their context (*highlighted in red*) e.g. grading information ("Tumorgraduierung"). Published with kind permission of ©Fraunhofer SCAI 2016. All Rights Reserved

The TNM Code extraction use case describes a solution that extracts the TNM codification automatically from free text and converts it into a structured representation. The specification of the data to be extracted form free text is in detail:

- (T)umour
- (N)ode
- (M)etastasis
- Pre- and suffixes of TNM code
- Number of revisions
- Tumour grading
- Residual tumour status

The anonymized German pathology reports were provided by two different hospitals. The corpus contains in total 4000 reports, one for each case/patient. The textual layout differs between each report and especially between the different departments and hospitals. The manually defined gold standard contains the correct (sub-) codifications of 400 pathology reports and was provided by both clinical partners in order to measure the performance of the extraction across the two hospitals. In addition, a guideline was developed for the correct extraction of relevant data objects.

The text mining service makes use of a set of regular expressions and machine learning in order to detect the different spelling variants (across different hospitals) of TNM codifications in text. A system of more than 15 different regular expressions (regEx) has been designed within the UIMA framework in order to identify the TNM codification in text. The system is flexible enough to identify full and partial TNM codifications, which is necessary to handle the various different spelling/codification variants across the hospitals. In a second step, each identified codification is analysed and separated into parts by further regular expressions. The algorithm extracts all relevant pre- and suffixes of the T, N, M and optional parts of the code. The regEx component was generated based on the expressions found in the training data and tested with the gold standard data.

The detection and interpretation of the tumour grading and the classification of the residual tumour is highly context specific (e.g. local vs. clinical) and there are no underlying, detectable rules that can be used for the definition of regular expression. This was the motivation for the development of a machine learning component in the UIMA text mining workflow supporting the extraction of context-specific information. We used Support Vector Machines (SVM) in combination with a special set of context dependent features to model this problem. The machine learning system was trained on the described manually annotated data and was evaluated by an independent test set afterwards by using tenfold cross validation.

With the developed regEx approach, the overall TNM codification term is identified with an F-score above 93% (see Table 2). Almost no false codes are extracted and the main recall error is the partial recognition of the TNM terms. The interpretation of the main TNM classes includes the correctly interpreted pre- and suffixes of each component. The developed machine learning system extracted the context specific data objects such as tumour grading and residual tumour classification. Table 3 shows the achieved performance for extracting these data points using the machine learning approach.

Table 2 Text mining results using the manually annotated data from two hospitals by the application of regEx

Data source	Data object to be extracted	F-Score
300 documents hospital 1	T	0.98
	N	0.93
	M	0.98
	Number of revision	0.95
100 documents hospital 2	T	0.98
	N	0.96
	M	0.99

Table 3 F-Score of tenfold cross validation of the machine learning algorithm using the manually annotated data two hospitals

Data source	Data object to be extracted	F-Score
300 documents hospital 1	Tumour grading	0.95
	Residual tumour classification	0.93
100 documents hospital 2	Tumour grading	1.0
	Residual tumour classification	0.96
All documents	Mean tumour grading	0.96
	Mean residual tumour classification	0.93

The overall level of performance is quite high also across both hospitals. The trained classifier is obviously able to generalize across the different terminologies, layouts, and spelling variants a hospital uses. The final text mining service was used to extract the structured TNM and grading information from all pathology reports available in the participating hospitals.

4 Discussion

Within cloud4health, Fraunhofer SCAI was responsible to set up a secure community cloud environment for information extraction of clinical records. In compliance with the German Federal Data Protection Act and the current BSI standard, multiple layers of security mechanisms were developed and implemented to ensure data protection and privacy. Although only anonymized records have been processed in cloud4health, we established the cloud environment in such a way that highest security levels could be achieved. With the growing data coming from translational medicine (e.g. with the integration of genomic data information into treatment decisions) the demand of cloud processing environments becomes more relevant. In patient care, the analysis of patient data is already dispersed over multiple locations and institutional bodies, which makes an external cloud environment a useful solution for data integration and analysis as well. Currently, in Germany there are no guidelines to set up secure cloud environments. The system and the security mechanisms introduced here could be used as foundation to set up a cloud environment for other uses cases with high demands on security. In addition to the introduced security mechanism, further general arrangements deemed to be necessary:

1. Regular updates in compliance with worldwide security standards
2. A trusted partner for hosting the cloud service
3. Establishment of standardized audit processes
4. Segmentation of the data processed in the cloud to reduce the harm in case of a data breach

Moreover, to increase data protection even further, further research to establish processes working on encrypted data is necessary.

In cloud4health, text mining services were established within the cloud environment to extract and structure information hidden in textual descriptions. The flexible and scalable solution presented here enables a lot of different large-scale extraction use cases not necessarily limited to the field of healthcare. The solution facilitates different biomedical applications coming from various institutions like public sector facilities, health insurance funds, industry and small and medium-sized businesses (medicine, biotechnology, and pharmaceutical industry), and public and private hospitals. Possible application scenarios in the healthcare sector are the verification of clinical guidelines, the monitoring of quality and costs, feasibility studies, the retrospective filling of registers and/or studies, the patient recruiting for studies and the structuring of huge data resources and migration of data across different tools and software solutions. To cover different application areas, we implemented several use cases within cloud4health. In this work, two use cases were described in detail which show that we can reach reasonable results with the established services. In case of the more complex endoprosthesis use case, acceptable performance rates between 0.82 and 0.89 F-score were reached. For the pathology application, high performance rates above 0.93 F-score were achieved and the system was used to structure TNM (Tumor Nodes Metastasis classification system) and grading information for huge data sets (>500.000).

The usage of the UIMA framework for text mining has the advantage that it supports a modular structure. A lot of different natural language processing tools are available to adapt the system to new use cases. In the applications presented here, it was necessary to develop new terminology resources and information extraction processes. For other use cases, the now existing terminology resources, recognizer and extraction processes can be applied easily. In these cases, the effort of the adaptation process is comparable low.

Another advantage in using an external cloud is the separation of responsibilities. The hospital as a client is responsible for the ETL process and the set-up of a transfer database. This process was integrated into a clinical portal implemented by the clinical partners in the cloud4health project. The described Cloud Gateway is integrated within this clinical portal. For new use cases, the text mining services are developed independently from the ETL process. After the development phase of a text mining service, the new workflow needs to be uploaded into the cloud store and is afterwards available without further adaptations within the hospital. By using this solution, the hospital has no installation overhead to establishment new use cases. On the other hand, the processing of data itself is completely independent from the developers of the text mining service. The data processing is controlled from the hospitals themselves and the developers have no access to the virtual machines invoked by the user.

In summary, we have established a secure cloud environment with a set of extraction services that ensures data privacy, processing performance as well as extraction quality. Although the infrastructure and the services have been established to solve specific use cases, both the cloud environment as well as the text mining services are established in a generic fashion. It is possible to use this cloud environment for various purposes. Moreover, the text mining services can be easily adapted for different extraction purposes.

Acknowledgements The project cloud4health has been funded by the German Federal Ministry of Economics and Technology in the funding program "Trusted Cloud" (FKZ 01MD11009).

Besides Fraunhofer SCAI, four other partners participated in and contributed to the project: Averbis GmbH, located in Freiburg, coordinated cloud4health, set up the UMIA based text mining environment in the cloud and developed text mining services as well. The Friedrich-Alexander-University Erlangen-Nuremberg and the RHÖN-KLINIKUM AG Bad Neustadt/Saale provided the clinical data and set up the clinical extraction workflow. Finally, TMF—Technology, Methods, and Infrastructure for Networked Medical Research, Berlin, was responsible for data protection related issues.

References

1. American Recovery and Reinvestment Act, Website, 2009. Online at https://www.washington. edu/research/gca/recovery/index.html, visited 18 Dec 2015
2. Berliner Forschungsplattform Gesundheit, Website. Online at http://medinfo.charite.de/ forschung/berliner_forschungsplattform_gesundheit/, visited 18 Dec 2015.
3. D. Carrell, A strategy for deploying secure cloud-based natural language processing systems for applied research involving clinical text, in *Proceedings of the 44th Hawaii International Conference on System Sciences, IEEE Computer Society*, pp. 1–11 (2011)
4. W.W. Chapman, P.M. Nadkarni, L. Hirschman, et al., Overcoming barriers to NLP for clinical text: the role of shared tasks and the need for additional creative solutions. J. Am. Med. Inform. Assoc. **18**, 540–543 (2011)
5. K. Chard, M. Russell, Y. Lussier, E. Mendonça, J. Silverstein, A cloud-based approach to medical NLP, in *AMIA Annual Symposium Proceedings, PMC, US National Library of Medicine, National Institutes of Health*, pp. 207–216 (2011)
6. J. Christoph, L. Griebel, I. Leb, et al., Secure secondary use of clinical data with cloud-based NLP services. Towards a highly scalable research infrastructure. Methods Inf. Med. **54**, 276–282 (2015)
7. C. Chute, J. Pathak, G. Savova, et al., The SHARPn project on secondary use of electronic medical record data: progress, plans, and possibilities, in *AMIA Annual Symposium Proceedings, PMC, US National Library of Medicine, National Institutes of Health*, pp. 248–256 (2011)
8. Cloud Standards Customer Council, Impact of Cloud Computing on Healthcare, Tech. Rep. Online at http://www.cloud-council.org/deliverables/CSCC-Impact-of-Cloud-Computing-on-Healthcare.pdf, visited 18 Dec 2015
9. cloud4health – Cloud Computing für Big-Data-Analysen in der Medizin, Website, 2013. Online at http://cloud4health.de/, visited 18 Dec 2015
10. F. Dankar, K. El Emam, A. Neisa, T. Roffey, Estimating the re-identification risk of clinical data sets. BMC Med. Inform. Decis. Mak. **12**, 66 (2012)
11. J.C. Denny, Chapter 13: Mining electronic health records in the genomics era, PLoS Comput. Biol. **8**, e1002823 (2012)
12. J.C. Denny, A. Spickard, R.A. Miller, et al., Identifying UMLS concepts from ECG impressions using KnowledgeMap, in *AMIA ... Annual Symposium proceedings/AMIA Symposium. AMIA Symposium*, pp. 196–200 (2005)
13. T.M. Deserno, V. Deserno, V. Lowitsch, et al., Aspekte des datenschutzgerechten Managements klinischer Forschungsdaten, in *Proceedings of the 2012 GI Jahrestagung*, pp. 1491–1505 (2012)
14. DROOLS – Business Rules Management System, Website, 2015. Online at http://www.drools. org, visited 18 Dec 2015
15. G.S. Dunham, M.G. Pacak, A.W. Pratt, Automatic indexing of pathology data. J. Am. Soc. Inf. Sci. **29**, 81–90 (1978)
16. Elektronische Fallakte, Website. Online at http://www.fallakte.de, visited 18 Dec 2015
17. P.L. Elkin, A.P. Ruggieri, S.H. Brown, et al., A randomized controlled trial of the accuracy of clinical record retrieval using SNOMED-RT as compared with ICD9-CM, in *Proceedings of the AMIA Symposium*, pp. 159–163 (2001)
18. Endoprothesenregister Deutschland, Website, 2014. Online at http://www.eprd.de, visited 18 Dec (2015)
19. Federal Data Protection Act, Website, 1990. Online at http://www.gesetze-im-internet.de/ englisch_bdsg/index.html, visited 18 Dec 2015
20. Federal Register/Vol. 78, No. 17 – Modifications to the Health Insurance Portability and Accountability Act, Website, 2013. Online at https://www.gpo.gov/fdsys/pkg/FR-2013-01-25/pdf/2013-01073.pdf, visited 18 Dec 2015
21. C. Friedman, L. Shagina, Y. Lussier, G. Hripcsak, Automated encoding of clinical documents based on natural language processing. J. Am. Med. Inform. Assoc. **11**, 392–402 (2004)

22. T. Ganslandt, S. Mate, K. Helbing, U. Sax, H.U. Prokosch, Unlocking data for clinical research – the German i2b2 experience. Appl. Clin. Inform. **2**, 116–127 (2011)
23. L. Griebel, H.-U. Prokosch, F. Köpcke, et al., A scoping review of cloud computing in healthcare, (2015). Online at http://bmcmedinformdecismak.biomedcentral.com/articles/10.1186/s12911-015-0145-7, visited 6 Nov 2015
24. H. Gurulingappa, B. Müller, M. Hofmann-Apitius, A semantic platform for information retrieval from E-health records, in *The Twentieth Text REtrieval Conference (TREC 2011) Proceedings* (2011)
25. D. Hanisch, K. Fundel, H.-T. Mevissen, R. Zimmer, J. Fluck, ProMiner: rule-based protein and gene entity recognition. BMC Bioinf. **6**, 1–9 (2005)
26. T. Hupperich, H. Löhr, A.-R. Sadeghi, M. Winandy, Flexible patient-controlled security for electronic health records, in *IHI '12: Proceedings of the 2Nd ACM SIGHIT International Health Informatics Symposium* (Association for Computing Machinery, New York, 2012), pp. 727–732
27. IT-Grundschutz webpages of the Federal Office for Information Security (BSI), Website, 2015. Online at https://www.bsi.bund.de/EN/Topics/ITGrundschutz/itgrundschutz_node.html, visited 18 Dec 2015
28. Is Open Source Software Insecure? An Introduction to the Issues, Website, 2013. Online at http://oss-watch.ac.uk/resources/securityintro, visited 18 Dec 2015
29. M. Li, S. Yu, Y. Zheng, K. Ren, W. Lou, Scalable and secure sharing of personal health records in cloud computing using attribute-based encryption, IEEE Trans. Parallel Distrib. Syst. **24**, 131–143 (2013)
30. S. Medicine, Recommendations and practices for using cloud computing in medical environments: Stanford Medicine, Information Resources and Technology, Tech. Rep. Online at https://med.stanford.edu/irt/security/cloud.html, visited 6 Nov 2015
31. C. Neuhaus, R. Wierschke, M.V. Löwis, A. Polze, Aspekte des datenschutzgerechten Managements klinischer Forschungsdaten, in *Proceedings of the 2011 GI Jahrestagung, Workshop "Zukunftsfähiges IT-Management im medizinischen Bereich"* (2011)
32. Operational Data Model, Website, 2015. Online at http://www.cdisc.org/odm, visited 18 Dec 2015
33. OpenNebula Project Webpages, Website, 2015. Online at http://opennebula.org/, visited 18 Dec 2015
34. OpenVPN, Website, 2015. Online at https://openvpn.net/, visited 18 Dec 2015
35. Open vSwitch – Production Quality, Multilayer Open Virtual Switch, Website, 2015. Online at http://openvswitch.org/, visited 18 Dec 2015
36. K. Pommerening, K. Helbing, T. Ganslandt, J. Drepper, Identitätsmanagement für Patienten in medizinischen Forschungsverbünden, in *Proceedings of the 2012 GI Jahrestagung*, pp. 1520–1529 (2012)
37. H.B. Rahmouni, T. Solomonides, M.C. Mont, S. Shiu, Privacy compliance and enforcement on European healthgrids: an approach through ontology. Philos. Trans. Math. Phys. Eng. Sci. **368**, 4057–4072 (2010)
38. S. Rea, C. Chute, J. Pathak, et al., Building a robust, scalable and standards-driven infrastructure for secondary use of EHR data: the SHARPn project. J. Biomed. Inform. **45**, 763–771 (2012)
39. Reform of EU Data Protection Rules, Website, 2015. Online at http://ec.europa.eu/justice/data-protection/reform/index_en.htm, visited 18 Dec 2015
40. N. Regola, N.V. Chawla, Storing and using health data in a virtual private cloud. J. Med. Internet Res. **15**, e63 (2013)
41. G.K. Savova, J.J. Masanz, P.V. Ogren, et al., Mayo clinical text analysis and knowledge extraction system (cTAKES): architecture, component evaluation and applications. J. Am. Med. Inform. Assoc. **17**, 507–513 (2010)
42. U.K. Schneider, Sekundärnutzung klinischer Daten – Rechtliche Rahmenbedingungen, in *Schriftenreihe der TMF, Band 12, Medizinisch Wissenschaftliche Verlagsgesellschaft* (2015)

43. SHARP – Strategic Health IT Advanced Research Projects, Website, 2013. Online at https://www.healthit.gov/policy-researchers-implementers/strategic-health-it-advanced-research-projects-sharp, visited 18 Dec 2015
44. SHARPn – Secondary Use of EHR Data, Website, 2013. Online at https://www.healthit.gov/policy-researchers-implementers/secondary-use-ehr-data, visited 18 Dec 2015
45. C.P. Shen, C. Jigjidsuren, S. Dorjgochoo, et al., A data-mining framework for transnational healthcare system. J. Med. Syst. **36**, 2565–2575 (2012)
46. Sicheres Cloud-basiertes Datenmanagement im Umfeld der klinischen Forschung: Project Webpages, Website, 2014. Online at http://www.cloudi-o.de, visited 18 Dec 2015
47. SkIDentity – Trusted Identities for the Cloud, Website, 2015. Online at https://www.skidentity.de/en/home/, visited 18 Dec 2015
48. K. Tomanek, P. Daumke, F. Enders, et al., An interactive de-identification-system, in *Proceedings of the GMDS 2013* (2013)
49. TREC Medical Records Track, Website, 2015. Online at https://www.i2b2.org/NLP/, visited 18 Dec 2015
50. TREC Medical Records Track, Website, 2015. Online at http://trec.nist.gov/data/medical.html, visited 18 Dec 2015
51. TRESOR – Trusted Ecosystem for Standardized and Open Cloud-Based Resources: Project Webpages, Website, 2015. Online at http://www.cloud-tresor.de, visited 18 Dec 2015
52. Trusted Cloud Projekt, Website, 2015. Online at https://www.trusted-cloud.de/projekt, visited 18 Dec 2015
53. UIMA – Unstructured Information Management Architecture, Website, 2015. Online at http://uima.apache.org/, visited 18 Dec 2015
54. O. Uzuner, I. Solti, E. Cadag, Extracting medication information from clinical text. J. Am Med. Inform. Assoc. **17**, 514–518 (2010)
55. Ö. Uzuner, B.R. South, S. Shen, S.L. DuVall, 2010 i2b2/VA challenge on concepts, assertions, and relations in clinical text. J. Am. Med. Inform. Assoc. **18**, 552–556 (2011)
56. E.M. Voorhees, W. Hersh, Overview of the TREC 2012 medical records track, in *The Twenty-First Text REtrieval Conference (TREC 2012) Proceedings*, ed. by E. Voorhees, L. Buckland (2012)
57. X. Wang, G. Hripcsak, M. Markatou, C. Friedman, Active computerized pharmacovigilance using natural language processing, statistics, and electronic health records: a feasibility study. J. Am. Med. Inform. Assoc. **16**, 328–337 (2009)
58. C. Wang, K. Ren, W. Lou, J. Li, Toward publicly auditable secure cloud data storage services. IEEE Netw. **24**, 19–24 (2010)
59. C. Wang, S.S.M. Chow, Q. Wang, K. Ren, W. Lou, Privacy-preserving public auditing for secure cloud storage. IEEE Trans. Comput. **62**, 362–375 (2013)
60. H. Xu, S.P. Stenner, S. Doan, et al., MedEx: a medication information extraction system for clinical narratives. J. Am. Med. Inform. Assoc. **17**, 19–24 (2010)

Dimensionality Reduction for the Analysis of Time Series Data from Wind Turbines

Jochen Garcke, Rodrigo Iza-Teran, Marvin Marks, Mandar Pathare, Dirk Schollbach, and Martin Stettner

1 Introduction

Wind energy is nowadays a very important component of the energy production, capable of meeting 11.4% of the EU's electricity demand in 2015 [3]. By 2010, more than 71,000 wind turbines were designed, built, commissioned, and are still operating in the EU.[1] We investigate two data analysis tasks arising in this domain. One is the analysis of data from numerical simulations of wind turbines during product design, certification, and commissioning, the other is the analysis of data arising during the condition monitoring of installed wind turbines.

First, the simulation of the non-linear dynamic response of numerical models of wind turbines is an essential element of design, modification, certification, and site-dependent load assessment [4]. In the course of design and optimization, hundreds of wind turbine configurations may be assessed. In order to evaluate a

---

[1] European Wind Energy Association www.ewea.org/wind-energy-basics/faq.

J. Garcke (✉)
Fraunhofer Institute for Algorithms and Scientific Computing SCAI, Schloss Birlinghoven, 53757 Sankt Augustin, Germany

Institute for Numerical Simulation, Rheinische Friedrich-Wilhelms-Universität Bonn, Wegelerstr. 6, 53115 Bonn, Germany
e-mail: jochen.garcke@scai.fraunhofer.de

R. Iza-Teran • M. Marks • M. Pathare
Fraunhofer Institute for Algorithms and Scientific Computing SCAI, Schloss Birlinghoven, 53757 Sankt Augustin, Germany

D. Schollbach
Weidmüller Monitoring Systems GmbH, Else-Sander-Straße 8, 01099 Dresden, Germany

M. Stettner
GE Global Research, Freisinger Landstraße 50, 85748 Garching bei München, Germany

---

© Springer International Publishing AG 2017
M. Griebel et al. (eds.), *Scientific Computing and Algorithms in Industrial Simulations*, DOI 10.1007/978-3-319-62458-7_16

single configuration, thousands of transient simulations of the model's response to turbulent and deterministic wind in normal operation, in a storm, under mechanical stresses, in fault situations, and during the loss of connection to the power grid—and various combinations thereof—may be conducted. A single simulation typically generates outputs for several hundreds software "sensors" extracting information like local loads, deflections, velocities, and wind inflow from the simulation model, in most cases for simulated 10 min real time intervals, sampled commonly at 10 Hz, resulting in 5–50 MB of data per simulation run. Mining and interpreting the gigabytes of data produced in this way represents a serious challenge.

Second, condition monitoring systems (CMS) rely on empiric models derived from physical sensor data. Applied to wind turbines, they have the goal of determining the health of the components, ensuring safety (e. g. by detecting ice build-ups on rotor blades), and reducing the wear of the mechanical parts. Diagnosis and fine tuning have become attainable thanks to the condition information provided by various sensors and collected by the turbine control station. Precise control strategies can be derived from this information, for example in order to reduce wear, hence prolonging a healthy state of the installation. The condition monitoring of rotating machine parts has a long tradition, whereas the monitoring of rotor blades with oscillation analysis has been studied far less deeply up until now. Therefore, we concentrate on the analysis of rotor blade oscillation data enriched with operational data of the wind turbine. The latter includes for example the meteorological conditions and the operational modes of the wind turbine. The sensor data is kept available for examinations at least as long as the wind turbine remains operational, which can be up to 20 years, to provide input for upcoming analysis and machine learning tasks. Therefore large data repositories can and need to be analysed to provide sustainable services to the wind turbine operators.

This article focuses on data analysis procedures which shall assist the engineer and simplify the overall workflow by providing novel, tailored meta data based on similarity and/or anomaly detection in raw output data. For the analysis of the arising data we investigate dimensionality reduction approaches which obtain a low dimensional embedding of the data. The standard dimensionality reduction method is principal component analysis (PCA), which gives a low-dimensional representation of data which are assumed to lie in a subspace. But if the data in fact resides in a nonlinear structure, PCA gives an inefficient representation with too many dimensions. Therefore, and in particular in recent years, algorithms for nonlinear dimensionality reduction, or manifold learning, have been introduced, see e.g. [11]. The goal of these approaches consists in obtaining a low-dimensional representation of the data which respects the intrinsic nonlinear geometry.

In case of numerical simulations, a goal is to facilitate interpretation of the system response as obtained from a vast number of simulations. We use nonlinear dimensionality reduction for time series obtained from the numerical simulations and aim for the detection of patterns, anomalies, or similar behavior. For the condition monitoring, one looks for a small number of significant key indicators which describe the health of a wind turbine. Here, using linear dimensionality reduction, we are able to compute a low-dimensional basis which represents a baseline, deviations from it indicate anomalies in the sensor data, this information can then be used for predictive maintenance.

In Sect. 2 we describe the two application domains in more detail. Section 3 contains our data analysis approach for time series from numerical simulations, while in Sect. 4 we present our approach for anomaly detection in sensor data.

2   Time Series Characteristics in Wind Energy

We consider two different situations where time series data arise in wind energy research. First, during the product development, process data stems from numerical simulations of wind turbines. Second, data is obtained during the condition monitoring of installed wind turbines. Other time series data in the wind energy domain, for example, concerns wind speed prediction or wind turbine control [10].

2.1   Numerical Simulations of Wind Turbines

Numerical simulation of wind turbines is an essential element in the design of wind turbine systems, the fine-tuning of installations, or the exploration of the effect of upgrades [4]. Input meta data encompass environmental and operational conditions, referred to as Design Load Cases, DLCs, as specified by certification bodies, as well as instantiations of the wind turbine's individual components such as its foundation, tower, nacelle, drive train, generator, hub, blade, and control system. This information is commonly compiled in files identified by or populated with the aid of this meta data. See Fig. 1 for an overview of the components of the analysis workflow.

The simulation tool itself contains commonly a condensed dynamic representation of the structure, with simplified "engineering models" for the effective description of wind, aerodynamics, or actuator dynamics, based on first principle physics, tuned with field data and engineering knowledge, permitting simulation times on desktop machines which are about one order of magnitude smaller than real time. High performance computing solutions employing 3D computational fluid dynamics solutions such as the Navier-Stokes equations and detailed structural models (such as non-linear FE solvers) might be used during the development of this wind turbine system simulation tool, or in case of root cause analysis, but are very rarely used during the design process of a wind turbine. Such approaches are usually unsuitable because of excessive run times, amplified by the already described need to execute vast amounts of simulations [9].

Raw data output of these simulations are time series of "sensors", for instance representing the response of the first bending mode in a direction normal to the blade's chord-wise extension, labeled as "first flap mode", or the bending moment in this direction as experienced at the root of a rotor blade. Sensor signals will exhibit broad-banded "noise" resulting from the structure's response to wind turbulence.

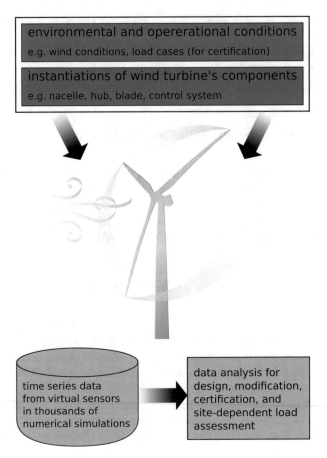

Fig. 1 In the design process of wind turbines, a large number of time series data is generated by numerical simulations, due to the needed investigations of different conditions and components affecting the behavior of the wind turbine in operation. Published with kind permission of ©Fraunhofer SCAI 2016. All Rights Reserved

For sensors on the rotating system (rotor and drive train), these signals will be superimposed by pronounced periodicity at multiple, varying frequencies.

Due to the complexity and volume of this raw data, automated post-processing is mandatory to assist the engineer in their analysis. The standard case requires extraction of extrema on several levels of simulation bundles, or accumulation of so-called Damage Equivalent Loads, DELs. This meta data is then used either to size components or to check margins to allowable values ensuring safe operation of the system.

In some cases, however, this type of meta data is insufficient, for instance as information on physical behavior leading to the identified extreme values is lost. A similar need occurs when trends with regard to the impact of input meta data must be identified. Resolving to raw output data is possible, but cumbersome and time consuming.

2.2 Condition Monitoring of Wind Turbines

A problem in the day-to-day operation of wind turbines is their downtime due to structural damage and other faults. In order to increase their output, one would want to minimize these outages. This goal can be achieved by implementing *condition monitoring* (CM) techniques. In contrast to conventional maintenance strategies (like run to breakdown), CM allows early detection and even prediction of damage. This is done by monitoring the current state of the turbines with the help of sensors [10].

With the use of CM it is possible to perform condition based maintenance, which has financial advantages to traditional strategies since, in particular, downtime and maintenance can be better planned. With the increasing number of installed wind turbines, the importance of CM will only grow in this field. Furthermore, with the higher efficiency of modern turbines, outages result in higher economical losses. Further advantages of CM are the prevention of secondary defects [12] caused by the failure of certain components, and total loss of the engine can be prevented by a timely shut-down.

Damage can occur in all components and control systems of the turbine. The annual frequency of these damage types varies and they have different impacts on the downtime of the turbine. While just 7% of all outages are caused by rotor blade failures, the average caused downtime of this damage type amounts to 4 days [5]. This motivates our investigation to improve CM methods in this field.

Vibration analysis is one of the main methods used in CM [14]. The idea behind this method is to analyze the change of vibrational properties which occur with the emergence of structural damage [1]. Therefore, we concentrate on the analysis of rotor blade oscillation data enriched with operational data of the wind turbine. Here, Fourier analysis is widely used to obtain the frequency spectrum of the measured vibrations. Often spectra of the undamaged case are taken as a point of reference for further measurements. Additionally, frequency bands which are associated with damage features can be easily isolated [1].

The analysis of the rotor blade oscillation data has to be based on robust behavioral models of the rotor blade. This means that these models ideally should be generally applicable for different wind turbine types, and should also be valid for a broad range of operational modes and meteorological conditions. Since in condition monitoring the models themselves are deduced from historical data, it is necessary to investigate the data from a vast amount of wind turbines over a broad range of operational conditions in order to validate these behavioral models. Various kinds of analysis methods are suitable for this task, especially signal analysis and statistical methods for regression, clustering, and optimization. Since the approach is empirical, the results are fraught with statistical uncertainty.

In particular, to detect changes in vibration behavior, machine learning methods [6] are used. In real applications there is often not enough data available to train supervised learning methods, so unsupervised learning algorithms are employed. Especially anomaly detection, also called novelty detection, is a method from this domain. Here, one builds a model from the existing data of the undamaged turbine. Additional data points, i.e. measured during the normal operation, are checked for conformity with this model. When such current data deviates from the model, the occurrence of damage is assumed and further investigations are triggered.

3 Exploration of Time Series Data from Numerical Simulations

For the analysis of time series data obtained from numerical simulations, we here introduce an approach based on nonlinear dimensionality reduction. We propose an interactive exploration tool assisting the engineer in the detection of similarities or anomalies in large bundles of numerical simulations of wind turbines. The specific situation we consider is the so-called storm load case, in which the engineer searches for abnormal system behavior. In this scenario strong winds have caused the wind turbine to shut down and to assume a survival configuration in which blades are rotated to cause minimum drag when the hub points directly into the wind. Furthermore, after shutdown, connection with the power grid is lost so that no controls are operational. This means in particular that the machine cannot react to changes in wind direction anymore. Numerical simulations now provide time series data of the behavior of the wind turbine in such a "parked" position under storm conditions.

3.1 Virtual Sensor Data from Wind Turbine Simulations

As a result from the numerical simulation, we obtain a time series $\mathcal{T}$ of given length D, where D is the length of the simulation time in seconds divided by the time step size for saving the sensor data, which can but does need to be the same as the time step of the numerical simulation. The impact of wind direction is studied by using wind directions $\alpha, \alpha = 1, \ldots, M$, where M typically is 36, i.e. the wind directions are changed in steps of $10°$. To address the turbulent nature of the wind, multiple (random) wind seeds are used, which we denote by $k, k = 1, \ldots, K$, where a typical value of K is 10. The average wind speed is given by regulations. For one wind turbine configuration we already have $M \times K$ simulations $\mathcal{T}_\alpha^k$, e.g. 360 for typical setups. In Fig. 2 we show some exemplary time series of a virtual sensor indicating the contribution of the so-called flapwise bending mode of one blade, which in this storm case represents bending within the plane of the rotor. The different quality of the response of the parked (standing still or idling) wind turbine as a function of the (now changed) wind direction is obvious to the human observer.

In order to quantify similarities between all bundles of simulation results, the first step in the analysis workflow is to extract the required data from all the simulations for the chosen sensor or sensors. If required, pre-processing steps are applied depending on the sensor/signal characteristics, e.g. scaling for data

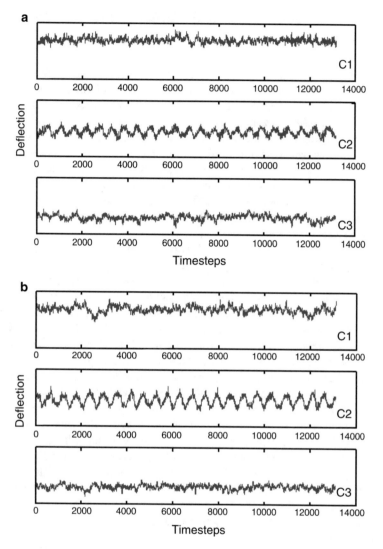

Fig. 2 Exemplary time series of a (virtual) flapwise mode sensor on a blade, for different wind directions, i.e. cases (**a**) and (**b**). In each case, the *first row* shows time series data from two simulations from cluster C1, *second row* from cluster C2, and *third row* from cluster C3, refer to Fig. 4 for details. Published with kind permission of ©Fraunhofer SCAI 2016. All Rights Reserved

normalization, transformation to polar coordinates for blade position and deflection values, transformation to frequency domain for periodic signals, peak detection, etc. Generally, the pre-processing steps depend on the analysis goal and the specific sensor. Here, we are interested in an initial anomaly detection and proceed with the raw time series data.

3.2   Nonlinear Dimensionality Reduction for Time Series Analysis

We aim for approaches which are capable of comparing several time series in a semi-automatic fashion, and allow the detection of similarities or anomalies in the results. The basis for such a comparison is a suitable concept of distances between the high-dimensional data points, which we then aim to preserve in a so-called nonlinear dimensionality reduction, where points which are nearby in the high-dimensional space are still nearby in the lower dimensional space [11]. For visualization purposes one often embeds the simulation data into three dimensions, i.e. each numerical simulation is represented as a point in this space. Datasets can now be efficiently explored to find similarities and differences, and the embedded space can be navigated to investigate the properties of simulated time series which are close to each other, i.e. similar [11].

In applications one needs a suitable notion of distance between different time series to allow such a dimensionality reduction. The Euclidean distance is commonly the first choice for such a measure. However, it is not suited for all conceivable applications, which may require completely different distance functions. For instance, the Euclidean distance will classify signals which are shifted in time, but are otherwise very similar in nature and magnitude, as dissimilar. But this is a situation typical for wind turbine simulation data; the time shift itself is due to variation in initial conditions and wind speed and seed, and is irrelevant when the dynamic characteristics of the wind turbine are investigated. One frequently used approach for signals which may vary in time or speed is called dynamic time warping (DTW) which consists of an optimal mapping between two signals implemented by a minimization of the pairwise distances under certain restrictions [8]. Lower distances indicate parts of the signal with high similarity. Evaluation has a computational complexity of $O(D^2)$, but under some assumptions it can be optimized by restricting the computation to only a suitable chosen part of the evaluation domain [8]. A variant of this algorithm is FastDTW, which uses a multilevel approach that recursively projects a solution from a coarser resolution and refines the projected solution, reducing the effort to $O(D)$ under suitable assumptions [15].

For applications in signal processing it is also necessary to be able to capture signal similarity at different time scales, specially in the presence of noise or locally influencing factors. Therefore, a way to decompose a signal into several components with varying degrees of detail is very useful. For example, one method that has been employed with good success for such a task is empirical mode decomposition [7]. The method does not use a set of fixed basis functions like in the case of Fourier or wavelet transforms, but it actually uses a spline interpolation between maximum and minimum in order to successively generate so called "intrinsic mode functions" (IMF). Once the IMFs are obtained, a nonlinear dimensionality reduction method can be applied based on these, albeit with the assumption that all signals generate compatible signal modes. Other such mathematically inspired pre-processing approaches can also be used before the eventual nonlinear dimensionality reduction.

The overall workflow of the studied data analysis procedure now consists of four steps.

1. In the first step of *data extraction* the specific data to be analyzed is extracted from the archived raw time series data. The selection of the data can depend on the wind turbine specifications, the load case, the wind directions, the chosen sensors, the analysis goal, etc.
2. During the *pre-processing* step the raw time series data is modified. For example transformations to the frequency domain can be applied, large variations in individual signals scales may be removed, or signals could be combined.
3. For the *dimensionality reduction*, computations are performed with the pre-processed data. For example, suitable distances are computed between all pairs of time series data, followed by a decomposition of the obtained similarity matrix.
4. The obtained low dimensional embedding is then used for an interactive *exploration* of the data. One can study the properties of nearby points in the embedding, which represent similar time series. The interactive study can be supported by computing the correlation, the mutual information, or similar quantities measuring relations between attributes, between the embedded dimensions and design parameters, or between quantities of interest in the output.

3.3 Diffusion Maps

In this work, we consider diffusion maps [2] as a nonlinear dimensionality reduction approach. Here, first a weighted graph $G = (\Omega, W)$ is constructed, where the vertices of the graph are the R observed data points $\mathcal{T}_\alpha^k \in \mathbb{R}^D$. The weight which is assigned to the edge between two data points $\mathbf{u} := \mathcal{T}_\alpha^k \in \Omega$ and $\mathbf{v} := \mathcal{T}_{\tilde\alpha}^{\tilde k} \in \Omega$ is given by $w(\mathbf{u}, \mathbf{v}) = e^{-\Delta^2(\mathbf{u},\mathbf{v})/\epsilon}$, where the value of ϵ controls the neighborhood size. The term $\Delta(\mathbf{u}, \mathbf{v})$ can in general be an application-specific, locally defined distance measure that defines the way signals are to be compared. In our application of time series data, we could use the DTW distance measure if the time shift between signals is irrelevant, or the Euclidean distance if capturing time shift is important. The employed Gaussian kernel $e^{-\Delta^2}$ is a common choice in many applications, and we apply it in our experiments. This stems from the relationship between heat diffusion and random walk Markov chains, the basic observation is that if one takes a random walk on the data, walking to a nearby data-point is more likely than walking to another one that is far away.

With this construction, one defines a Markov process on the graph where the transition probability for going from $\mathbf{u}$ to $\mathbf{v}$ is given by $p(\mathbf{u}, \mathbf{v}) = w(\mathbf{u}, \mathbf{v}) / \sum_{z \in \Omega} w(\mathbf{u}, \mathbf{z})$. Hence, if the points are similar, then we have a high transition probability, and if they are not, then the transition probability is low. The Markov chain can be iterated for q time steps and a so called "diffusion distance" D_q can be defined in a natural way [2].

Diffusion maps now embed high-dimensional data into the low-dimensional space such that the diffusion distances D_q are (approx.) preserved. It has been shown in [2] that this is accomplished by the map $\Psi_q : v \rightarrow [\lambda_1^q \psi_1(v), \lambda_2^q \psi_2(v), \ldots, \lambda_s^q \psi_s(v)]$, where λ_j and ψ_j are the eigenvalues and right eigenvectors of the similarity matrix $P = p(\mathbf{u}, \mathbf{v})_{\mathbf{u},\mathbf{v}}$. Furthermore, one can approximate the distance between transition probabilities using the Euclidean distance in the reduced space. The number of terms used for the embedding, i.e. the dimension s of the low-dimensional space, depends on the weight w and the value of ϵ, besides the properties of the data set itself. In case of a spectral gap, i.e. when only the first few eigenvalues are significant and the remaining ones are small, one can obtain a good approximation of the diffusion distance with only a few terms. For details about the theoretical background of the approach we refer to [2].

3.4   Numerical Results

To evaluate the proposed data analysis procedure, we study a generic utility sized 1.5 MW power wind turbine model with three blades of roughly 40 m length. We consider numerical simulations of a storm load case with turbulent wind with average speed of 39 m/s. All (external) controls are off due to loss of connection to the grid. The wind direction is varied in 10° steps from 0° to 360°, with 10 wind seeds per wind directions, resulting in 360 individual simulations. The time window for the wind turbine simulation is roughly 11 min, where the sensor data is sampled at 0.05 s intervals for 1360 sensors. The dimension of each time series is equal to the number of data samples, i.e. it is 13107 in this case. We concentrate in the following on the analysis of sensor data for rotor speed, rotor position, blade deflection (flapwise and edgewise), tower deflection, and tower torsion only.

We first consider data from the rotor speed sensor with no pre-processing, and use the Euclidean distance. The dimensionality is reduced to three for visual purposes, while capturing the main behavior of the data. Figure 3 (left) shows the low dimensional representation. There are 360 points in the displayed embedding plot, each point in three dimensions represents one simulation. Simulations in the same cluster in such an embedding have similar behavior in regard to the rotor speed sensor.

Since each simulation corresponds to a particular wind direction and wind seed, we can also visualize the results with a polar plot (see Fig. 3, right) by using the first three coordinates from the low-dimensional embedding as RGB values to derive a color value for each simulation. Each cell in the plot represents the time series from one simulation, while the position of each cell is based on the wind direction (azimuthal) and the wind seed (radial). For example, the 36 simulations with seed A form the innermost circle, where the wind direction in degrees is marked on the boundary of the outer most cell. Simulations with similar behavior are close to each other in the low dimensional coordinates, and are therefore shown in similar color in the polar plot. For example, one easily observes that the wind directions marked

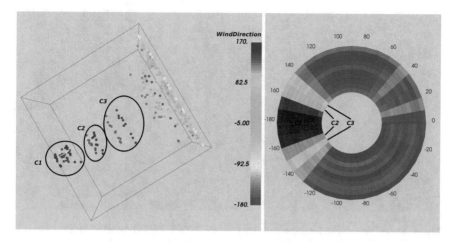

Fig. 3 Embedding results for rotor speed sensor data, using diffusion maps with the Euclidean distance. For simulations in clusters C3 to C1, wind turbine is observed to be rotating with increasing speed, i.e. simulations in cluster C1 have highest rotor speed. Published with kind permission of ©Fraunhofer SCAI 2016. All Rights Reserved

by C1, C2, and C3 in the right plot form clusters in the adjoining embedding plot on the left of Fig. 3. Further investigation of the raw data confirms that the rotor is rotating while in the idle configuration for the simulations in these three clusters, but for the other wind directions it is not. Additionally, the rotor speed in cluster C1 is higher than in C2 and C3, and the rotor speed in simulations from the C2 cluster is again higher than in C3. This is also apparent from the location of these clusters along one dimension of the low-dimensional space, which appears to be dominated by the rotor's rotational speed sensor. No information can be obtained on the quality of rotor rotation, however.

As a second example we consider the flapmode sensor for blade one, again no pre-processing was performed for the sensor. The embedding plot in Fig. 4 also shows three different clusters (note that C1-C3 are different from the cluster shown previously in Fig. 3). Further investigations of the raw data time series confirm that the rotor is in a still position for the simulations in cluster C1 and C3, and the position of blade 1 for simulations in cluster C1 is different from that in cluster C3. Simulations in cluster C2 have a rotating rotor. In simulations with wind direction of −40° the rotor did rotate for half of the simulation time, and is inconsistent with the behavior with simulations having adjacent wind directions of −30° and −50°. This anomaly is very relevant from an engineering perspective and can be easily identified with our data analysis procedure.

Using DTW for this data provides slightly more details in cluster C2. We considered only the range of wind directions from this cluster, where the rotor is actually turning. We performed an embedding using both the Euclidean distance and the distance obtained from DTW, as seen in Fig. 5. While the Euclidean distance incorporates differences from different wind seeds, initial conditions, and thus time

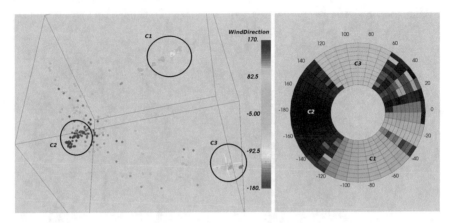

Fig. 4 Embedding results for flapwise mode sensor data from blade 1, using diffusion maps with the Euclidean distance. Similar behavior is observed among simulations in each cluster and an anomaly is detected for simulations with wind direction −40°. Published with kind permission of ©Fraunhofer SCAI 2016. All Rights Reserved

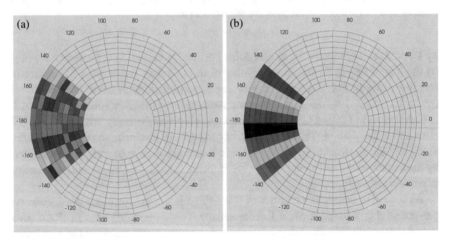

Fig. 5 Embedding results for flapwise mode sensor data from blade 1, using different distance measures for Δ in diffusion maps. In difference to the Euclidean distance, DTW allows a clear distinction for the different wind directions. Published with kind permission of ©Fraunhofer SCAI 2016. All Rights Reserved. (**a**) Embedding with Euclidean distance. (**b**) Embedding with DTW

shifts in signals, DTW effectively removes this dependency permitting a clear distinction for the different wind directions.

These examples show the possibilities of the nonlinear dimensionality reduction to allow a quick identification of similarly behaving simulations and the detection of anomalies in certain simulations.

4 Anomaly Detection Based on Linear Dimensionality Reduction for Condition Monitoring Sensor Data from Wind Turbines

For the detection of anomalies in the sensor data of wind turbines we now propose an approach based on dimensionality reduction. The method utilizes principal component analysis (PCA) on vibrational sensor measurements of a wind turbine, given in the form of frequency spectra (computed from the time series data), to obtain a baseline by way of the low-dimensional data representation. Further measurement samples are projected into this basis to detect deviation of the coefficients from those of the baseline. Our approach is evaluated with real life data which is provided by Weidmüller Monitoring Systems (WMS). Note that initial investigations of nonlinear dimensionality reduction approaches for this real life data did not show advantages for anomaly detection against the employed linear approach.

4.1 Sensor Data from Rotor Blades

The vibrational data from the rotor blades are collected by two acceleration sensors per blade, mounted at different angles. The *flap*-sensors align with the flat side of the blade, and the *edge*-sensors are mounted towards the edge of the blade, see Fig. 6. Instead as raw time series data, the information is stored in the form of frequency spectra and also contains environmental and operational parameters. The different orientation of the edge and flap sensors cause differences in the kind of vibrations they are monitoring. In particular, flap sensors undergo stronger vibrations because they monitor vibrations in the wind direction.

The data points stem from hourly measurements. In each measurement cycle the vibration from the sensors is captured over the course of 100 s and afterwards transformed into the frequency domain. Additionally, the following operational

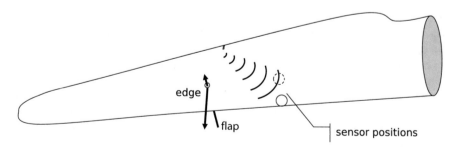

Fig. 6 Blade with edge/flap sensor positions. Published with kind permission of ©Weidmüller Monitoring Systems GmbH 2016. All Rights Reserved

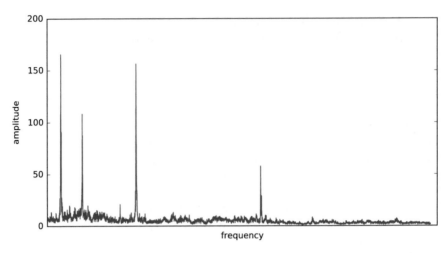

Fig. 7 A typical frequency spectrum, taken from a single sensor during one measurement cycle. Published with kind permission of ©Fraunhofer SCAI 2016. All Rights Reserved

parameters are saved: Time stamp of measurement, rotational speed, power output of the turbine, wind velocity, pitch angle of the rotor blades, and environmental temperature, where the instationary values are averaged over the time window.

A typical frequency spectrum is shown in Fig. 7, it can be interpreted as a vector of length D containing real numbers. To visualize different spectra at the same time, we arrange them into a spectrogram as seen in Fig. 8. This allows an overview of all spectra from one sensor, and events like shifts of peaks can be easily detected. For example, Fig. 8 shows clearly some differences between certain measurement points, i.e. the blue lines are from measurements during the downtime of the turbine, therefore having lower amplitudes than the other spectra.

4.1.1   Pre-processing

Before one analyzes the data, one needs to perform some pre-processing steps. For example, the data points are divided into different classes according to their operational parameters. These classes represent different operational modes, e.g. grouped by rotational speed or similar wind velocities. In the next step the variables to be examined are selected; these are certain frequency bands which are associated with damage features.

Experiments have shown that rotational speed is one of the main contributors to the change in the observed frequency spectra. Especially since the data sets originate from turbines which are essentially operated with constant rotational speeds (see Fig. 9), it is advised to separate data classes by rotational speed.

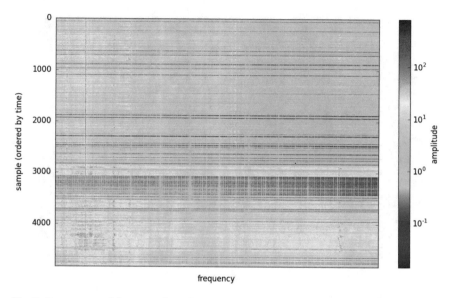

Fig. 8 Spectrogram of frequency data of a single sensor, spectra are sorted by time parameter *from top to bottom*. Published with kind permission of ©Fraunhofer SCAI 2016. All Rights Reserved

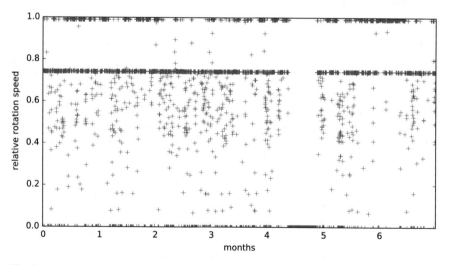

Fig. 9 Rotational speed over time of one turbine. The *two horizontal lines* on which most of the points accumulate represent the two operational speeds at which the turbine is operated. Published with kind permission of ©Fraunhofer SCAI 2016. All Rights Reserved

Figure 10 shows the spectrogram of measurements taken under the same rotational speed. Compared to Fig. 8, it is clearly seen that these frequency spectra are more similar to each other.

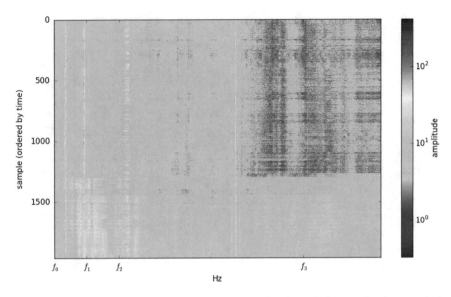

Fig. 10 Spectrogram of samples taken under same rotational speed. Frequencies f_i are marked to illustrate ways to select variables described in Sect. 4.1.1. Published with kind permission of ©Fraunhofer SCAI 2016. All Rights Reserved

Instead of processing the frequency spectra as a whole, smaller frequency bands can be selected for analysis. The underlying idea behind analyzing smaller portions of the spectrum separately is that each band can contribute to the damage signal on its own. This is favorable when damage manifests locally in some frequency bands which are dominated in variance by other non-relevant frequencies. On the downside, this approach can result in a higher false positive rate. Furthermore, one might know from engineering principles in which parts of the spectrum certain damage features manifest themselves, and therefore one can select these bands for detection of these specific features. Additionally, it saves computational time when several narrow frequency bands are analyzed since fewer variables are needed.

In the following, we assume that $\mathcal{Y} = \{\mathbf{y}_1, \ldots, \mathbf{y}_m\} \subset \mathbb{R}^D$ is the set of m samples of dimension D which remain after the described pre-processing steps are taken, in particular this means that the vectors $\mathbf{y}_i$ do not contain the whole spectrum, but the selected frequency band. In our case the vectors $\mathbf{y}_i \in \mathbb{R}^D$ can be interpreted as discrete frequency spectra as seen in Fig. 7.

4.2 Anomaly Detection in Sensor Data

4.2.1 Model of Undamaged State

One common approach for anomaly detection is to create a baseline model of the undamaged condition of the wind turbine. To achieve this, we select a point in time

t until which the turbine is assumed to be free of damage. Let

$$\mathcal{Y}_t := \{\mathbf{y}_1, \dots, \mathbf{y}_t\} \subset \mathcal{Y}$$

be the set of data points gathered until this point in time.

From $\mathcal{Y}_t$, we calculate the baseline model using principal component analysis (PCA), which is a widely used method for linear dimensionality reduction. The core principle of PCA is to find a linear projection of the original data set into a new orthogonal basis by computing an eigenvalue decomposition. In this basis the individual components are statistically independent, allowing a distinction of relevant and redundant information in the data set. By discarding all but the first few principal components corresponding to latent variables with the highest variance, a low dimensional representation of the data is found which keeps most of the relevant information. Note that the use of PCA implies a further pre-processing step, namely the centering and scaling of the data $\mathcal{Y}_t$.

We collect the data samples $\mathcal{Y}_t$ into a matrix $\mathbf{Y}_t := [\mathbf{y}_1, \dots, \mathbf{y}_t] \in \mathbb{R}^{D \times t}$. Let $\mathbf{C} := \frac{1}{t} \mathbf{Y}_t \mathbf{Y}_t^T \in \mathbb{R}^{D \times D}$ be the sample covariance matrix and $\mathbf{C} = \mathbf{V} \mathbf{\Lambda} \mathbf{V}^T$ its eigenvalue decomposition, where $\mathbf{\Lambda} = \mathrm{diag}(\lambda_1, \dots, \lambda_D)$ consists of eigenvalues ordered by magnitude and $\mathbf{V}$ contains the according eigenvectors. The projection matrix $\mathbf{P} \in \mathbb{R}^{S \times D}$ is given by

$$\mathbf{P} := \mathbf{I}_{S \times D} \mathbf{V}^T. \tag{1}$$

The samples $\mathbf{y}_i \in \mathbb{R}^D$ are mapped to the first S principal components by $\mathbf{P}\mathbf{y}_i = \mathbf{x}_i \in \mathbb{R}^S$, with $S < D$. Deviations of the remaining points $\mathcal{Y} \setminus \mathcal{Y}_t$ from this baseline are measured in a suitable fashion. Note that a similar approach is described in [13].

4.2.2 Deviation from the Undamaged State

The PCA can be used to detect anomalous behavior in a process or system by measuring the variation of the coefficients of the samples in the PCA model. For this purpose the Q-statistic and T^2-statistic are frequently used tools. In order to apply them, we have to assume that the underlying process is normally distributed. The Hotelling's T^2-statistic is a generalization of the Student's t-statistic and is based on the scores of the projection of a sample.

As stated above, $\mathbf{P}$ is the transformation matrix from the PCA-model of $\mathcal{Y}_t$ and $\mathbf{\Lambda} = \mathrm{diag}(\lambda_1, \dots, \lambda_S)$ is the covariance matrix of the latent variables, obtained by the PCA algorithm. The T^2-statistic of sample $\mathbf{y}_i$ is calculated by

$$T_i^2 := \mathbf{P}\mathbf{y}_i \mathbf{\Lambda}^{-1} \mathbf{y}_i^T \mathbf{P}^T = \sum_{j=1}^{S} \frac{x_{ij}^2}{\lambda_j}. \tag{2}$$

The T^2-statistic detects variation in the subspace of the first S (not discarded) principal components, measuring the variation of samples within the PCA model. We then define a tolerance threshold m_T by:

$$m_T := \max_{\mathbf{y} \in \mathcal{Y}_t} |\mathbf{P}\mathbf{y}\mathbf{\Lambda}^{-1}\mathbf{y}^T\mathbf{P}^T|. \tag{3}$$

If the values from the T^2-statistic of the remaining data points exceed m_T, the presence of an anomaly, potentially indicating a damage or fault, is assumed.

An alternative would be the Q-statistic, which is a measure for the (residual) difference between a sample and its projection onto the model. Note that in our experiments we observed essentially the same results for the T^2-statistic and the Q-statistic.

4.3 Methodology for Damage Detection

The pre-processing steps and the calculation of the baseline parameters are now collected into a damage detection method. This method analyzes data $\mathcal{Y} = \{\mathbf{y}_1, \ldots, \mathbf{y}_m\} \subset \mathbb{R}^D$ from a single sensor of the turbine and detects deviations from the undamaged state and consists of the following steps:

1. pre-processing
   – divide data points into classes determined by operational modes
   – select frequency ranges to be analyzed
2. select data points $\mathcal{Y}_t$ describing the undamaged state of the turbine
3. compute baseline model from $\mathcal{Y}_t$ using PCA
4. set thresholds for T^2- or Q-statistics
5. use the statistics to measure deviations from the baseline model

Let $\mathbf{Y} \subset \mathcal{Y}$ be the class of data points which satisfy the parameters of a selected operational mode, e.g. with the same rotational speed. Furthermore let $j, k \in \{1, 2, \ldots, D\}$ be the lower and upper bound of a selected frequency band. Let t be the point in time until which the installation is assumed to be without damage. Algorithm 5 returns the baseline computed from the frequency spectra describing the undamaged state. To cover the biggest possible portion of the initial data $\mathcal{Y}$, several classes of operational modes have to be chosen accordingly. Additionally, when l many frequency bands are selected for analysis, we iterate the above algorithm over these selections yielding l different baseline models.

For a given collection of frequency spectra and a baseline model, Algorithm 6 returns the corresponding damage signal $\mathbf{s}$. This output can be interpreted as a one dimensional time series. When l many frequency bands are selected, we obtain i different damage signal time series in the form of a vector $[\mathbf{s}_1, \ldots, \mathbf{s}_l]$. To obtain one signal value, we sum them up entrywise in our analysis $\mathbf{s}[i] := \sum_{k=1}^{l} \mathbf{s}_k[i]$. Alternatively one can take the maximum over the signal entries $\mathbf{s}_k[i]$.

Algorithm 5: Calculate Baseline Model for a Sensor

Data: $\mathbf{Y} \subset \mathcal{Y}$ frequency spectra describing undamaged state, j, k frequency range
Result: baseline model $\mathbf{P}, \Lambda, m_T$
$n := |\mathbf{Y}|, i := 1$
for $\mathbf{y} = (y_1, \ldots, y_D) \in \mathbf{Y}$ **do**
   $\mathbf{Y}_t[i] \leftarrow (y_j, \ldots, y_k)^T$ ▷ selection of frequency variables
   $i \leftarrow i + 1$
calculate matrices $\mathbf{P}, \Lambda$ for baseline model after (1) using $\mathbf{Y}_t$
calculate threshold m_T from $\mathbf{Y}_t$ according to (3)
return $\mathbf{P}, \Lambda, m_T$

Algorithm 6: Calculate Damage Signal

Data: $\mathbf{Y} \subset \mathcal{Y}$ frequency spectra, j, k frequency range, baseline model $\mathbf{P}, \Lambda, m_T$
Result: damage signal $\mathbf{s} \in \mathbb{R}^n$
$n := |\mathbf{Y}|, i := 1$
for $\mathbf{y} = (y_1, \ldots, y_D) \in \mathbf{Y}$ **do**
   $\mathbf{y} \leftarrow (y_j, \ldots, y_k)$ ▷ selection of frequency variables
   compute T_i^2 after (2)   ▷ compute T^2-statistic for each $\mathbf{y}$
   $s[i] \leftarrow \begin{cases} 0 & \text{if } T_i^2 \leq m_T \\ T_i^2/m_T & \text{else} \end{cases}$ ▷ compute damage signal for each $\mathbf{y}$
   $i \leftarrow i + 1$
return s

The data points occurring after time stamp t can be interpreted as a continuous data feed in real time. Therefore, after the computation of a baseline model with Algorithm 5, this data feed can be directly evaluated by Algorithm 6, and a continuous signal feed is generated in real time.

4.4 Numerical Results

In this section we present the application of the proposed method on real wind turbine data as described in Sect. 4.1. Each data point consists of a frequency spectrum from each sensor and operational parameters taken during the measurement cycle. The data set contains around 4000 data points taken over the course of 7 months. Data from the first month was used to compute the baseline model. The occurrence of damage in the given time frame is known, and its date was determined by a proprietary method of WMS.

During pre-processing, the selection of data points by environmental parameters takes place. Because the turbine at hand is operated under constant rotational speed, the data points originating from the first rotational mode are selected for analysis. This leaves us with around 2000 data points which cover the time domain of the data quite well. Additionally, we perform selection of frequency bands. Here, we distinguish between two different approaches:

- **(F1)** analysis of the whole frequency band between the frequencies f_2 and f_3 in Fig. 10
- **(F2)** the whole spectrum is divided into several smaller frequency bands of the same width, i.e. intervals of length $f_1 - f_0$ in Fig. 10, which are analyzed separately, and their results are summarized afterwards

Based on the decay of the eigenvalues for this data, we keep the first three principal components for the computation of the baseline model. The data points are centered and scaled by the mean and standard deviation of the baseline data points.

The following figures illustrate the results of the method, which was applied to the data from the three *flap*-sensors. The three plots in one figure show the values of the damage signal (computed by the described algorithm) over time. The damage signal is an indicator for deviation from the baseline model, and big values can be associated with the occurrence of damage. The left vertical line indicates the point in time for which the baseline was selected. The dashed line represents the date t_d at which the damage is first detected by the proprietary method of WMS evaluated for the stored sensor data. Note that the method of WMS was not available for live monitoring of the wind turbine at that time.

In Fig. 11 we show results for which frequency band F1 was evaluated. The first rotor blade does not show significant deviations, the second blade on the other hand

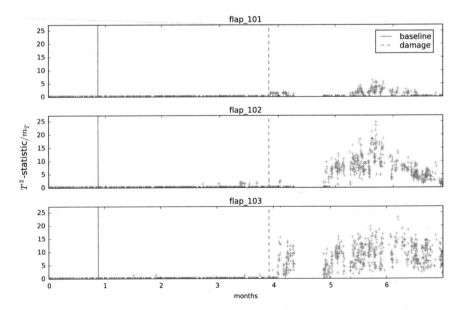

Fig. 11 Damage signal from evaluation of frequency band **F1**. The signal from rotor blade 3 increases after date t_d, the assumed date of the damage occurrence. Published with kind permission of ©Fraunhofer SCAI 2016. All Rights Reserved

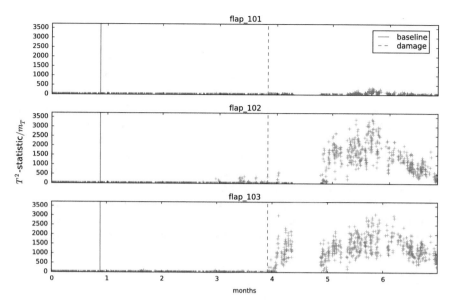

Fig. 12 Sum of damage signals from evaluation of frequency bands **F2**. The signal from rotor blade 3 increases after date t_d, the assumed date of the damage occurrence. Published with kind permission of ©Fraunhofer SCAI 2016. All Rights Reserved

shows deviations before the assumed date of damage occurrence t_d. If this has to be attributed to anomalies which were not included into the baseline model or if it is an indicator for damage is unknown. One can see clearly that the damage signal from rotor blade 3 increases after date t_d, so the purely data-driven procedure obtains results very similar to the original approach.

After the fault, the turbine was shut down during month 4, visible by the gap between month 4 and 5. After re-activation of the turbine in month 5, the deviations persist and even increase in blade 1 and 2, which indicates a change of operational or environmental conditions due to the repair of the rotor blades and other maintenance efforts. For the purpose of condition monitoring, a new baseline would need to be derived in such a situation.

Figure 12 shows the sum of damage signals from frequency bands **F2**. The results are essentially the same, showing that the anomaly detection procedure is, at least somewhat, independent from the frequency bands which are chosen for the evaluation.

We also applied our approach on sensor data from a second wind turbine with a similar fault and observed results consistent to those presented here, i.e. the fault can be detected at more or less the same time as the proprietary method.

5   Conclusions

We described how linear and nonlinear dimensionality reduction approaches can be used for analyzing time series data in the wind energy domain. For the analysis of time series data stemming from numerical simulations in the virtual product development process, we introduced an approach which allows the quick identification of similarly behaving simulations and the detection of anomalies in simulation results. Furthermore, we proposed an approach based on linear dimensionality reduction which is able to detect changes in vibrational properties of rotor blades and which can be associated with the occurrence of faults.

A natural extension of this work would be to evaluate the data of more than one sensor at a time. For analyzing multiple sensors concurrently, suitable concepts of distances have to be introduced in this setting, and one has also to investigate, e.g., correlations in time between different sensors. Generally, other distance measures suitable for time series have to be investigated, in particular those which can be interpreted to observe certain physical principles and are invariant to certain transformations.

Acknowledgements The authors would like to thank the German Federal Ministry of Education and Research (BMBF) for the opportunity to do research in the VAVID project under grant 01IS14005. We cordially thank Henning Lang and Tobias Tesch for their assistance with the numerical experiments.

References

1. I. Antoniadou, Accounting for Nonstationarity in the Condition Monitoring of Wind Turbine Gearboxes, PhD thesis, University of Sheffield, 2013
2. R. Coifman, F. Lafon, Diffusion maps. Appl. Comput. Harmon. Anal. **21**, 5–30 (2006)
3. European Wind Energy Association (EWEA), Wind in power - 2015 European statistics, tech. report, European Wind Energy Association, February 2016
4. R. Gasch, J. Twele (eds.), *Wind Power Plants* (Springer, Berlin, 2012)
5. B. Hahn, M. Durstewitz, K. Rohrig, Reliability of wind turbines, in *Wind Energy*, ed. by J. Peinke, P. Schaumann, S. Barth (Springer, Berlin, 2007), pp. 329–332
6. T. Hastie, R. Tibshirani, J. Friedman, *The Elements of Statistical Learning*, 2nd edn. (Springer, Berlin, 2009)
7. N.E. Huang, Z. Shen, S. Long, et al., The empirical mode decomposition and the Hilbert spectrum for nonlinear and non-stationary time series analysis. Proc. R. Soc. Lond. Ser. A Math. Phys. Eng. Sci. **454**, 903–995 (1998)
8. F. Itakura, Minimum prediction residual principle applied to speech recognition. IEEE Trans. Acoust. Speech Signal Process. **23**, 67–72 (1975)
9. J.M. Jonkman, The new modularization framework for the FAST wind turbine CAE tool, in *Proceedings of the 51st AIAA Aerospace Sciences Meeting, 2013*. also Tech. RepNREL/CP-5000-57228, National Renewable Energy Laboratory, Golden, CO.
10. A. Kusiak, Z. Zhang, A. Verma, Prediction, operations, and condition monitoring in wind energy. Energy **60**, 1–12 (2013)
11. J.A. Lee, M. Verleysen, *Nonlinear Dimensionality Reduction* (Springer, Berlin, 2007)

12. W. Lu, F. Chu, Condition monitoring and fault diagnostics of wind turbines, in *Prognostics and Health Management Conference, 2010*. PHM '10., Jan. 2010, pp. 1–11

13. L. Mujica, J. Rodellar, A. Fernández, A. Güemes, Q-statistic and T^2-statistic PCA-based measures for damage assessment in structures. Struct. Health Monit. **10**, 539–553 (2011)

14. R.B. Randall, *Vibration-based Condition Monitoring: Industrial, Automotive and Aerospace Applications* (Wiley, Hoboken, 2011)

15. S. Salvador, P. Chan, FastDTW: Toward accurate dynamic time warping in linear time and space. Intell. Data Anal. **11**, 561–580 (2007)

Energy-Efficiency and Performance Comparison of Aerosol Optical Depth Retrieval on Distributed Embedded SoC Architectures

Dustin Feld, Jochen Garcke, Jia Liu, Eric Schricker, Thomas Soddemann, and Yong Xue

1 Introduction

Atmospheric aerosols are liquid or solid particles suspended in the air from natural and anthropogenic origin. They scatter and absorb solar radiation, and, to a lesser extent, scatter, absorb and emit terrestrial radiation (*direct effects*). Additionally, aerosols acting as cloud condensation nuclei and ice nuclei are referred to as *indirect effects*. A consequence of the direct effect (caused by absorbing aerosols) which changes cloud properties is called the semi-direct effect [5]. Besides the fact that aerosols affect the air quality and human health [8], such aerosol effects are currently considered one of the largest uncertainties in global radiative forcing [11]. These aspects have made the characterization of atmospheric aerosols a great concern in recent years.

The Aerosol Optical Depth (AOD), a measure of light extinction by aerosols in the atmospheric column above the earth's surface, is a comprehensive variable to remotely assess the aerosol burden in the atmosphere [14]. AOD data can be used by applications like the atmospheric correction of remotely sensed surface features, monitoring of sources and sinks of aerosols, radiative transfer models etc.

D. Feld (✉) • E. Schricker • T. Soddemann
Fraunhofer Institute for Algorithms and Scientific Computing SCAI, Schloss Birlinghoven, 53757 Sankt Augustin, Germany
e-mail: dustin.feld@scai.fraunhofer.de

J. Garcke
Fraunhofer Institute for Algorithms and Scientific Computing SCAI, Schloss Birlinghoven, 53757 Sankt Augustin, Germany

Institute for Numerical Simulation, Rheinische Friedrich-Wilhelms-Universität Bonn, Wegelerstr. 6, 53115 Bonn, Germany
e-mail: jochen.garcke@scai.fraunhofer.de

J. Liu • Y. Xue
Institute of Remote Sensing and Digital Earth, Beijing, China

© Springer International Publishing AG 2017
M. Griebel et al. (eds.), *Scientific Computing and Algorithms in Industrial Simulations*, DOI 10.1007/978-3-319-62458-7_17

Compared with ground measurements, satellite remote sensing provides an effective method for monitoring spatial distribution and temporal variation of aerosols. Many approaches have been developed for the retrieval of AOD using satellite remote sensing observations, including the use of advanced very-high-resolution radiometer (AVHRR), medium-resolution imaging spectrometer (MERIS), moderate-resolution imaging spectroradiometer (MODIS), multi-angle imaging spectroradiometer (MISR) and others [6].

A wide range of datasets have been published since the operation of MODIS sensor on TERRA and AQUA satellites. The broad swath of 2330 km enables MODIS to provide global coverage with near daily frequency [14]. AOD datasets derived from MODIS observations have been used to estimate surface particulate matter [13], construct global climatology [10] etc. One commonly used AOD dataset can be obtained from the *National Aeronautics and Space Administration* (NASA) based on the DarkTarget and DeepBlue method [2]. The work in this paper adopts the synergetic retrieval of aerosol properties (SRAP) algorithm developed by in [17] to generate AOD datasets over China.

2 Method

Although many approaches have been developed to retrieve AOD, it is still a difficult task to retrieve AOD over land because it is not easy to separate aerosols' signals from the land surface contributions. This paper takes the SRAP-MODIS algorithm as the study case. It has been developed to solve the aerosol retrieval problem over bright land surfaces. The algorithm utilizes the high-frequency multi-temporal and multi-spectral information from MODIS data aboard both TERRA and AQUA satellites to produce the AOD results.

The SRAP-MODIS algorithm is a simple but practical algorithm introduced in [16] on an operational bi-angle approach model for retrieving AOD and the earth surface reflectance [12]. More details can be found in [17].

In the algorithm, the ground surface reflectance $A_{i,j}$ is expressed by

$$A_{i,j} = \frac{(aA'_{i,j} - b_i) + b_i(1 - A'_{i,j})e^{\varepsilon(b_i-a)(0.00879\lambda_j^{-4.09}+\beta_i\lambda_j^{-\alpha})b'_i}}{(aA'_{i,j} - b_i) + a(1 - A'_{i,j})e^{\varepsilon(b_i-a)(0.00879\lambda_j^{-4.09}+\beta_i\lambda_j^{-\alpha})b'_i}}, \tag{1}$$

where $i = 1, 2$ represent the observations of TERRA MODIS and AQUA MODIS respectively, and $j = 1, 2, 3$ stand for three visible spectral bands at central wavelengths of 470, 550, 660 nm.

The symbols in Eq. (1) include both, *known* variables extracted from MODIS hierarchical data format (HDF) information such as $A'_{i,j}$ (the reflectance on the *Top Of Atmosphere* (TOA)) and *unknown* variables to be solved for, the symbols in Eq. (1) are listed and explained in Table 1.

Table 1 Symbols and applied values used in the SRAP algorithm

	Symbol	Implication
Input	$A'_{i,j}$	Apparent reflectance (reflectance on the top of atmosphere)
	a	$a = 2$ (cf. [12] and [16])
	b_i	$b_i = sec\theta_i$, with θ_i the solar zenith angle for two satellite observations
	ε	$\varepsilon = 0.1$
	λ_i	Wavelengths for three visible bands
	b'_i	$b'_i = sec\theta'_i$, with θ'_i the sensor zenith angle for two satellite observations
Output	$A_{i,j}$	Ground surface reflectance
	β_i	Ångstrom's turbidity coefficient
	α	Wavelength exponent

By assumption, the ground surface bidirectional reflectance properties and aerosol types and properties do not change for two MODIS observations within short time intervals between TERRA and AQUA overpass. Thus, the wavelength exponent α is invariant for two observations and three visible bands. Ångstrom's turbidity coefficient, which represents the concentration of aerosol particles, may change for two overpass TERRA and AQUA observations; hence we have β_1 and β_2 for two overpass times.

The ground surface reflectance $A_{i,j}$ can be approximated by the variation in the wavelength and the variation in the geometry [1]. Under this assumption, the ratio of two views' ground surface reflectance K_j for the wavelength j can be formulated as

$$K_j = \frac{A_{1,j}}{A_{2,j}}, \tag{2}$$

where $A_{1,j}$ and $A_{2,j}$ are the surface reflectances for the TERRA MODIS and AQUA MODIS.

Since aerosol extinction decreases rapidly with the wavelength, the AOD at $2.12\,\mu m$ is very small compared to that at the visible spectra bands. The atmospheric contribution at $2.12\,\mu m$ is relatively small, hence $K_{\lambda=2.12\mu m}$ can be approximated as the ratio between the reflectance on the TOA. Besides, since K_j is assumed independent of the wavelength, the value at $2.12\,\mu m$ can be used for the visible bands. Thus, Eq. (3) is serving as the constraint between two ground surface reflectances for each visible band:

$$\frac{A_{1,j}}{A_{2,j}} = K_j = K_{\lambda=2.12\mu m} = \frac{A'_{1,\lambda=2.12\mu m}}{A'_{2,\lambda=2.12\mu m}}. \tag{3}$$

As a result, the unknown variables reduce to three, i.e. Ångstrom's turbidity coefficient β_i ($i = 1, 2$) for two overpass observations and the wavelength exponent α. Three equations containing three unknown variables to be solved are formed.

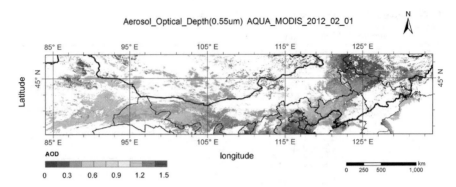

Fig. 1 AOD result map for AQUA MODIS at 550 nm band. Published with kind permission of ©Fraunhofer SCAI 2016. All Rights Reserved

In our implementation, the three variables β_1, β_2 and α from the obtained nonlinear equations are solved using the Broyden quasi-Newton method in the implementation from the 'Numerical Recipes in C' [9]. The known variables in Table 1 are taken as the input data for the method while it outputs the results for β_i and α. These values are then used to calculate the resulting ground surface reflectances $A_{i,j}$ and to further calculate the AOD τ_A according to Ångstrom's turbidity formula

$$\tau_A = \beta_i \lambda_j^{-\alpha}. \tag{4}$$

Figure 1 shows an example of an AOD result map for the observed benchmark region.

3 Implementation

For an appropriate AOD retrieval using the SRAP algorithm, pre-processing procedures such as cloud masking, absorption and geometric corrections, interpolations etc. need to be implemented and applied in addition to the final AOD model solving. We profiled the runtime of all serial procedures for the retrieval workflow on a workstation system in an earlier publication [4]. The profiling results showed that the final AOD model solving procedure takes up almost 50% of the total workflow runtime and, thus, is the most time-consuming part. It is therefore in the exclusive focus of the study in this paper. The pseudo code for the AOD model solving part is shown in Algorithm 1. It mainly consists of *data input and output (I/O)* procedures, i.e. the image data reading and writing from or to hard disk steps and the *computation* performing the solving of nonlinear equations and AOD calculation according to Ångstrom's turbidity formula addressed in Sect. 2. The reading and writing of image data from and to disc were implemented using

---

Algorithm 1: Pseudo code for the AOD model solving.

Data: corrected reflectance, sensor zenith, solar zenith, initial values
Result: AOD results
read-in the scene's image data (with GDAL)
create an image representation of each spectral band
if *calculate DISTRIBUTED* **then**
   | split the scene (each representation) in #nodes blocks
   | distribute the data via message passing (MPI)

OnEach *Node* **do in parallel**
   if *use CUDA GPU* **then**
     | copy image representations to the accelerator's memory

   for *each pixel p in the scene* **do in parallel**
     | use $\mathbf{x}_p$ to solve (1) with (3) for
     | α (wavelength exponent) and
     | β (Ångstrom's turbidity coefficient)
     | calculate (4)
     | τ_A (AOD according to Ångstrom's turbidity formula)

   if *use CUDA GPU* **then**
     | copy back results from the accelerator's memory

if *calculate DISTRIBUTED* **then**
   | collect results from all nodes
write resulting images to files

---

the Geospatial Data Abstraction Library (GDAL) [15]. These functions are not parallelized and are therefore executed on a single core of the host CPU for all versions of the implementation. The computation procedure has a pixel-based nature in the operations without communication across the pixels. Thus, the solving of nonlinear equations and the AOD calculation for each pixel can be assigned to an individual parallel thread without any explicit or implicit synchronization.

For multi-core processors, the loop in Algorithm 1 was parallelized using an OpenMP directive '#pragma omp parallel for' with the advanced scheduling strategy 'schedule (dynamic)'. This choice of a dynamic scheduling strategy is crucial for a good resulting performance as the time needed for the individual pixels' calculation varies depending on the input vector. This is based on varying convergence speeds for different inputs on the one hand and on the fact that different pixels follow different control flows in the calculation kernel on the other hand. For example, the calculation for a pixel over sea finishes way faster than one over land. If a static scheduling is applied, coherent regions over sea are typically mapped to the same core and this core would therefore become idle long before other 'over-land' cores finish their calculation. As the calculation time for each pixel is relatively long, the overhead for dynamic scheduling pays off. Alternatively, a static scheduling with a chunk-size of 1 behaves similar. This influence was intensively studied and verified in an earlier publication [3].

For the GPU implementation, the Compute Unified Device Architecture C (*CUDA-C*) was used. A respective AOD kernel was designed and implemented so that each thread corresponds to the calculation for one pixel. The pixels are

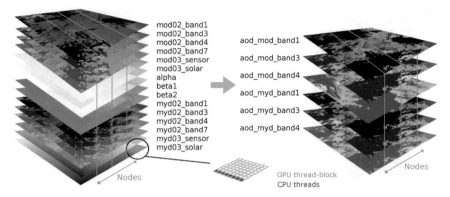

Fig. 2 Multi-level parallelization on the spectral bands' data hypercube. Published with kind permission of ©Fraunhofer SCAI 2016. All Rights Reserved

distributed on a *grid-block* of parallel threads. After reading the images from disc, the input images' data are copied from the CPU host's main memory to the GPU device's global memory at once, and, after computation of the AOD result, data are copied back to the CPU host and written to output files on disc. The thread-block size was configured to 8×8 parallel threads per thread-block and the dimensions of the grid of thread-blocks were set dynamically corresponding to the image size to cover the whole image with thread-blocks.

For distributed computing, the Message Passing Interface (*MPI*) was used to combine the computing power of multiple nodes. The input images were uniformly split into approximately identically sized pieces according to the number of nodes. Each resulting part of the input image is distributed to its node and AOD is computed there either on the multi-core processors or the GPU. Because of the static splitting, the workload on the different nodes may be unbalanced but due to the relatively slow interconnect with gigabit ethernet, a splitting into smaller portions combined with a dynamic distribution on request is not investigated for this architecture. It could nevertheless improve the multi-node performance furthermore. After the computations on all nodes have finished, the AOD results are finally gathered from all nodes and combined. Algorithm 1 and Fig. 2 depict the overall parallel procedure of the AOD retrieval ($\mathbf{x}_p$ represents the vector containing observations for the different spectral bands for pixel p).

4 Embedded Low-Energy System

We set up an embedded system combined of four NVIDIA Tegra K1 (*TK1*) boards as shown in Fig. 3. These boards are designed to provide relatively high computing power especially for GPU applications while consuming only a relatively low amount of energy. They are primarily targeted for mobile applications. The TK1

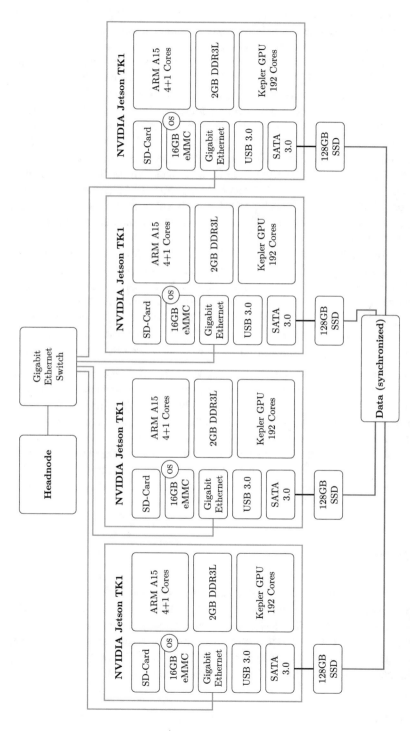

Fig. 3 Jetson TK1 cluster configuration. Published with kind permission of ©Fraunhofer SCAI 2016. All Rights Reserved

builds on the same NVIDIA Kepler architecture that drives high performance graphic compute units like the Tesla K20, but with much fewer cores (192 compared to 2496), slower and less memory (2 GB *DDR3* memory shared with the CPU compared to 5 GB of exclusive *GDDR5* memory) and other restrictions. Each board is driven by a 4 + 1 ARM Cortex A15 with four fast cores and one ultra low-energy core. As a consequence, a whole TK1 board has a power intake of not more than 10 W while a high performance graphics card may take in up to 225 W. In the middle of such 'extreme' GPUs, there is a wide variety of CUDA capable GPU accelerators, like the NVIDIA GTX 680 we use in our benchmarks. This card contains 1536 CUDA cores, 2 GB exclusive GDDR5 memory and consumes up to 200 W.

Undoubtedly, such high performance GPU devices like the K20 provide computational power that is of a higher order of magnitude (3.52 Tflops [peak single-precision]) than the TK1's GPU chip (326 Gflops [peak single-precision]).[1] As well, a high performance Intel Xeon CPU computes way faster than the equipped ARM cores on the TK1.

Certainly, the best performing hardware solution with a fast CPU and a fast GPU composed of the aforementioned hardware would be a system combining Intel Xeon CPUs with NVIDIA Tesla GPUs. If hard energy constraints similar to automotive or on-board satellite situations exist, a low-energy constellation like on the TK1 may be chosen. The question to be answered in this paper is, what to choose if the *energy efficiency* of such different configurations is the criterion of choice. This can be formalized by the $\frac{\text{pixels}}{\text{joule}} = \frac{\text{pixels}}{\text{wattsecond}}$ ratio. Real-time computations producing new input data every M minutes combined with the demand to save energy are typical scenarios. If new data arrives every M minutes, a solution with lower performance and lower energy consumption may be preferable to high performance solutions as long as it processes the data set in less than M minutes to meet the time constraint.

In the following benchmarks, each hardware constellation is used to compute the AOD for the same earth region so that the number of pixels is constant. Therefore, the overall energy consumption can be used to compare the energy efficiency of different hardware constellations.

5   Benchmarks

The benchmarks' input data are extracted and pre-processed from MODIS HDF data, which can be downloaded from the *Level-1 and Atmosphere Archive and Distribution System* (LAADS Web) [7]. The data was selected randomly from February 1, 2012 and it covers 84–134°E, 38–48°N, which corresponds to 5000 × 1000 pixels. The spatial resolution of each pixel is 1 km. In the following benchmarks, calculations are performed with single-precision floating point data.

---

[1]http://www.nvidia.com.

Table 2 Benchmark system configurations

		CPU	GPU
Workstation	Type	Intel® Xeon® E3-1275 V2	NVIDIA® GTX 680
(Scientific Linux 6.6)	#Cores	4 (8 with HT)	1536
	Clock-Speed	3,5 GHz	1058 MHz
	Memory	8 GB DDR3	2 GB GDDR5
Tegra cluster	Type	ARM Cortex A15	NVIDIA® Kepler™
(Gentoo Linux)	#Cores	4(+1) quad-core	192 CUDA cores
	Clock-Speed	Up to 2.3 GHz	852 MHz
	Memory	2 GB DDR3 (shared memory space)	

5.1 Benchmark Environment

We used two benchmark systems in our experiments: the aforementioned low-energy system based on four NVIDIA Tegra K1 boards from Sect. 4, each equipped with an ARM 4(+1) quad-core CPU and an NVIDIA Kepler GPU, and a workstation equipped with an Intel Xeon E3-1275 V2 CPU and an NVIDIA GTX 680 GPU. Details are shown in Table 2.

First we compare the performance of the two hardware systems and their different processor types in Sect. 5.2, second we show how they behave in terms of energy consumption and efficiency in Sect. 5.3. As the TK1 contains not only the quad-core CPU and a GPU but also an additional low power core, we as well analyze the potential of this core in both counts (energy consumption and efficiency). Regarding this, the four powerful cores are named *HPCores*, the additional low power core is named *LPCore*. The TK1 '*System on Chip*' boards are also referred to as *SoC* and the workstation as *XeonWS*.

All runs were performed 10 times, the maximal and minimal runtime and energy values were neglected and the average of the remaining 8 runs is reported. The relative standard deviation of the runtimes of all single node runs was below 1% in our benchmarks while the one of the multi-node runs with four TK1 boards was still below 2%. The respective ranges of power intake are shown by the shaded areas in Figs. 7, 8, 9, 10, 11, 12, 13, 14.

Whenever only one TK1 system is used, the results are reported in green, if more than one TK1 boards are used, it is shown in purple and for the Xeon workstation in gray.

5.2 Performance Benchmarks

Table 3 contains the calculation runtimes for all constellations while Fig. 4 only shows the most relevant times as a bar chart. The results for one board (1xSoC) reveal that using one of the faster four cores instead of the low energy core improves

Table 3 Calculation runtime, HPCo≙High Power Core, LPCo≙Low Power Core, T≙Threads

Runtime	1xSoC					4xSoC	XeonWS		
	CPU			GPU		MPI	CPU		GPU
	1 LPCo	1 HPCo	4 HPCo	1 LPCo	1 HPCo	1 HPCo+GPU	1T	4T	
Calculation	2914.83	2861.37	717.37	47.14	46.93	18.13	192.15	49.28	3.92
I/O	4.78	4.08	3.94	4.87	4.24	4.23	1.81	1.44	1.44
Overall	2919.62	2865.45	721.31	52.01	51.16	22.36	193.97	50.72	5.36

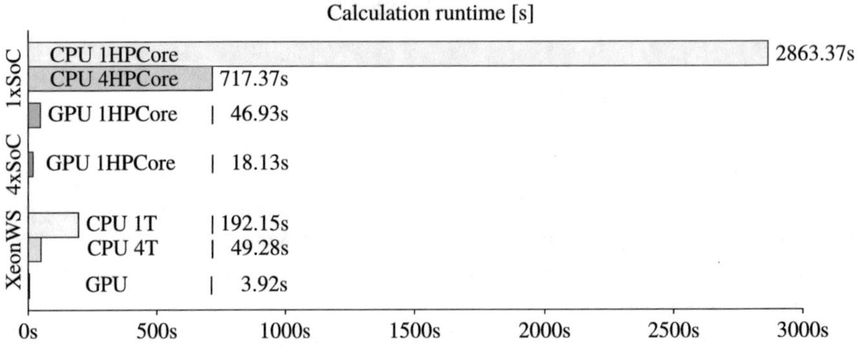

Fig. 4 Calculation runtime, HPCore≙High Power Core, T≙Threads. Published with kind permission of ©Fraunhofer SCAI 2016. All Rights Reserved

the runtime by only about 2%. Using all four fast cores with multi-threading instead of one of those leads to a near-ideal overall speedup of 3.97×. Executing the code on one SoC on the GPU, while the serial I/O is performed on one of the faster ARM CPU cores, is 14× faster than on the four main ARM cores. Activating the low power core instead of a normal one leads to a 15% performance loss in the CPU based I/O routines while it does not significantly influence the GPU based calculation time. Therefore, the best performing single SoC configuration is to use one of the faster ARM cores for the I/O routines along with the GPU for the calculation part.

Distributing the work on all four boards with this configuration leads to an additional speedup of 2.3×. Figure 5 shows how the runtime decreases the more boards are used to process the whole work. It as well shows how the speedup stagnates from 1.63× when taking two boards, 2.07× for three and, finally, 2.6× for all four boards. This confirms that the interconnect via gigabit ethernet, which is used for the MPI communication, can become a de facto bottleneck in such a system configuration.

Comparing those runtimes to the workstation shows that multi-threading on the Xeon CPU as well scales near-optimal with a speedup of 3.82× on four cores and that the Xeon CPU performs about 15× better than one ARM core. The resulting

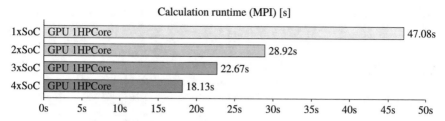

Calculation runtime (MPI) [s]

1xSoC	GPU 1HPCore — 47.08s
2xSoC	GPU 1HPCore — 28.92s
3xSoC	GPU 1HPCore — 22.67s
4xSoC	GPU 1HPCore — 18.13s

0s 5s 10s 15s 20s 25s 30s 35s 40s 45s 50s

Fig. 5 Calculation runtime (MPI), HPCore≙High Power Core, T≙Threads. Published with kind permission of ©Fraunhofer SCAI 2016. All Rights Reserved

performance on all Xeon cores is comparable to the performance of one TK1 with the GPU. The GPU in the workstation expectedly reaches the best overall performance.

5.3 Energy Benchmarks

Table 4 contains the respective energy consumptions for all configurations. The power intake was measured with an ISO certified digital multimeter 'Voltcraft VC870' at the maximal clock-rate of one measurement per second. This introduces an inaccuracy for small runtimes that should be equalized by averaging the consumption of multiple repeated runs. As in the previous section, Fig. 6 shows the most relevant results as a bar chart. As the system allows it, we deactivated all cores that are not used in any of the TK1 measurements.

Executing the code on four cores instead of only one on the CPU consumes only about 3650.93 Ws compared to 12309.49 Ws and is therefore much more energy efficient as the four core version is 3.97× faster and only has a 18% higher average power intake (~5.1 W cp. to 4.3 W—see Figs. 7 and 8). This is based on a relatively high power intake for all the periphery in the idle system. The best energy efficiency on one TK1 board is reached when using the GPU.

As shown in Sect. 5.2, exhausting the compute power of all four TK1 boards leads to the best TK1 performance but not with an ideal scaling due to the inter-board communication via ethernet. As the power intake grows linearly when using multiple identical boards simultaneously (plus some extra power for the ethernet switch), using multiple boards does not improve the energy efficiency.

Comparing those results to the ones of the workstation, the TK1 system is for every configuration (single-core CPU, multi-core CPU, GPU) more energy efficient than its pendant on the workstation. Nevertheless, the high power intake of the workstation's components is almost equalized by the likewise faster computing. The biggest advantage in terms of energy efficiency can be seen in the GPU benchmarks. While the workstation's GPU consumes 880.95 Ws for the whole calculation, a single TK1 board settles this task with consuming only 339.61 Ws.

Table 4 Energy consumption, HPCo≘High Power Core, LPCo≘Low Power Core, T≘Threads

	1xSoC					4xSoC
	CPU			GPU		MPI
Power/energy	1 LPCo	1 HPCo	4 HPCo	1 LPCo	1 HPCo	1 HPCo+GPU
Ø Power intake [W]	4.14	4.30	5.06	6.48	6.64	27.32
Energy consumption [Ws]	12086.11	12309.49	3650.93	337.20	339.61	610.88
Pixels/Ws	414	406	1370	14828	14723	8185

	XeonWS		
	CPU		GPU
Power/energy	1T	4T	
Ø Power intake [W]	79.07	103.88	164.36
Energy consumption [Ws]	15336.89	5268.56	880.95
Pixels/Ws	326	949	5676

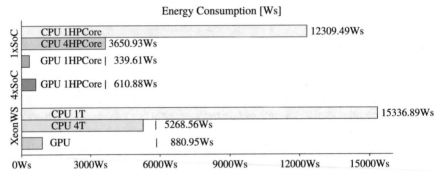

Fig. 6 Energy consumption, HPCore≘High Power Core, T≘Threads. Published with kind permission of ©Fraunhofer SCAI 2016. All Rights Reserved

For better comparison, Table 4 additionally contains the $\frac{pixels}{Ws}$ ratio which quantifies how many pixels can averagely be calculated with the energy of 1 Ws (resp. 1 J).

Figures 7, 8, 9, 10, 11, 12, 13, 14 illustrate the power intake over time for the different runs. The shaded areas represent the range of variation in the repeated runs. It can be concluded that the power intake on the TK1s is generally very stable over time (cp. especially Fig. 14). On the workstation, there is more deviation in the intake, especially for the GPU runs, but this was probably caused by the very short runtimes with only few measurement points.

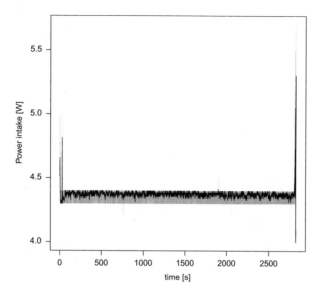

Fig. 7 Power intake 1xSoC CPU 1HPCore. Published with kind permission of ©Fraunhofer SCAI 2016. All Rights Reserved

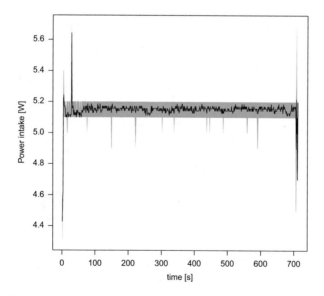

Fig. 8 Power intake 1xSoC CPU 4HPCore. Published with kind permission of ©Fraunhofer SCAI 2016. All Rights Reserved

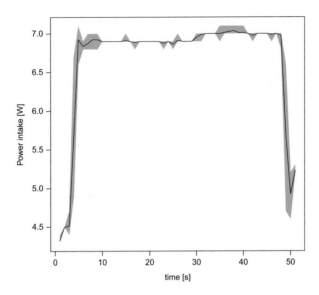

Fig. 9 Power intake 1xSoC GPU 1HPCore. Published with kind permission of ©Fraunhofer SCAI 2016. All Rights Reserved. The ranges of power intake among the 10 repeated runs are shown by the *shaded areas*

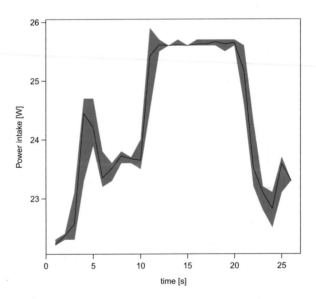

Fig. 10 Power intake 4xSoC GPU 1HPCore. Published with kind permission of ©Fraunhofer SCAI 2016. All Rights Reserved. The ranges of power intake among the 10 repeated runs are shown by the *shaded areas*

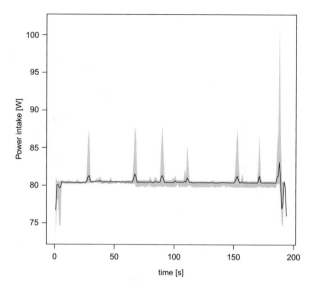

Fig. 11 Power intake XeonWS CPU 1T. Published with kind permission of ©Fraunhofer SCAI 2016. All Rights Reserved. The ranges of power intake among the 10 repeated runs are shown by the *shaded areas*

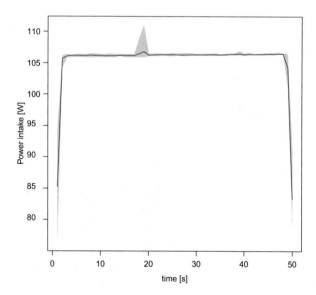

Fig. 12 Power intake XeonWS CPU 4T. Published with kind permission of ©Fraunhofer SCAI 2016. All Rights Reserved. The ranges of power intake among the 10 repeated runs are shown by the *shaded areas*

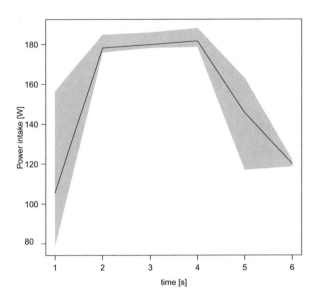

Fig. 13 Power intake XeonWS GPU. Published with kind permission of ©Fraunhofer SCAI 2016. All Rights Reserved. The ranges of power intake among the 10 repeated runs are shown by the *shaded areas*

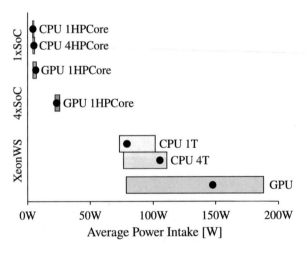

Fig. 14 Power intake range (● ≙ Average). Published with kind permission of ©Fraunhofer SCAI 2016. All Rights Reserved. The ranges of power intake among the 10 repeated runs are shown by the *shaded areas*

6 Discussion

Concerning the suitability for parallel execution and the applicability for different parallel architectures, the results from Sect. 5 show that the code scales very well in a multi-threaded shared memory environment and is exceptionally well suited for GPU computing. As a result, a GPU on a TK1 board can achieve a performance comparable to the Xeon CPU while consuming only a fraction of the energy. The difference in both, energy consumption and performance, of the low power ARM core on the TK1 to one of the 'normal' cores was rather small in our benchmarks. Due to the existent but not ideal multi-node scaling, the usage of multiple boards improves performance on the one hand but diminishes energy-efficiency on the other hand.

In an earlier publication [3], we investigated the performance and energy efficiency of the retrieval code in a server environment equipped with a high performance Kepler K20 GPU. The calculation of the same scene took 2.61 s on the GPU consuming 621.91 Ws (approximately 60% is consumed by the GPU, 40% by the remaining system) while processing 8040 pixels per Ws. Thus, the energy efficiency of this high-end server system is in between the one of the GPU in the benchmarked workstation system and the one on a TK1 board.

Which configuration to choose therefore strongly depends on the environmental constraints. The most energy efficient solution for executing the presented code is to use the GPU of one TK1 board. If the energy restrictions are a bit looser, a compound of several such boards can further improve the performance while keeping the energy intake relatively low. Such a system outperforms our Xeon CPU on four threads by a factor of 2.7× while the Xeon system consumes more than 8.6× times the energy. It has to be noted that the idling GPU power intake is included in such CPU measurements on the workstation.

When looking purely for highest computational performance with no energy constraints, the workstation GPU clearly wins.

Considering all benchmarks and both goals, performance and energy-efficiency, a very interesting workstation constellation for the future will be a low energy CPU along with a high-end GPU.

7 Outlook

In this paper, we investigated the potential of an embedded SoC architecture equipped with a multi-core ARM processor and an NVIDIA Kepler GPU in terms of both, computational performance and energy efficiency for the retrieval of AOD. The results show that embedded SoC boards like the NVIDIA Tegra K1 provide a relatively high computational power paired with a low power intake, especially if a code can make efficient use of the GPU.

In the future, we plan to port more codes that are used in real-time scenarios to such architectures to save energy in scenarios where either the energy constraints are naturally tight or where the provided computational power of such systems is simply sufficient.

Acknowledgements This work was partially funded by the German Ministry for Education and Research (BMBF) under project grant 01IS13016A within the ITEA2-Project MACH.

References

1. R.J. Flowerdew, J.D. Haigh, An approximation to improve accuracy in the derivation of surface reflectances from multi-look satellite radiometers. Geophys. Res. Lett. **22**, 1693–1696 (1995)
2. R.C. Levy, S. Mattoo, L.A. Munchak, et al., The Collection 6 MODIS aerosol products over land and ocean. Atmos. Meas. Tech. **6**, 2989–3034 (2013)
3. J. Liu, D. Feld, Y. Xue, J. Garcke, T. Soddemann, Multicore processors and graphics processing unit accelerators for parallel retrieval of aerosol optical depth from satellite data: implementation, performance, and energy efficiency. IEEE J. Sel. Top. Appl. Earth Obs. Remote Sens. **8**, 2306–2317 (2015)
4. J. Liu, D. Feld, Y. Xue, et al., An efficient geosciences workflow on multi-core processors and GPUs: a case study for aerosol optical depth retrieval from MODIS satellite data. Int. J. Digital Earth **9**, 748–765 (2016)
5. U. Lohmann, J. Feichter, Global indirect aerosol effects: a review. Atmos. Chem. Phys. **5**, 715–737 (2005)
6. L. Mei, Y. Xue, H. Xu, et al., Validation and analysis of aerosol optical thickness retrieval over land. Int. J. Remote Sens. **33**, 781–803 (2012)
7. NASA, Level 1 and Atmosphere Archive and Distribution System (LAADS Web). Website, 2015. Online at http://ladsweb.nascom.nasa.gov/, visited 20 Oct 2015
8. U. Pöschl, Atmospheric aerosols: composition, transformation, climate and health effects. Angew. Chem. Int. Ed. **44**, 7520–7540 (2005)
9. W.H. Press, S.A. Teukolsky, W.T. Vetterling, B.P. Flannery, *Numerical Recipes in C: The Art of Scientific Computing*, 2nd edn. (Cambridge University Press, New York, 1992)
10. L.A. Remer, R.G. Kleidman, R.C. Levy, et al., Global aerosol climatology from the MODIS satellite sensors. J. Geophys. Res. Atmos. **113**, D14S07 (2008)
11. T.F. Stocker, D. Qin, G.-K. Plattner, et al., Technical summary, in *Climate Change 2013: The Physical Science Basis. Contribution of Working Group I to the Fifth Assessment Report of the Intergovernmental Panel on Climate Change*, ed. by T.F. Stocker, D. Qin, G.-K. Plattner, et al. (Cambridge University Press, Cambridge/New York, 2013), book section TS, pp. 33–115
12. J. Tang, Y. Xue, T. Yu, Y. Guan, Aerosol optical thickness determination by exploiting the synergy of TERRA and AQUA MODIS. Remote Sens. Environ. **94**, 327–334 (2005)
13. T.-C. Tsai, Y.-J. Jeng, D.A. Chu, J.-P. Chen, S.-C. Chang, Analysis of the relationship between MODIS aerosol optical depth and particulate matter from 2006 to 2008. Atmos. Environ. **45**, 4777–4788 (2011)
14. Y. Wang, Y. Xue, Y. Li, et al., Prior knowledge-supported aerosol optical depth retrieval over land surfaces at 500m spatial resolution with MODIS data. Int. J. Remote Sens. **33**, 674–691 (2012)
15. F. Warmerdam, GDAL - Geospatial Data Abstraction Library. Website, 2015. Online at http://www.gdal.org/, visited 20 Oct 2015
16. Y. Xue, A.P. Cracknell, Operational bi-angle approach to retrieve the earth surface albedo from AVHRR data in the visible band. Int. J. Remote Sens. **16**, 417–429 (1995)
17. Y. Xue, X. He, H. Xu, et al., China Collection 2.0: the aerosol optical depth dataset from the synergetic retrieval of aerosol properties algorithm. Atmos. Environ. **95**, 45–58 (2014)

Part IV
A Short History

The Fraunhofer Institute for Algorithms and Scientific Computing SCAI

Ulrich Trottenberg and Anton Schüller

The Fraunhofer Institute for Algorithms and Scientific Computing SCAI, together with its forerunner institutes, will be 50 years old in 2018. This means that a large part of the development of information technology has been accompanied by SCAI. And at least on the software and the algorithmic side, SCAI has contributed substantially to this development.

Before SCAI became a Fraunhofer institute in 2001, SCAI had been one of the eight institutes of the GMD, originally the *Gesellschaft für Mathematik und Datenverarbeitung mbH*, and renamed to *GMD—German National Research Center for Information Technology* in 1995.

We distinguish five phases of the institute's history:

1. foundation of the GMD, the first decade (1968–1977),
2. numerical simulation, multigrid and parallel computing (1978–1991),
3. SCAI—algorithms and scientific computing (1992–2001),
4. SCAI as a Fraunhofer institute—the first years (2002–2009),
5. new fields of research and new business departments (2010–2016).

Before we take a closer look into SCAI's history since 1968, we would like to give an impression of the dynamic development of information technology in the past decades.

Figure 1 shows how drastically the computing times for a typical problem in scientific computing were reduced from 1980 to 2000. The illustration clarifies that this progress is due to both, algorithmic and hardware developments. Altogether, these developments reduced the computing times by a factor of 20 millions. This example illustrates why the role of computer science and applied mathematics has

U. Trottenberg (✉) • A. Schüller
Fraunhofer Institute for Algorithms and Scientific Computing SCAI, Schloss Birlinghoven, 53757 Sankt Augustin, Germany
e-mail: ulrich.trottenberg@scai.fraunhofer.de

© Springer International Publishing AG 2017
M. Griebel et al. (eds.), *Scientific Computing and Algorithms in Industrial Simulations*, DOI 10.1007/978-3-319-62458-7_18

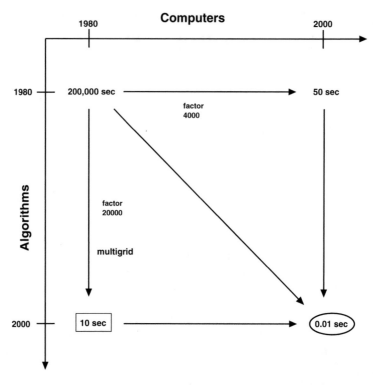

Fig. 1 Progress by faster algorithms and faster computers from 1980 to 2005 for the solution of the discrete Poisson equation (see also [34]): The algorithmic progress was achieved by the development of multigrid methods compared to a classical relaxation type method. This algorithmic progress is not just an exponential speed up, but a *disruptive* change in complexity from $O(N^2)$ to $O(N)$, where N denotes the number of unknowns. The hardware progress includes the change from vector computing to parallel computing. The fact that the hardware and software speed-up factors are not just multiplied is due to the fact that the parallel efficiency of multilevel methods for a problem of fixed size decreases with the number of processors, such that some loss has to be accepted. Published with kind permission of ©Fraunhofer SCAI 2016. All Rights Reserved

been so important for the scientific progress in the last decades. Computations which can be performed within seconds today would have required several months of computing time 30 years ago. (We just mention, that, of course, nowadays much larger numerical simulations are run regularly on today's computers.)

These developments have to be kept in mind when considering the quite involved history of SCAI. We do not describe all details of this history, but restrict ourselves to the main milestones and highlights which still have some relevance today.

1 Foundation of the GMD, the First Decade (1968–1977)

The GMD itself was founded on April 23, 1968 as the *Gesellschaft für Mathematik und Datenverarbeitung mbH*. The shareholders of the GMD were the *Federal Republic of Germany* holding 90% and the *Federal State of North Rhine-Westphalia* holding 10%. The foundation was politically motivated. The idea was to establish a research center for applied mathematics and data processing that structurally followed the concept of the Nuclear Research Centers Jülich and Karlsruhe (*Kernforschungsanlage Jülich* and *Kernforschungszentrum Karlsruhe*). For this purpose, the *Institut für Instrumentelle Mathematik (IIM)* (headed by Heinz Unger) at the University of Bonn was extended and moved into laboratory buildings belonging to Schloss Birlinghoven near Bonn.

The initiators and first managing directors of the GMD were Ernst Peschl and Heinz Unger, both professors of mathematics at the University of Bonn. At the end of 1969 the GMD had established three mathematical and two computer science oriented institutes,

- the *Institut für Reine Mathematik* directed by Ernst Peschl,
- the *Institut für Angewandte Mathematik* directed by Heinz Unger,
- the *Institut für Numerische Datenverarbeitung* directed by Fritz Krückeberg,
- the *Institut für Theorie der Automaten und Schaltnetzwerke* directed by Karl Heinz Böhling,
- the *Institut für Informationssystemforschung* (ISF) directed by Carl Adam Petri,

and, additionally, a computing center and several independent departments.

Already in 1970, in the context of the so called *software crisis* (the finding that the software had become more expensive and more complex than the hardware, leading to the introduction of *software engineering*) and the setup of courses of study for information technology at German universities, a new focus was laid on engineering disciplines and software technology. Fritz Krückeberg became chairman of the executive board of the GMD, and—in order to strengthen the *practical* numerical work of the GMD, Johann Schröder, a renowned numerical analyst, was offered to lead the mathematical activities of the GMD. Johann Schröder, however, did not accept the offer.

The extremely rapid pace of development of information technology in the 1970s was reflected by corresponding changes and extensions of the research topics of the GMD and by several new institutes working in information technology. Consequently, mathematics slowly drifted into the background of GMD's research priorities. In order to concentrate the mathematical work of the GMD, the three mathematical institutes were integrated into one, the *Institut für Mathematik (IMA)*, in 1973 (see Fig. 2). The prevalence of information technology was also reflected by a reformulation in the articles of association of the GMD in 1975, which reduced the mathematical research of the GMD to those areas which were of particular importance for computer science.

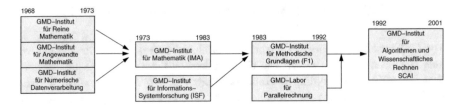

Fig. 2 History of the institute SCAI between 1968 and 2001. Published with kind permission of ©Fraunhofer SCAI 2016. All Rights Reserved

Following the demands of its shareholders, the GMD focused on a few research areas enforcing more collaboration among its institutes and intensified cooperations with industrial partners in 1978. In this situation, Krückeberg invited Ulrich Trottenberg, at that time professor of applied mathematics at the University of Bonn, to join the institute IMA as a consultant and to build up an application oriented numerical group.

We will not describe the history of the GMD systematically, but restrict ourselves to issues of direct relevance for SCAI. We only want to mention that over the years (1978–2000), the GMD was restructured several times, finally consisting of eight research institutes, four of which were located in Sankt Augustin, two in Berlin and two in Darmstadt.

2   Numerical Simulation, Multigrid and Parallel Computing (1978–1991)

Soon after Ulrich Trottenberg had started his work for the institute IMA, he was able to hire several young highly qualified numerical scientists (from the universities of Bonn and Cologne), among them Klaus Stüben who became the first head of the numerical group in 1978. The numerical group started its research and development work on *fast elliptic solvers*, in particular on the *Total Reduction* method [31, 32]. (Fast elliptic solvers are algorithms which efficiently solve the large systems of equations which arise when discretizing Poisson-like partial differential equations.) This approach is (formally) related to the *multigrid* principle [3, 16, 46]. Therefore, Achi Brandt, professor of Applied Mathematics at the Weizmann Institute at Rehovot, Israel, and one of the (re-)inventors of multigrid, visited the institute IMA in order to learn more about Total Reduction and to start a cooperation. As a first result of this cooperation the *MGR solvers* [25], a combination of Total Reduction and multigrid, were conceived and implemented. And indeed, the *MGR-CH-1* algorithm [6] a bit later turned out to be the fastest *Poisson Solver* worldwide.

Stüben and Trottenberg soon recognized that the multigrid principle was both, much more general than *and* nearly as efficient as Total Reduction. From then on, the group concentrated its methodological work on MGR-like methods and multigrid, with all its generalizations and applications including geometric and algebraic multigrid [26, 39] methods. Soon the numerical GMD group became one of the most renowned multigrid groups in the world. A very intensive and fruitful cooperation between Achi Brandt's group and the numerical GMD group was established which lasted for several decades. This cooperation was also reflected by the fact that Trottenberg later was one of the initiators of the foundation of the *Minerva Carl F. Gauss-Center "Scientific Computation"* in Rehovot. From 1993 to 2003 Brandt was its director and Trottenberg the chairman of its supervisory board.

SIHEM: Already in 1978, the numerical IMA group started a first industrial project, solving the following practical problem for the steel industry (Mannesmann AG): Before steel blocks are rolled, they have to be heated up to a certain fixed temperature (e.g. 1500 K) in such a way that the final temperature distribution is as homogeneous as possible. For the numerical simulation and energy minimization of the heating process, a (nonlinear parabolic) heat conduction equation had to be solved. For that purpose the interactive program SIHEM was developed, numerically making use of Newton linearization, Crank-Nicolson implicit discretization and fast elliptic solvers for each time step. Together with a graphical interface, SIHEM was presented at the Hannover Fair 1979—and considered as a big success.

Soon after the results of the SIHEM project had been demonstrated, the Volkswagen AG asked the institute for a cooperation in order to develop a multigrid solver for the numerical simulation of in-cylinder turbulent flows in combustion engines [27, 38]. This was the first contractual work of the institute for an industrial partner, financing two scientists for 2 years.

In spite of these quite spectacular successes, the board of directors of the GMD decided to make *information technology for organizations* the working focus of the whole GMD in 1980—which would had left little space for numerics and engineering oriented mathematics.

In order to stabilize the work of the numerical group and to obtain more political and financial support for the important and promising fields *Numerical Simulation and Computational Engineering*, Trottenberg directly contacted Andreas von Bülow, at that time the Federal Minister of Research and Technology (BMFT). As a result, the chairman of the supervisory board of the GMD, Fritz-Rudolf Güntsch, approved a reasonable and enduring budget for the numerical activities.

A period of a lasting upswing of the GMD began when Norbert Szyperski (1931–2016) became chairman of the executive board of the GMD in 1981. He implemented a reorganization of the GMD institutes and—after discussions with the shareholders—enforced significantly better conditions for collaborative projects. In particular, he strongly supported the numerical activities and the work with Petri nets [37] from the very beginning.

In 1983, Ulrich Trottenberg accepted the offer to become director of IMA, first as an extraofficial director, and in a joint appointment as a full professor of applied mathematics and scientific computing at the University of Cologne since 1984. In 1983, the institutes IMA and ISF were merged and became the new institute F1 (see Fig. 2), jointly directed by Krückeberg (1928–2012), Petri (1926–2010) and Trottenberg.

> **SUPRENUM:** In 1983, in view of Japan's 5th Generation Program and in order to stay competitive, the German federal research centers expressed the urgent need for concentrated research in two central fields:
>
> 1. innovative supercomputer architectures and
> 2. fast methods for numerical simulation.
>
> At that time several computer architecture groups in Germany had developed new concepts for (parallel) supercomputers. A joint proposal by Wolfgang Giloi, director of the GMD Institute for Computer Architecture and Software Technology (FIRST) in Berlin, (in charge of the computer architecture) and Trottenberg (in charge of the numerical software) was selected by an international committee of experts. The corresponding SUPRENUM project (SUPerREchner für NUMerische Anwendungen) was funded from 1984 to 1990.
>
> The SUPRENUM system, an MIMD architecture with up to 256 computing nodes, together with system and application software, was developed by a consortium of 13 partners from industry, federal research centers and university institutes. A small system consisting of 32 computing nodes was presented at the Hannover Fair 1989 with a lot of national and international resonance. The full system was shipped in 1990; with a measured performance of 3.2 GigaFlops for a matrix multiplication, it was the worldwide fastest MIMD parallel computer in 1991.
>
> Within the SUPRENUM development, the role of the institute F1 was to develop fast numerical *parallel* algorithms, exemplarily for all fields of numerical simulation and optimization. One also theoretically important result was that fast multigrid methods could be parallelized and implemented efficiently on parallel computers. This was not trivial since multigrid has essential sequential features—so that there had been a lot of doubts whether multigrid could be efficiently parallelized at all.

An immediate follow-up project of SUPRENUM was *GENESIS* [18] funded by the EU (1989–1992). Originally planned as a European parallel supercomputer development, the companies BULL, INMOS (Transputer) and PALLAS (a spin-off of the institute F1) as coordinator of GENESIS soon decided to concentrate on software environments for high performance numerical computing on distributed memory architectures.

In the course of the GENESIS project, portability became a major issue and the PARMACS message-passing library [4, 17], a joint development of the institute F1 and the Argonne National Lab, was chosen as the common central programming model. The broad experience of the institute F1 with the PARMACS significantly influenced the definition of the Message Passing Interface (MPI) standard.

> **HLRZ** (*Höchstleistungsrechenzentrum*, **High Performance Computing Center**): Additionally, the Institute F1 was strongly involved in the conception and the foundation of the HLRZ, a joint venture of GMD, KFA (today FZJ) and DESY. At the end of 1995, on behalf of Minister Riesenhuber, Trottenberg formed and coordinated a scientific committee, which worked out a detailed concept for such a supercomputer environment. This *HLRZ-concept* was realized in 1997, with specific roles for the FZJ and the GMD.
>
> In the FZJ, the highest performance computers, i.e. the production workhorses, were installed—vector computers at first, and parallel computers later. In the GMD part of the HLRZ, the institute F1 was responsible for the experimental use of innovative parallel computer architectures and ran a variety of parallel machines. These machines were used internally and externally for methodical developments, for the test of application software and tools, as well as for benchmarking and optimization.
>
> Later, in 1992, when the diversity of architectures for parallel machines began to consolidate, the computer park at the GMD was reduced to a single architecture (first a Connection Machine CM-5 of Thinking Machines, later an IBM SP2). The GMD terminated its membership in the HLRZ in 1998, but continued its research activities in high performance computing.

Summarizing the most successful research activities and highlights of the institute during this early period (1978–1991), we would like to emphasize three fields:

- the *development of algebraic and geometric multigrid methods* (see [41, 46]),
- the *efficient parallelization of numerical algorithms within the SUPRENUM project* [43, 44] and
- the *Petri nets theory* [37].

In recognition of his merits for informatics, Carl Adam Petri received the Konrad-Zuse-Medaille, the reputable German lifetime achievement award in computer science, in 1993. In 1996 he was honored with the Werner-von-Siemens-Ring, one of the highest awards for technical sciences in Germany.

3　SCAI: Algorithms and Scientific Computing (1992–2001)

When Petri retired in 1991, Thomas Lengauer became his successor, in a joint appointment as a full professor of computer science at the University of Bonn. In his part of the institute, named *Discrete Algorithms*, Lengauer started essentially two new fields of research in SCAI: *cutting and packing optimization*—based on Lengauer's discrete and combinatorial optimization background—and *bioinformatics*, a new field in informatics. Apart from the new topics in Lengauer's part of SCAI, the Petri group continued its work on Petri nets and computer algebra.

After the definition of an overall new research agenda for the institute, the institute F1 was renamed to *Institute for Algorithms and Scientific Computing (SCAI)* in 1994.

Lengauer's bioinformatics department soon became one of the leading groups in this field. SCAI was the computer science coordinator for the formation of the BMFT initiative *Molecular Bioinformatics*. This initiative was funded with 23 million DM from 1993 to 1997 supporting eight research clusters formed by universities, research institutes, and industry. Moreover, a priority program of the German Science Foundation (DFG) on *Efficient Algorithms for Discrete Problems and their Applications* was coordinated by SCAI and started its work in January 1995.

One of the most prominent research topics of SCAI's bioinformatics department was the development of the protein-ligand docking tool FlexX [24]. This development has been honored by several scientific awards and was the motive for Lengauer and his coworkers Michael Rarey, Ralf Zimmer and Christian Lemmen to establish the BioSolveIT GmbH, Sankt Augustin, in 2001.

But not only SCAI's bioinformatics activities were a real breakthrough, also the cutting and packing optimization research of Lengauer and his coworker Ralf Heckmann soon became highly successful and industrially relevant. The corresponding research results were the basis for the software products AutoNester (for the optimized 2D cutting of fabrics, sheet metal or wood) and PackAssistant (for the optimal 3D packing of identical parts in standard containers). Today, these two software products are the economically most successful software developments of SCAI.

In Trottenberg's part of the institute, the traditional topics like multigrid and the development of efficient parallel algorithms were successfully continued and generalized in the early 1990s.

The SUPRENUM and GENESIS projects had shown that fast numerics and efficient parallelism were compatible. Large industrial codes, however, were not yet available for parallel computers. In particular, the problem of parallelizing *existing software for numerical simulation* was not solved: The great multitude of commercial industrial simulation codes was—so far—running only on sequential computers. These fundamental practical problems were addressed in the 1990s in national and international programs and projects.

The HPSC initiative: The basis for the German program was the *HPSC initiative* (High Performance Scientific Computing). In this initiative experts from meteorology and climate research, physics, chemistry and molecular biology, fluid and structural mechanics, computer science, applied mathematics and finally from the German supercomputer centers formed seven expert groups. The overall initiative and the work of these expert groups was coordinated by the institute F1/SCAI, namely by Ulrich Trottenberg (spokesman) and Johannes Linden (reporting). The final report [45] was submitted to the BMFT ministry in spring 1992. As a consequence, the ministry set up the HPSC program funded with 30 million DM from 1994 to 1996. Within this program, SCAI was in charge of several projects with significant impact on industry. We just mention the projects POPINDA (with the German aircraft industry) [33], KALCRASH [23], AUTOBENCH and AUTO-OPT (with the German automotive industry) [22] and BEMOLPA (with the German pharma industry) [36].

Independent of HPSC, but in the same spirit, was the project SESIS with the German shipbuilding industry, funded by the German Federal Ministry for Economic Affairs and Energy (BMWi) some years later. In SESIS, SCAI developed a structured, modular and flexible platform for the early stage ship design [15, 19, 21].

An HPSC-like EU funded program was based on the *HPCN initiative* (High Performance Computing and Networking). Here, the focus was on the parallelization of *existing software packages*. Prominent EU-projects of SCAI in this context were EUROPORT [5, 40] and TTN [20].

Another important SCAI activity in this period was the development of ADAPTOR [1, 2], a compilation system for the transformation of Fortran programs using high-level data parallelism to message-passing programs. ADAPTOR was awarded the IBM European Parallel Computing Consortium (ECCO) prize for the best parallel software development tool in 1994. Later on, ADAPTOR was used in many projects to develop and evaluate High Performance Fortran compilation techniques.

Various parallelization projects of SCAI were honored by international and national awards, too (ECCO 1993, 1994; Mannheim SuParCup 1993, 1995).

Parallelization of the IFS (Integrative Forecasting System): The most spectacular success probably was the efficient parallelization of the IFS code— the operative software permanently used by the ECMWF (European Center for Medium Range Weather Forecast, Reading, UK) for the daily European weather prediction—in 1993 [12–14]. This first parallelization of a production code consisting of several hundred thousand lines of code could be achieved by a project team consisting of three SCAI scientists within 6 months and proved to be highly efficient. This result was a breakthrough for parallel computing in meteorology worldwide. Furthermore, this work demonstrated that the parallelization of such large codes was feasible in a reasonable amount of time and effort.

The outstanding role of SCAI in parallel and high performance computing was also recognized by industry. So, SCAI signed cooperation agreements with IBM and NEC. NEC opened a branch lab in the TechnoPark of the GMD in Sankt Augustin in order to foster the intensive cooperation with SCAI.

Daimler-Benz financed the development of the programming environment for parallel computing TRAPPER. TRAPPER indeed was a very advanced project targeting the development of driverless cars. The suitability of TRAPPER was demonstrated in practical tests of an automatic steering car already in 1997. To avoid collisions, the car had been equipped with 18 cameras which were able to detect traffic signs, street lanes and obstacles like cars or persons in real-time. The controlling software consisted of 180 distributed processes, which ran on 60 parallel processors residing in the trunk of the test vehicle. The test car was autonomously able to keep track, to hold safe distance to cars in the front and to manage overtaking maneuvers [28–30].

In 2000, the political decision to merge the GMD and the Fraunhofer Gesellschaft was jointly made by the German Ministry of Education and Research (BMBF) and the presidents of the Fraunhofer Gesellschaft, Hans-Jürgen Warnecke, and of the GMD, Dennis Tsichritzis. The integration process turned out to be quite complicated. The plan was neither welcomed by the employees nor by most of the institute directors of the GMD. During the tangled preparation of the merger, the SCAI directors received offers from the Max-Planck Society (Lengauer) and from the German Aerospace Center (DLR) (Trottenberg). The DLR was also interested in integrating essential parts of SCAI.

Finally, Lengauer accepted the offer to become one of the directors of the Max-Planck Institute for Informatics in Saarbrücken and left SCAI in 2001. Some of his department heads and coworkers from the bioinformatics group accepted calls of various universities and left SCAI, too (Ralf Zimmer, University of Munich; Matthias Rarey, University of Hamburg; Christel Marian, University of Düsseldorf; Joachim Selbig, now University of Potsdam). Trottenberg decided to stay with SCAI as its director, but additionally accepted an offer to become a part time director in the DLR.

Before the integration of the GMD institutes into the existing Fraunhofer Information and Communication Technology (ICT) Group, which consists of all Fraunhofer institutes primarily engaged in the fields of mathematics, computer science and communication technology, two *technical integration committees* were formed, headed by Jose Encarnacao for the six "old" and Trottenberg for the eight "new" institutes. The task of these committees was to check possible overlaps in the research fields of the institutes and to look for topics of cooperation.

4 SCAI as a Fraunhofer Institute, the First Years (2001–2009)

Since its integration into the Fraunhofer-Gesellschaft in July 2001, SCAI has been the *Fraunhofer Institute for Algorithms and Scientific Computing*, briefly *Fraunhofer SCAI*. This integration meant a big challenge for the institute, in particular with respect to the financial model of the Fraunhofer-Gesellschaft. In the GMD, the success criteria had been a mix of different requirements: publications, third-party-funding, qualified graduations (in particular, dissertations) and foundations of spin-off companies. According to these criteria, SCAI had been one of the most successful institutes in the GMD. Whereas making 30% third-party-funding of the institute's budget was regarded as a success at the GMD, 70% third-party funding of the institute's budget is regarded as a minimum for a Fraunhofer institute. Even more important for the Fraunhofer model is that the direct industrial income should be at least 40% of the overall budget.

In order to fulfill these new financial conditions within 5 years, the institute had to set up a thorough strategic planning.

- The brain drain in the bioinformatics department mentioned above was a severe problem. For some time after 2001, it was not sure whether the bioinformatics activities of Lengauer's part of the institute could be continued at all. In 2002 Martin Hofmann (now: Hofmann-Apitius) could be engaged as a new department head so that the work was continued, now with a strong emphasis on *text mining*, mainly for biology, medicine and pharmacology.
- There was no doubt that SCAI would only be able to fulfill the new financial conditions within 5 years if the revenues were increased and, in addition, the expenses were reduced significantly.
- As a result, this meant that the institute had to abandon the work on several scientific topics, for which significant financial revenues could not be expected in the next years. This included the work on Petri nets, but also the development of efficient geometric multigrid methods, both research fields, in which SCAI had been among the worldwide leading institutes since the 1980s.
- Accordingly, the number of employees (without students and Ph.D. students) decreased steadily from 85 full time equivalent employees in 2001 to 65 at the end of 2005.
- Corresponding to these developments, the divisional structure of SCAI was subjected to several adaptations. At the end of 2004, SCAI consisted of the departments Bioinformatics (BIO, headed by Martin Hofmann-Apitius), Optimization (OPT, headed by Ralf Heckmann), Simulation Engineering (SIAN, headed by Johannes Linden) and Numerical Software (NUSO, headed by Clemens-August Thole).
- The most important strategic decision of the institute, however, was to not only develop innovative, highly efficient algorithms but—using such developments as a basis—to go a step further and to *generate and market* corresponding *software*

products, being well aware of the increased demands for robustness and user-friendliness of the software.

Within a few years, these strategic actions proved to be effective and SCAI became eventually successful according to the Fraunhofer criteria. The industrial revenues increased continuously to 20% in 2004 and to more than 40% in 2007. The overall third-party funding increased to more than 45% in 2004 and to nearly 75% in 2007. This means that 2007 was the first year in which SCAI could meet the financial conditions of a Fraunhofer institute. Between 40 and 50% of the SCAI budget has been generated by SCAI software products and by direct industrial contract work since then every year, with the exception of the crisis year 2009.

The first and to date most important software products deal with

- the optimized automatic placement of markers on fabrics, sheet metal, wood or other materials (AutoNester) [7],
- the optimized packing of identical parts in standard containers (PackAssistant) [9],
- the efficient solution of very large systems of linear equations (SAMG) [42],
- the solution of multidisciplinary problems (MpCCI) [47].

For the PackAssistant development, Fraunhofer SCAI was awarded the innovation prize for Climate and Environment in 2010 in the category of environmentally friendly products and services. The award is donated by the German Federal Ministry for the Environment, Nature Conservation and Nuclear Safety (BMU) and the Federation of German Industries (BDI).

For more information on these and further software products, we refer to the corresponding contributions in this book and to SCAI's product web pages [10] .

In 2002, the University of Bonn, the RWTH Aachen, the Hochschule Bonn-Rhein-Sieg and the Fraunhofer-Gesellschaft founded the Bonn-Aachen International Center for Information Technology (B-IT). The B-IT is financed by a foundation which got means from the Bonn-Berlin compensation, project funds of the German Federal Ministry of Education and Research and complementary funds of the Federal State of North Rhine-Westphalia. The B-IT offers international master programs in computer science and applied IT. SCAI has been particularly involved with regard to the M.Sc. in Life Science Informatics.

From 2006 to 2012, SCAI maintained a branch lab at the Mathematical Institute of the University of Cologne which cooperated closely with Caren Tischendorf, who was a professor of applied mathematics at the University of Cologne in that period.

In Germany, the Year of Science 2008 was dedicated to mathematics. SCAI used this opportunity to increase awareness of the importance of applied mathematics for modern society. In fact, Germany's Federal President Horst Köhler and his wife Eva Luise Köhler honoured SCAI's achievements with a visit in August 2008. Apart from the Arithmeum in Bonn, SCAI was the only mathematical institute in Germany honoured by a Presidential visit during the Year of Mathematics 2008.

Together with the German National Ministry of Education and Research and with the Fraunhofer Institute for Industrial Mathematics ITWM, SCAI organized a large congress in 2009. The objective of this congress was to elaborate the relevance of applied mathematics for today's industrialized society and to give impulses for a modern realistic mathematical education at school. In a joint activity, the Mathematical Institute of the University of Cologne and SCAI developed exemplary teaching modules which take up applications such as weather prediction, mp3, or the prediction of traffic jams, and work out the mathematics behind (see [8, 35] for details).

In order to give interested high school students the opportunity to learn more about modern algorithmic and applied mathematics, SCAI has organized 2-week summer schools regularly since 2009.

SCAI had become more and more successful in marketing and selling its software products in the early 2000s. But since this kind of work is no genuine task of a research institute, SCAI founded the scapos AG as a spin-off to outsource this work in 2009. The Fraunhofer-Gesellschaft became one of the shareholders of the scapos AG.

5 New Fields of Research and New Business Areas (2010–2016)

In 2009, SCAI could win Michael Griebel, full professor of applied mathematics and director of the *Institute for Numerical Simulation* at the University of Bonn, as a second director of SCAI. From 2010 to 2012 Trottenberg and Griebel directed the institute jointly. Since Trottenberg's retirement in 2012, Griebel has been the only director of SCAI.

Griebel immediately started to build a strong bridge between Fraunhofer SCAI and the Institute for Numerical Simulation at the University of Bonn. Correspondingly, SCAI has maintained a branch lab at the Institute for Numerical Simulation since 2010.

As mentioned above, SCAI had consisted of the departments BIO, NUSO, OPT and SIAN since 2004. Of course, the adaptation to the requirements and the developments of the industrial markets required adjustments in the research topics of the departments. For example, OPT extended its research activities and now covers a much larger scope of different application areas in discrete optimization. Correspondingly, BIO established a strong focus on dementia diseases in recent years and contributes its expertise in text mining in projects with the pharmaceutical, the biotech, and the life science software industry.

Since 2010, in addition to its department structure, SCAI has introduced ten business areas in order to present its work in a more customer oriented way. We first describe this process for the four existing departments and then report on newly formed business areas.

Up to 2010, SIAN had established three business areas: *Multiphysics* (headed by Klaus Wolf), *High Performance Computing* (headed by Thomas Soddemann) and *Computational Chemical Engineering* (headed by Dirk Reith and Anton Schüller).

NUSO had built up three business areas on fast solvers, data compression and robust design. In 2011, the robust design group became the new department *High Performance Analytics*, which also has a focus on the simulation and optimization of energy networks, and Tanja Clees was appointed its head. In 2012, Thole decided to found the software company *SIDACT GmbH*, based on a license agreement with SCAI. The main activities of this company are the development of compression algorithms for numerical simulation results (*FEMZIP*) and the analysis of such results in order to obtain new insights, e.g. in crash simulation (*DIFFCRASH*). The remaining part of NUSO (headed by Klaus Stüben until his retirement in summer 2016) continued its work in the business area *Fast Solvers*.

Moreover, several fully new business areas have been established at SCAI, beginning with *Virtual Material Design* (headed by Jan Hamaekers), which integrated most of the activities of the computational chemical engineering group, in 2010. Further new business areas have been *Numerical Data-Driven Prediction* (since 2011, headed by Jochen Garcke), *Computational Finance* (since 2013, headed by Thomas Gerstner), and *Meshfree Multiscale Methods* (since 2013, headed by Marc Alexander Schweitzer who also leads the business area *Fast Solvers* after Stüben's retirement). Correspondingly, the business areas of SCAI are now:

- Bioinformatics,
- Fast Solvers,
- High Performance Computing,
- Multiphysics,
- Optimization,
- Computational Finance,
- High Performance Analytics,
- Meshfree Multiscale Methods,
- Numerical Data Driven Prediction,
- Virtual Material Design.

We do not address the quite large number of projects of SCAI in recent years here, but refer to the detailed contributions in this book and to SCAI's project web pages [11].

Including students and Ph.D. students, SCAI employed about 146 people in 2016. The budget in 2016 was about 11 million euros.

References

1. T. Brandes, Exploiting advanced task parallelism in High Performance Fortran via a task library, in *Proceedings of EuroPar 99*, ed. by P. Amestoy, et al. Lecture Notes in Computer Science, vol. 1685 (Springer, Berlin/Heidelberg, 1999), pp. 833–844
2. T. Brandes, F. Zimmermann, ADAPTOR – a transformation tool for HPF programs, in *Programming Environments for Massively Parallel Distributed Systems* (Birkhäuser, Basel, 1994), pp. 91–96.
3. A. Brandt, Multi-level adaptive solutions to boundary-value problems. Math. Comput. **31**, 333–390 (1977)
4. R. Calkin, R. Hempel, H.-C. Hoppe, P. Wypior, Portable programming with the PARMACS message-passing library. Parallel Comput. **20**, 615–632 (1994)
5. J. Elliott, K. Stüben (eds.), EUROPORT: industrial high-performance computing, in *Arbeitspapiere der GMD 905* (Gesellschaft für Mathematik und Datenverarbeitung, Sankt Augustin, 1995)

6. H. Foerster, K. Stüben, U. Trottenberg, Non-standard multigrid techniques using checkered relaxation and intermediate grids, in *Elliptic Problem Solvers*, ed. by M.H. Schultz, (Academic Press, New York, 1981), pp. 285–300.
7. Fraunhofer Institute for Algorithms and Scientific Computing SCAI, AutoNester-T. Website, 2016. Online at https://www.scai.fraunhofer.de/en/business-research-areas/optimization/products/autonester-t.html. Accessed 16 Dec 2016
8. Fraunhofer Institute for Algorithms and Scientific Computing SCAI, Mathematik für die Praxis. Website, 2016. Online at https://www.scai.fraunhofer.de/mathematik-fuer-die-praxis.html. Accessed 16 Dec 2016
9. Fraunhofer Institute for Algorithms and Scientific Computing SCAI, PackAssistant. Website, 2016. Online at https://www.scai.fraunhofer.de/en/business-research-areas/optimization/products/packassistant.html. Accessed 16 Dec 2016
10. Fraunhofer Institute for Algorithms and Scientific Computing SCAI, Products. Website, 2016. Online at https://www.scai.fraunhofer.de/en/products.html. Accessed 16 Dec 2016
11. Fraunhofer Institute for Algorithms and Scientific Computing SCAI, *Projects*. Website, 2016. Online at https://www.scai.fraunhofer.de/en/projects.html. Accessed 16 Dec 2016
12. U. Gärtel, W. Joppich, A. Schüller, First results with a parallelized 3D weather prediction code. Parallel Comput. **19**, 1427–1429 (1993)
13. U. Gärtel, W. Joppich, A. Schüller, Parallel computing for weather prediction, in *High-Performance Computing and Networking I*, ed. by W. Gentzsch, U. Harms. Lecture Notes in Physics, vol. 196 (Springer, Berlin, 1994)
14. U. Gärtel, W. Joppich, A. Schüller, Portable parallelization of the ECMWF's weather forecast program – technical documentation and results, in *Arbeitspapiere der GMD 820* (Gesellschaft für Mathematik und Datenverarbeitung, Sankt Augustin, 1994)
15. T. Gosch, W. Abels, E. Esins, et al., SESIS - Entwicklung eines integrierten schiffbaulichen Entwurfs- und Simulationssystems. in Bundesministerium für Wirtschaft und Technologie - BMWi-, Bonn; Projektträger Jülich -PTJ-: Statustagung Schifffahrt und Meerestechnik 2008. Tagungsband : 11. Dezember 2008, Rostock-Warnemünde Bonn: BMWI, 2008
16. W. Hackbusch, *Multi-Grid Methods and Applications* (Springer, Berlin, 1985)
17. R. Hempel, L. Bomans, D. Roose, The Argonne/GMD macros in FORTRAN for portable parallel programming and their implementation on the Intel iPSC/2. Parallel Comput. **15**, 119–132 (1990)
18. T. Hey, O.A. McBryan (eds.), Suprenum and GENESIS. Special Issue of Parallel Comput. **20**, 10–11 (1994)
19. O. Kraemer-Fuhrmann, Y. Raekow, Business experiment ship building, in *Grid and Cloud Computing: A Business Perspective on Technology and Applications*, ed. by K. Stanoevska-Slabeva, T. Wozniak, S. Ristol (Springer, Berlin, 2010), pp. 159–172
20. O. Krämer-Fuhrmann, The HPCN-TTN network of technology transfer nodes; take-up and transfer of HPCN technology in non-traditional industrial sectors, in *15th IMACS World Congress on Scientific Computation, Modelling and Applied Mathematics*, ed. by A. Sydow (1997)
21. O. Krämer-Fuhrmann, RCE - Reconfigurable Computing Environment. *NAFEMS Seminar "Simulation Data Management - Integration into the Product Development Process"* (2009)
22. A. Kuhlmann, C.-A. Thole, U. Trottenberg, AUTOBENCH/AUTO-OPT: Towards an integrated construction environment for virtual prototyping in the automotive industry, in *Recent Advances in Parallel Virtual Machine and Message Passing Interface*, ed. by J. Dongarra, D. Laforenza, S. Orlando (Springer, Berlin, 2003), pp. 686–690
23. G. Lonsdale, O. Kolp, Ein neues Verfahren für die parallele Behandlung der Contact-Impact Algorithmen bei Crashworthiness Simulationen, in *Statustagung des BMBF HPSC 95*, ed. by G. Wolf, R. Krahl, Projektträger des BMBF für Informationstechnik bei der DLR e.V. (DLR, Berlin, 1995), pp. 105–114
24. M. Rarey, B. Kramer, T. Lengauer, G. Klebe, A fast flexible docking method using an incremental construction algorithm. J. Mol. Biol. **261**, 470–489 (1996)

25. M. Ries, U. Trottenberg, G. Winter, A note on MGR methods. Linear Algebra Appl. **49**, 1–26 (1983)
26. J.W. Ruge, K. Stüben, Algebraic multigrid. Multigrid Methods **3**, 73–130 (1987)
27. B. Ruttmann, K. Solchenbach, A multigrid solver for the computation of in-cylinder turbulent flows in engines, in *Efficient Solution of Elliptic Systems*, ed. by W. Hackbusch. Notes on Numerical Fluid Mechanics, vol. 10 (Vieweg, Braunschweig/Wiesbaden, 1984), pp. 87–108
28. L. Schäfers, C. Scheidler, O. Krämer-Fuhrmann, TRAPPER: A graphical programming environment for parallel systems. Futur. Gener. Comput. Syst. **11**, 351–361 (1995)
29. L. Schäfers, C. Scheidler, O. Krämer-Fuhrmann, TT-TRAPPER: A software engineering environment for time-triggered embedded systems, in *IEEE International Workshop on Embedded Fault-Tolerant Systems 1996 Proceedings* (IEEE Press, Piscataway, 1996)
30. C. Scheidler, L. Schäfers, O. Krämer-Fuhrmann, Rapid prototyping with TRAPPER in the PROMETHEUS project 'Collision Avoidance', in *Rapid Prototyping in the Automotive Industries*, ed. by J. Soliman, D. Roller 1995, pp. 247–260; Proceedings for the 28th ISATA, Croydon
31. J. Schröder, U. Trottenberg, Reduktionsverfahren für Differenzenverfahren bei Randwertaufgaben I. Numer. Math. **22**, 37–68 (1973)
32. J. Schröder, U. Trottenberg, H. Reutersberg, Reduktionsverfahren für Differenzenverfahren bei Randwertaufgaben II. Numer. Math. **26**, 429–459 (1976)
33. A. Schüller (ed.), in *Portable Parallelization of Industrial Aerodynamic Applications (POPINDA)*. Notes on Numerical Fluid Mechanics, vol. 71 (Vieweg, Braunschweig/Wiesbaden, 1999)
34. A. Schüller, U. Trottenberg, R. Wienands, Schnelle Lösung großer Gleichungssysteme. Der Mathematik-Unterricht, Jahrgang 57, Heft 5, pp. 16–29 (2011)
35. A. Schüller, U. Trottenberg, R. Wienands, et al. Der Mathematik-Unterricht, Jahrgang 57, Heft 5, Oktober 2011
36. H. Schwichtenberg, G. Winter, H. Wallmeier, Acceleration of molecular mechanic simulation by parallelization and fast multipole techniques. Parallel Comput. **25**, 535–546 (1999)
37. E. Smith, Carl Adam Petri - Eine Biographie (Springer, Berlin, 2014)
38. K. Solchenbach, B. Steckel, in *Numerical Simulation of the Flow in 3d-Cylindrical Combustion Chambers Using Multigrid Methods*. Arbeitspapiere der GMD 216 (Gesellschaft für Mathematik und Datenverarbeitung, Sankt Augustin, 1986)
39. K. Stüben, Algebraic multigrid (AMG): experiences and comparisons. Appl. Math. Comput. **13**, 419–451 (1983)
40. K. Stüben, Europort-D: Commercial benefits of using parallel technology, in *Parallel Computing: Fundamentals, Applications and New Directions*, ed. by E. D'Hollander, G. Joubert, F. Peters, U. Trottenberg (Elsevier Science B.V., North-Holland, 1998), pp. 61–78.
41. K. Stüben, An introduction to algebraic multigrid, in *Multigrid* (Academic Press, New York, 2001). Appendix A in [46]
42. K. Stüben, J.W. Ruge, T. Clees, S. Gries, H.-J. Plum, Algebraic multigrid – from academia to industry, in *Scientific Computing and Algorithms in Industrial Simulations*, ed. by M. Griebel et al. (Springer International Publishing, Cham, 2017). doi:10.1007/978-3-319-62458-7_16
43. U. Trottenberg, Some remarks on the SUPRENUM project. Parallel Comput. **20**, 1397–1406 (1994)
44. U. Trottenberg, Das Superrechnerprojekt SUPRENUM (1984–1989), in *Informatikforschung in Deutschland*, ed. by B. Reuse, R. Vollmar (Springer, Berlin, 2008), pp. 176–187
45. U. Trottenberg, J. Linden, Situation und Erfordernisse des wissenschaftlichen Höchstleistungsrechnens in Deutschland. Informatik-Spektrum **15**, 218–220 (1992)
46. U. Trottenberg, C.W. Oosterlee, A. Schüller, *Multigrid*, (Academic Press, New York, 2001)
47. K. Wolf, P. Bayrasy, C. Brodbeck, et al., MpCCI – neutral interfaces for multiphysics simulation, in *Scientific Computing and Algorithms in Industrial Simulations*, ed. by M. Griebel et al. (Springer International Publishing, Cham, 2017). doi:10.1007/978-3-319-62458-7_16

Printed in the United States
By Bookmasters